Laboratory Manual

Human Biology

Fifteenth Edition

Sylvia S. Mader

Contributor

Jason Carlson

St. Cloud Technical and Community College

LABORATORY MANUAL TO HUMAN BIOLOGY, FIFTEENTH EDITION

Published by McGraw-Hill Education, 2 Penn Plaza, New York, NY 10121.

This book is printed on acid-free paper.

2 3 4 5 6 7 8 9 LMN 21 20 19 18

ISBN 978-1-259-93370-7
MHID 1-259-93370-9

Chief Product Officer, SVP Products & Markets: *G. Scott Virkler*
Vice President, General Manager, Products & Markets: *Marty Lange*
Vice President, Content Production & Technology Services: *Betsy Whalen*
Managing Director: *Lynn M. Breithaupt*
Executive Brand Manager: *Michelle Vogler*
Director of Development: *Rose M. Koos*
Senior Product Developer: *Anne Winch*
Digital Product Analyst: *Christine Carlson*
Marketing Manager: *Britney Ross*
Director, Content Production: *Linda Meehan-Avenarius*
Program Manager: *Angie Fitzpatrick*
Content Project Manager: *Mary Jane Lampe*
Senior Buyer: *Sandy Ludovissy*
Senior Designer: *David Hash*
Senior Content Licensing Specialist: *Lori Hancock*
Cover Image: *©mihtiander/iStock/Getty Images Plus*
Compositor: *Aptara® Inc.*
Typeface: *11/13 STIX MathJax Main*
Printer: *LSC Communications - Menasha*

All credits appearing on page or at the end of the book are considered to be an extension of the copyright page.

Some of the laboratory experiments included in this text may be hazardous if materials are handled improperly or if procedures are conducted incorrectly. Safety precautions are necessary when you are working with chemicals, glass test tubes, hot water baths, sharp instruments, and the like, or for any procedures that generally require caution. Your school may have set regulations regarding safety procedures that your instructor will explain to you. Should you have any problems with materials or procedures, please ask your instructor for help.

mheducation.com/highered

Contents

Applications for Daily Living

Preface

To the Instructor

The laboratory exercises in this manual are coordinated with *Human Biology*, a text that has two primary functions: (1) to achieve an understanding of how the human body works, and (2) to show the relationship of humans to other living things in the biosphere.

This manual may also be used in coordination with other biology texts. There are a sufficient number of laboratories and exercises within each lab to tailor the laboratory experience as desired. Then, too, many exercises may be performed as demonstrations rather than as student activities, thereby shortening the time required to cover a particular concept.

The Exercises

All exercises have been tested for student interest, preparation time, time of completion, and feasibility. The following features are particularly appreciated by adopters:

Integrated Opening. Integrated Opening. Each laboratory begins with a list of Learning Outcomes organized according to the major sections of the laboratory. The major sections of the laboratory are numbered in the laboratory text material.

Prelab Questions. The Learning Outcomes contain one prelab question per major section. These questions are meant to encourage students to read the laboratory before they start work. Instructors can require that the prelab questions be answered and turned in on the day of the lab and/or instructors can select a few to include in a laboratory quiz.

Self-Contained Content. Each laboratory gives all the background information necessary to understand the concepts being studied and to answer the questions asked. This feature will reduce student frustration and increase learning.

Human Emphasis. This manual accompanies *Human Biology*, so it is fitting for the laboratories to have a human emphasis. This emphasis means that it is not necessary to include "Human" in the title of the laboratories because they all have a human orientation.

Scientific Process. This manual provides opportunities for students to gain an appreciation of the scientific method. The first laboratory of the manual explicitly explains the steps of the scientific method and gives students an opportunity to use them.

Student Activities. Sequentially numbered steps guide students as they perform each activity. Color bars are used to designate whether an activity is an Observation or whether it is an Experimental Procedure. A boxed statement with a clock icon is added at the start of the day's work to call attention to any exercises that require more time than other exercises. A clock also appears within the procedure whenever time is required for the results to occur.

Application for Daily Living. Relevancy is of primary importance in the *Human Biology Lab Manual*. Relevancy is particularly apparent due to a feature called *Application for Daily Living*. These short boxes at the end of the lab tell students how the day's work applies in a practical way to a current health or environmental concern.

Laboratory Safety. Laboratory safety is of prime importance, and the listing now online at **http://connect.mheducation.com**, will assist instructors in making the laboratory experience a safe one. Warning notices occur throughout the manual as is appropriate to the particular procedure.

The Fifteenth Edition

Each Laboratory Review section is completely new for this edition. Fourteen short answer questions now help students identify key vocabulary and basic concepts critical to the topic of the exercise. Questions are organized to match the exercise's order of presentation, facilitating students' ability to identify an appropriate response. Each Laboratory Review also contains three new critical thinking questions that challenge students to make connections between the concepts presented, or to apply information from the exercise to real world examples.

Each laboratory exercise was carefully revised to ensure that directions to students are clear and follow common laboratory standards and practices. For example, the manual now directs students to use inexpensive plastic transfer pipets to measure volume rather than marking off test tubes and filling to a cm.

Lab 4: Chemical Composition of Cells
This laboratory was reorganized and written for added clarity and to matching learning objectives.

Lab 6: Body Tissues
A section was added to this lab to provide an overview of body tissues and a summary activity was added to review each type of body tissue.

Lab 8: Organization of the Body
Anatomy of the human torso was moved to the end of the lab after dissection of the fetal pig. Greater detail was added to dissection of the fetal pig and art showing internal anatomy has been added to supplement and clarify pictures.

Customized Editions

The 19 laboratories in this manual are now available as individual "lab separates," so instructors can custom-tailor the manual to their particular course needs. Find more information at **http://create.mheducation.com.**

Laboratory Resource Guide

The *Laboratory Resource Guide,* an essential aid for instructors and laboratory assistants, free to adopters of the *Laboratory Manual,* is online at **http://connect.mheducation.com.** The answers to all the laboratory questions are in the *Resource Guide.*

To the Student

Special care has been taken in preparing *Human Biology Laboratory Manual* so that you will **enjoy** the laboratory experience as you **learn** from it. The instructions and discussions are written clearly so that you can understand the concepts presented and the procedures you are doing. Student learning aids discussed next are designed to help you focus on important aspects of each exercise.

Student Learning Aids

Students will want to be aware of these learning aids:

Learning Outcomes set the goals for each laboratory session. The major sections of each laboratory are numbered, and the Learning Outcomes are grouped according to these topics. The Learning Outcomes will help you review the material for a laboratory practical or any other kind of exam.

Prelab Questions are a part of the Learning Outcomes; when you can answer the Prelab Questions, you are prepared for a meaningful laboratory experience. These questions will guide your study of the concepts and the student activities in a meaningful way.

The Introduction on the same page as the Learning Outcomes reviews background information necessary for comprehending the upcoming experiments.

Color bars call out exercises that require your active participation. Different colors are used to highlight Observations and Experimental Procedures. If you see a clock in an experimental procedure, it means that a certain amount of time is required before you can observe your results.

Icons are used to get your attention. A safety icon tells you when to be particularly careful about the use of a chemical; a clock icon at the start of the day's work tells you when you might want to set up a procedure ahead of time.

Tables and fill-in-the-blanks allow you to record the results of investigations and experiments. Be sure to answer each question carefully. These allow you to understand the procedure. In addition, each laboratory ends with a set of review questions covering the day's work.

Laboratory Preparation

Read the Learning Outcomes and be sure to answer the Prelab Questions. Also ***study*** the introductory material and the procedures before the day you have lab. To obtain an even better understanding of a topic, consult your text.

Explanations and Conclusions

Throughout a laboratory, you are often asked to formulate explanations or conclusions. To do so, you will need to synthesize information from a variety of sources, including the following:

1. Your experimental results and/or the results of other groups in the class. If your data are different from those of other groups in your class, do not erase your answer; add the other answers in parentheses.
2. Your knowledge of underlying principles. Obtain this information from the laboratory Introduction or the appropriate section of the laboratory and from the corresponding chapter of your text.
3. Your understanding of how the experiment was conducted and/or the materials used. *Note:* Ingredients can be contaminated or procedures incorrectly followed, resulting in reactions that seem inappropriate. If this occurs, consult with other students and your instructor to see if you should repeat the experiment.

In the end, be sure you are truly writing an explanation or conclusion and not just restating the observations made.

Color Bar and Icon

Throughout each laboratory, a color bar designates either an Observation or an Experimental Procedure.

Observation: An activity in which models, slides, and preserved or live organisms are observed to achieve a learning outcome.

Experimental Procedure: An activity in which a series of steps uses laboratory equipment to gather data and come to a conclusion.

Time: An icon is used to designate when time is needed before you can determine your results. A boxed statement at the beginning of the laboratory calls attention to any experimental procedure requiring a considerable amount of time. Start these activities at the beginning of the laboratory, proceed to other activities, and return to this experimental procedure when appropriate.

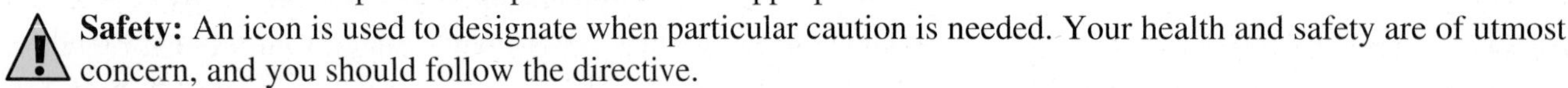

Safety: An icon is used to designate when particular caution is needed. Your health and safety are of utmost concern, and you should follow the directive.

A boxed statement at the beginning of the laboratory calls attention to any experimental procedure requiring a considerable amount of time. Start these activities at the beginning of the laboratory, proceed to other activities. An icon is used to designate when time is needed before you can determine your results.

Laboratory Review

Each laboratory ends with a number of short-answer questions that will help you determine whether you have accomplished the outcomes for the laboratory.

LABORATORY

1

Scientific Method

Learning Outcomes

1.1 Using the Scientific Method

- Outline the steps of the scientific method.
- Distinguish among observations, hypotheses, conclusions, and theories.

Prelab Question: Explain what a scientist means by the term *theory*.

1.2 Observations Concerning Heart Rate

- Observe and describe the rhythmic beating of the heart.
- Read the data presented by the laboratory manual about the heartbeat.

Prelab Question: How do scientists carry out the first step of the scientific method?

1.3 Formulating Hypotheses

- Formulate a hypothesis based on the observations.

Prelab Question: Why is a hypothesis called a tentative explanation?

1.4 Performing an Experiment and Coming to a Conclusion

- Design an experiment that can be repeated by others.
- Reach a conclusion based on observation and experimentation.

Prelab Question: In science, a conclusion is based on the data. Explain.

***Application for Daily Living:* The Stress Test**

Introduction

In everyday life we are often called upon to make observations and use our past experiences to come to a hypothesis that can be tested. For example, suppose you flipped the switch, but the light didn't come on. Most likely you would hypothesize that the bulb has burnt out, the socket isn't working, or there has been a power outage. Each of these possibilities could be tested until you have come to a conclusion that allows the problem to be fixed. In today's laboratory, we want to experience firsthand the manner in which a scientist uses these same steps to come to a conclusion.

1.1 Using the Scientific Method

Science differs from other human ways of knowing and learning by its process, which can be varied because it can be adjusted to where and how a study is being conducted. Still, the scientific process often involves the use of the scientific method. In today's laboratory, we will use the scientific method to come to a conclusion about the heartbeat rate. As depicted in Figure 1.1, the scientific method involves making observations, formulating hypotheses, doing experiments, and coming to a conclusion. Many conclusions on the same topic allow scientists to develop a scientific theory.

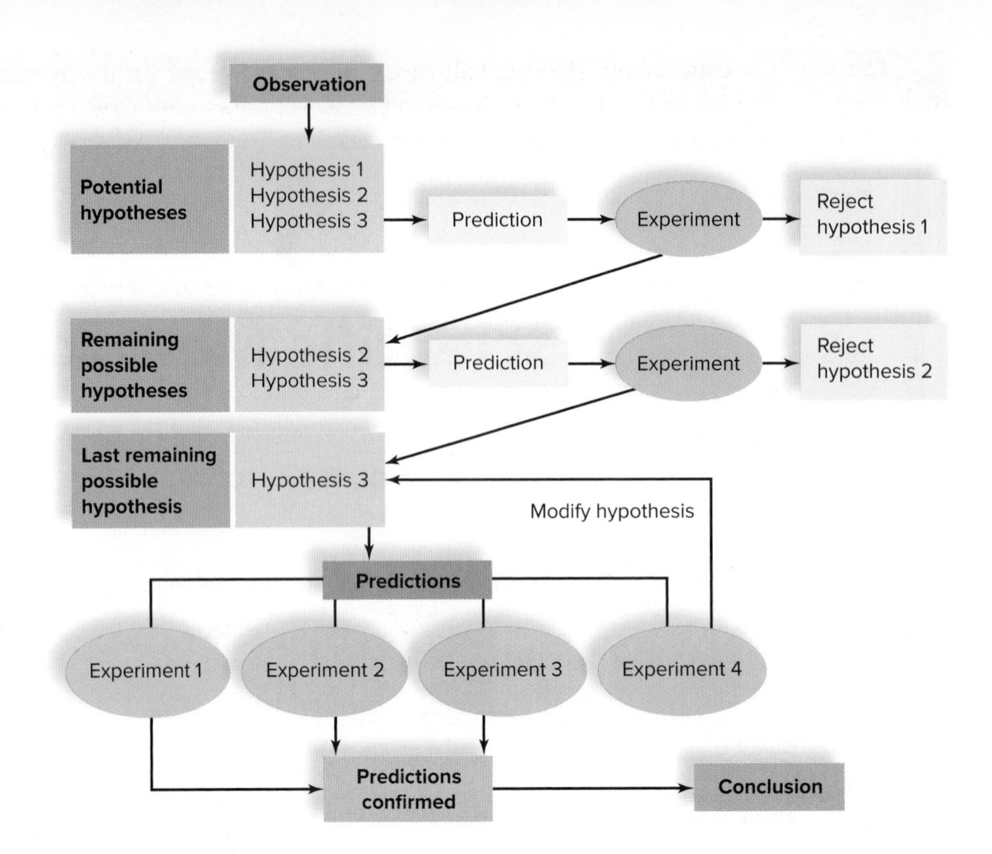

Figure 1.1 Flow diagram for the scientific method. On the basis of new and/or previous observations, a scientist formulates a hypothesis. The hypothesis is tested by further observations and/or experiments, and new data either support or do not support the hypothesis. The return arrow from experiment 4 indicates that a scientist often chooses to retest the same hypothesis or to test a related hypothesis. Conclusions from many different but related experiments may lead to the development of a scientific theory. For example, studies pertaining to development, anatomy, and fossil remains all support the theory of evolution.

Making observations: As a first step in their study of a topic, scientists use all their senses and also instruments, such as a microscope, to make **observations.** Observations can also include something you read, such as the upcoming information about the beating of the heart. If you want to learn more about the heartbeat rate, you could do a search of the literature or talk to a nurse or doctor about the heartbeat. Scientists keep notes about their observations, as you will be asked to do. Why is it helpful for a scientist to begin by making observations? ____

____

Formulating a hypothesis: Based on their observations, scientists come to a tentative explanation, called a **hypothesis,** about what it is they are investigating. Having observed the beating of the heart and having done a little literary research, you will be asked to formulate a hypothesis about the influence of exercise on the heartbeat rate. Why is a hypothesis sometimes called an "educated guess"?

____

Testing your hypothesis: After formulating a hypothesis, a scientist performs an **experiment,** which tests the hypothesis. A well-designed experiment must have a **control,** a sample or event that is not exposed to the testing procedure. If the control and the test produce the same results, either the procedure is flawed or the hypothesis is wrong. What is the purpose of a control? ____________________________________

____

The results of your experiment are called **data.** Data are any factual information that come to light because of your experiment or perhaps further observations you have made. It is important for a scientist to keep accurate records of all their data. When another person repeats the experiment, their data should be the same or something is amiss. Why must a scientist keep complete records of an experiment?

____

Coming to a conclusion: The data will either support or not support the hypothesis. A scientist can come to a **conclusion** that the hypothesis has been shown to be false but does not say that a hypothesis has been shown to be true. After all, some fault might be found with the method of collecting data. Why don't scientists say they have proven their hypothesis true? ____

____

Arriving at a scientific theory: A **theory** in science is an encompassing conclusion based on many individual conclusions in the same field. For example, the cell theory states that all organisms are composed of cells. The cell theory is based on observations made by many scientists over many years who have observed a countless variety of living things. We will not be developing any theories today. How is a scientific theory different from a conclusion? ____

____

1.2 Observations Concerning Heart Rate

You will follow the procedure for the scientific method outlined in Figure 1.1. The first step in this procedure is to make observations about the topic you are proposing to study.

Observation: The Human Heart

1. Your instructor will show a short video of the heart beating. Alternatively, feel your pulse on your neck or wrist. Do your observations suggest that blood moves through your body in a constant flow or a rhythmic flow? ____
2. In his text *Functional Human Anatomy,* James E. Crouch writes

   The heart is a pump which is of primary importance in the maintenance of the flow and pressure of the blood in the closed cardiovascular system. Its contractions start in the early days of embryonic development and must continue for the

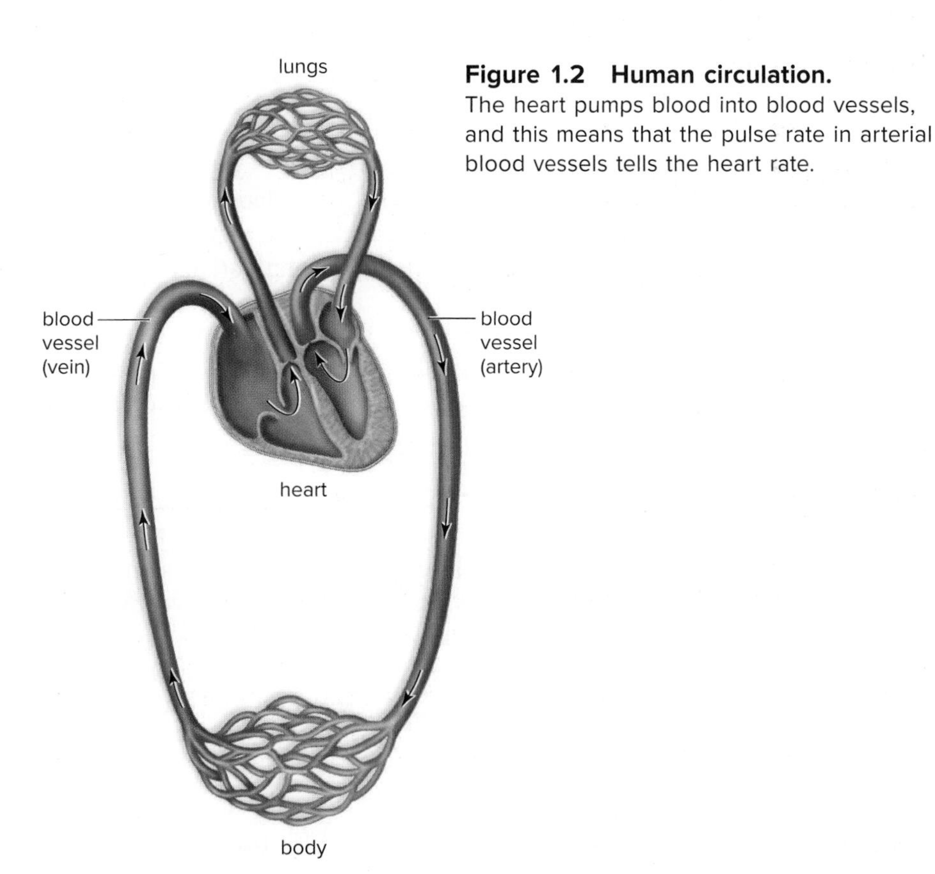

Figure 1.2 Human circulation. The heart pumps blood into blood vessels, and this means that the pulse rate in arterial blood vessels tells the heart rate.

duration of the individual's life. For the contractions to cease even for a few minutes would cause serious damage, or would be fatal. The heart beat occurs at an average rate of 72/min (72 times per minute) and adds up to about 100,000 times a day or about 2,600,000,000 beats in a lifetime of seventy years. If you clench and open your fist alternately at the rate of 72/min, the muscles involved will "feel" tired after about 2 minutes....[1]

3. When you take a spinning or kickboxing class, your instructor may ask you to periodically take your pulse rate. The pulse rate indicates how fast the heart is beating. In your experience, how does exercise affect the heart rate and pulse rate? ____________________

1.3 Formulating Hypotheses

Now that you have made some observations of the beating heart, had some literary input, and recalled past experiences, it is time to formulate a hypothesis about the heart rate and exercise.

1. Think of an exercise a subject could do in the laboratory at a slow speed and a fast speed.
   Record your choice here: ____________________
2. After a class discussion and input from your instructor, the class will have the subject do the same type exercise for the same length of time during the experiment. Together the class will decide what is meant by a "slow speed" and "fast speed" exercise and the time limit for each. Three minutes of exercise is suggested. Record the class decision about type of slow/fast exercise to be done and for how long:

   Slow exercise/time: ____________________ ***Fast exercise/time:*** ____________________

3. Formulate a hypothesis here that tells how you expect the heart rate to be affected by the chosen slow and fast exercise per length of time:
   Hypothesis: The heart rate will ____________________

1.4 Performing an Experiment and Coming to a Conclusion

Variables

1. Variables are differences in your experimental procedure or results. Two variables of value are the experimental variable and the dependent variable. The **experimental variable** is what you purposely change to get your results. What is your experimental variable?

   ____

   The **dependent variables** are the variables that the hypothesis predicts will be influenced by the experimental variable. What will be a dependent variable in your experiment?

   ____

2. The experimenter tries to control any other variables that could affect the results. For example, if a scientist were doing an experiment with plants, s/he would always use the same type plant and the same type pots and would subject the plants to the same treatment under the same circumstances. **Controlled variables** are aspects of an experiment the scientist makes consistent to prevent them from affecting the results. Otherwise, the experiment has too many variables and the results might be due to differences in treatment other than the one being tested. What are some possible variables that should be controlled or they could affect the reliability of your results? ____________________

[1]Crouch, James E. 1985. *Functional Human Anatomy,* 4th ed. Philadelphia. Lea & Febiger, p. 399.

3. Subject variability is often a concern when doing experiments with humans. Just as all the people signed up for an exercise class are different (Fig. 1.3), so are the people taking a human biology course. Then, too, we have to consider that some subjects are used to doing exercise and some are not. The hope is that if subjects are randomly chosen, and the results are averaged, then differences between subjects may cancel themselves out. We also have to realize that in general the larger the sampling size, the more confidence there is in the results. The Framingham heart study, on which much of our common knowledge regarding heart disease is now based, began with 5,209 subjects. The best we can do under the present circumstances is to pool our results as a class while realizing that even this sample size is probably inadequate. Skill of the experimenter is another variable that we have to consider, and all students should practice taking the pulse rate in the manner described next.

Figure 1.3 An aerobics class.
Just as the individuals taking spinning class are varied in their appearance, so participants in a scientific study are different in their physiological characteristics.

Experimental Design

1. It's time to consider what the experimenter is to do. The experimenters will take pulse rate measurements at the wrist using the radial artery, which pulses every time the heart beats.
   a. **Location of the radial artery.** The radial artery is located on the thumb side of the wrist, a little below the base of the thumb. If you are right-handed, use the tips of your index and third finger of your right hand to find the radial artery on your left wrist. (Left-handed people should do the opposite.) You should be able to feel the pulsing of the artery.
   b. **Taking a measurement.** Choose a timepiece to use, preferably a stopwatch, but if this is not available, use a watch with a second hand as a way to tell the number of seconds. To take a measurement, use only the tips of your index and third fingers. Holding gently, begin counting the pulse with "zero," and count for 10 seconds. Multiply by six, and the answer is your pulse rate per minute. Record your results:

      counts per 10 seconds ________________ × 6 = ________________ pulse rate per minute.
   c. When taking the pulse rate after exercising, begin as soon as the subject stops. If you let time pass, the pulse rate will slow, and your count will not be accurate.
2. As mentioned, every experiment needs to have a control. For this experiment, (a) the control should be a student who does not experience exercise. However, (b) the experiment could use a student who does exercise as long as the student fully recovers in between slow and fast exercise. Why is the first method better than the second method of providing a control? ____

   ____

   ____

   To generate enough data, the experiment will use the second method and assume that pulse rate before slow exercise and again, after recovery and before fast exercise, is the control sample (see Table 1.2).

Table 1.1 Physical Fitness Evaluation Chart for the Canadian Home Fitness Test

	Ten-Second Pulse Rate	
Age (yr)	**After First 3 Minutes of Exercise**	**After Second 3 Minutes of Exercise**
15–19	If 30 or more, stop exercising.	If 26 or less, you have the recommended personal fitness level.
20–29	If 29 or more, stop exercising.	If 25 or less, you have the recommended personal fitness level.
30–39	If 28 or more, stop exercising.	If 24 or less, you have the recommended personal fitness level.
40–49	If 26 or more, stop exercising.	If 23 or less, you have the recommended personal fitness level.
50–59	If 25 or more, stop exercising.	If 22 or less, you have the recommended personal fitness level.
60–69	If 24 or more, stop exercising.	If 22 or less, you have the recommended personal fitness level.

3. Table 1.1 gives the 10-second pulse rate that tells various ages when they should stop exercising. Use the following grid to create a line graph using the data from the first and second columns of Table 1.1. *Label the x-axis (horizontal) "Age of subject." Use only the first age of each category. Label the y-axis (vertical) "Maximum 10 sec pulse rate/3 mins exercise."*

The graph tells us that as age __________, the acceptable 10-second pulse rate after 3 minutes of exercise __________.

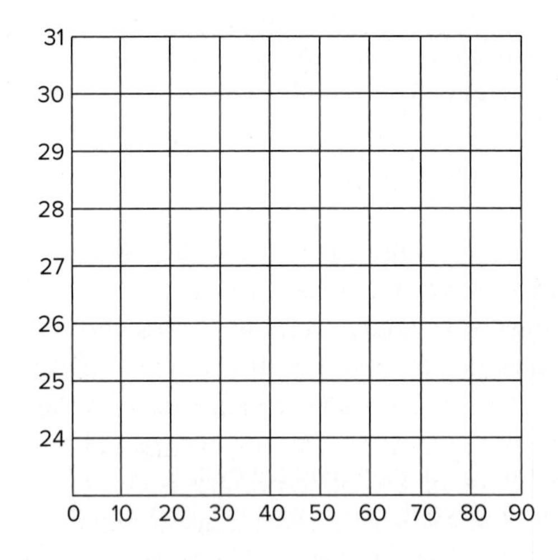

4. As scientists do, before you begin your experiment, summarize your experimental design. Include information concerning the subjects; the experimental variable (what the subjects will do); and how the results, called the data, will be gathered.

The Experiment

1. The members of your class will work in groups. Each group will contain at least one experimenter and one subject. You may switch off and be the experimenter for one round of the experiment and the subject for another round of the experiment. Record your results in Table 1.2

Table 1.2 Increase in Heart Rate Following Slow and Fast Exercise per Group

	Pulse Rate/min Before Exercise	Pulse Rate/min Slow Exercise	Pulse Rate/min Before Exercise	Pulse Rate/min Fast Exercise
First subject				
Second subject				

2. Your instructor or designated members of the class will collect the data, average the results, and tell you what data to record in Table 1.3.

Table 1.3 Increase in Heart Rate Following Slow and Fast Exercise per Class

	Pulse Rate/min Before Exercise	Pulse Rate/min Slow Exercise	Pulse Rate/min Before Exercise	Pulse Rate/min Fast Exercise
Class average				

Conclusion: Heart Rate Following Exercise

- Did the results of your experiment support the hypothesis, per your group? ______________________
  per class? ________________ Why do you say so? ______________________________________
  ____

Application for Daily Living

The Stress Test

A high pulse rate could possibly be an indication of heart irregularities. But a cardiac stress test, which examines a series of ECGs as the patient exercises (Fig. 1.4), is more reliable than taking the pulse rate after exercise. An abnormal ECG, which measures the electrical activity of the heart, is suggestive of an impending heart attack. One out of every five deaths in the United States is due to a heart attack, and there are over 1 million new and recurrent heart attacks per year. Heart attack is a fairly common occurrence, so it is often a good idea to determine the chances of having one.

During a stress test, the level of exercise is increased in 3-minute stages of ever-increasing incline and speed. The American Heart Association recommends a stress test as the first choice for patients with medium risk of coronary heart disease based on risk factors, such as smoking and family history of cardiac disorders. Over the past 20 years, more modern methods have been devised to test whether the heart is functioning properly, but they are more expensive and more invasive than a stress test.

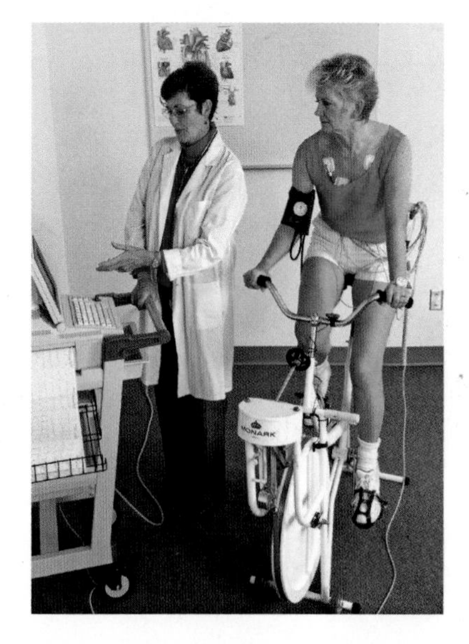

Figure 1.4 Cardiac stress test.

Human Laboratory Review 1

________________ **1.** Which process involves making observations, formulating hypotheses, doing experiments, and coming to a conclusion?

________________ **2.** What is a tentative explanation of observed phenomena?

________________ **3.** What is performed to test a hypothesis?

________________ **4.** What step in the scientific method follows experiments and observations?

________________ **5.** What do you call a sample that goes through all the steps of an experiment, but is not exposed to the experimental variable?

________________ **6.** What do you call the information scientists collect when doing experiments and making observations?

________________ **7.** What is an encompassing conclusion based on many individual conclusions?

Indicate whether statements 8 and 9 are hypotheses, conclusions, or scientific theories.

________________ **8.** The data show that vaccines protect people from disease.

________________ **9.** All living things are made of cells.

________________ **10.** Where is heart rate measured on the body?

________________ **11.** Heart rate is typically represented as beats per ______.

________________ **12.** Which variable is purposely changed in an experiment?

________________ **13.** Which variable is predicted by the hypothesis?

________________ **14.** How does exercise affect heart rate?

Thought Questions

15. Why is a theory more comprehensive than a conclusion?

16. Why is it important to have a control for an experiment?

17. Why is it important to have many subjects in an experiment?

LABORATORY

2

Measuring with Metric

Learning Outcomes

2.1 Length

- Compare and contrast the metric units for length: meter (m), centimeter (cm), millimeter (mm), micrometer (μm), and nanometer (nm).
- Know the abbreviation for each of these units.
- Convert the metric units for length from one type of unit to another.

Prelab Question: Is nanometer, centimeter, or micrometer the larger measurement unit for length?

2.2 Weight

- Compare and contrast the metric units for weight: kilogram (kg), gram (g), and milligram (mg).
- Know the abbreviation for each of these units.
- Convert the metric units for weight from one type of unit to another.

Prelab Question: Would you expect the weight of a penny to be given in terms of kilograms or grams?

2.3 Volume

- Compare and contrast the metric units for volume: liter (l) and milliliter (ml).
- Know the abbreviation for each of these units.
- Convert the metric units for volume from one type of unit to another.

Prelab Question: What units of measurement appear on a graduated cylinder?

2.4 Temperature

- Compare and contrast the Fahrenheit (F) and Celsius (C) temperature scales.
- Know the abbreviation for each of these units.
- Convert one type of temperature into another using a provided equation.

Prelab Question: Water boils at 100° in which scale?

Introduction

The metric system is the standard system of measurement in the sciences, including biology, chemistry, and physics (Fig. 2.1). It has tremendous advantages because all conversions, whether for volume, mass (weight), or length, are in units of ten. This base-ten system is similar to our monetary system, in which 10 cents equals a dime, 10 dimes equals a dollar, and so on. In this laboratory, you will get experience making measurements of length, volume, mass, and temperature.

Figure 2.1
The metric system is the system of measurement used in scientific laboratories.

2.1 Length

Metric units of length measurement include the **meter (m), centimeter (cm), millimeter (mm), micrometer (μm),** and **nanometer (nm)** (Table 2.1). The prefixes milli- (10^{-3}), micro- (10^{-6}), and nano (10^{-9}) are used with length, weight, and volume.

Table 2.1 Metric Units of Length Measurement

Unit	Meters	Centimeters	Millimeters	Relative Size
Meter (m)	1 m	100 cm	1,000 mm	Largest
Centimeter (cm)	0.01 (10^{-2}) m	1 cm	10 mm	
Millimeter (mm)	0.001 (10^{-3}) m	0.1 cm	1.0 mm	
Micrometer (μm)	0.000001 (10^{-6}) m	0.0001 (10^{-4}) cm	0.001 (10^{-3}) mm	
Nanometer (nm)	0.000000001 (10^{-9}) m	0.0000001 (10^{-7}) cm	0.000001 (10^{-6}) mm	Smallest

Experimental Procedure: Length

1. Obtain a small ruler marked in centimeters and millimeters. How many centimeters are represented? __________ One centimeter equals how many millimeters? __________ To express the size of small objects, such as cell contents, biologists use even smaller units of the metric system than those on the ruler. These units are the micrometer (μm) and the nanometer (nm).

   According to Table 2.1, 1 μm = __________ mm, and 1 nm = __________ mm.

   Therefore, 1 mm = __________ μm = __________ nm.

2. Measure the diameter (a line passing through the center with endpoints touching the circle) of the circle shown below to the nearest millimeter. This circle's diameter is

   __________ mm = __________ μm = __________ nm.

For example, to convert mm to μm:

$$\_\_\_\_\_\_\_\_ \text{ mm} \times \frac{1{,}000\ \mu\text{m}}{\text{mm}} = \_\_\_\_\_\_\_\_\ \mu\text{m}$$

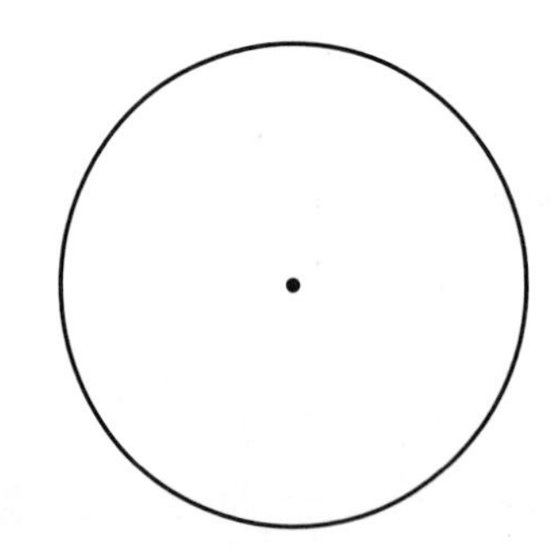

3. Obtain a meter stick. On one side, find the numbers 1 through 39, which denote inches. One meter equals 39.37 inches; therefore, 1 meter is roughly equivalent to 1 yard. Turn the meter stick over, and observe the metric subdivisions. How many centimeters are in a meter? __________ How many millimeters are in a meter? __________ The prefix *milli* means __________.
4. Use the meter stick and the method shown in Figure 2.2 to measure the length of two long bones from a disarticulated human skeleton. Lay the meter stick flat on the lab table. Place a long bone next to the meter stick between two pieces of cardboard (each about 10 cm × 30 cm), held upright at right angles to the stick. The narrow end of each piece of cardboard should touch the meter stick. The length between the cards is the length of the bone in centimeters. For example, if the bone measures from the 22 cm mark to the 50 cm mark, the length of the bone is __________ cm. If the bone measures from the 22 cm mark to midway between the 50 cm and 51 cm marks, its length is __________ mm, or __________ cm.
5. Record the length of two bones. First bone: __________ cm =__________ mm. Second bone: __________ cm = __________ mm.

Figure 2.2 Measurement of a long bone.
How to measure a long bone using a meter stick.

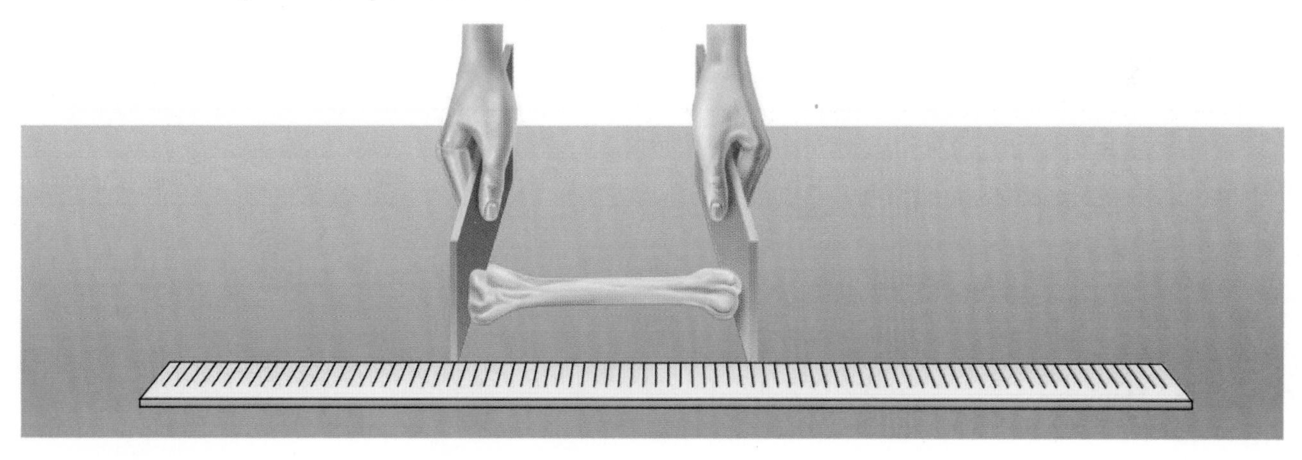

2.2 Weight

Two metric units of weight are the **gram (g)** and the **milligram (mg).** A paper clip weighs about 1 g, which equals 1,000 mg. 2 g = __________ mg; 0.2 g = __________ mg; and 2 mg = __________ g.

Experimental Procedure: Weight

1. Use a balance scale to measure the weight of a wooden block small enough to hold in the palm of your hand.
2. Measure the weight of the block to the tenth of a gram. The weight of the wooden block is __________ g = __________ mg.
3. Measure the weight of an item small enough to fit inside the opening of a 50 ml graduated cylinder. The item, a(n) __________, is __________ g = __________ mg.

2.3 Volume

Two metric units of volume are the **liter (l)** and the **milliliter (ml).** One liter = 1,000 ml.

Experimental Procedure: Volume

1. Volume measurements can be related to those of length. For example, use a millimeter ruler to measure the wooden block used in the previous Experimental Procedure to get its length, width, and depth.

   length = __________ cm; width = __________ cm; depth = __________ cm
   The volume, or space, occupied by the wooden block can be expressed in cubic centimeters (cc or cm^3) by multiplying: length × width × depth = __________ cm^3. For purposes of this Experimental Procedure, 1 cubic centimeter equals 1 milliliter; therefore, the wooden block has a volume of __________ ml.
2. In the biology laboratory, liquid volume is usually measured directly in liters or milliliters with appropriate measuring devices. For example, use a 50 ml graduated cylinder to add 20 ml of water to a test tube. First, fill the graduated cylinder to the 20 ml mark. To do this properly, you have to make sure that the lowest margin of the water level, or the **meniscus** (Fig. 2.3), is at the 20 ml mark. Place your eye directly parallel to the level of the meniscus, and add water until the meniscus is at the 20 ml mark. (Having a dropper bottle filled with water on hand can help you do this.) A large, blank, white index card held behind the cylinder can also help you see the scale more clearly. Now pour the 20 ml of water into the test tube.
3. Hypothesize how you could find the total volume of the test tube. ______________________________

   ____

   ____

   What is the test tube's total volume? ______________________________

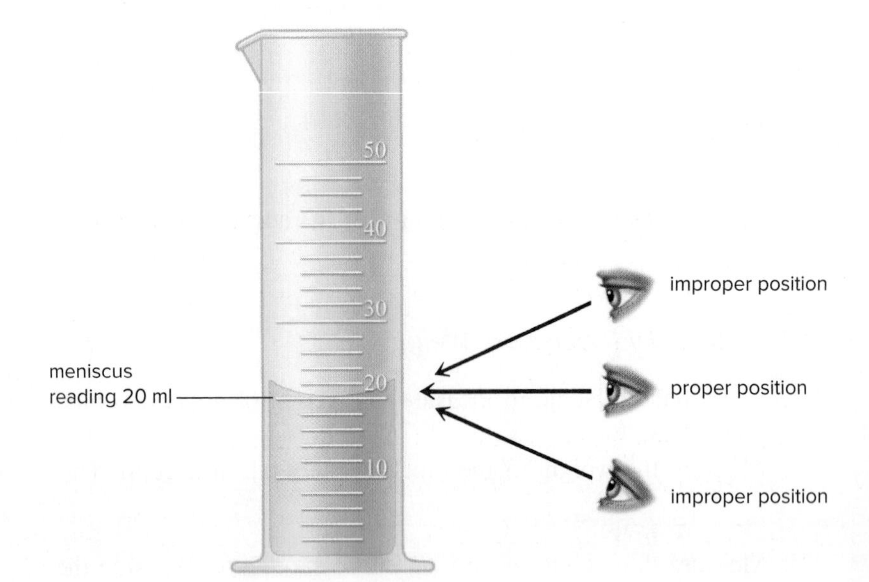

Figure 2.3 Meniscus.
The proper way to view the meniscus.

4. Fill a 50 ml graduated cylinder with water to about the 20 ml mark. Hypothesize how you could use this setup to calculate the volume of an object. ____________________

____

Now perform the operation you suggested. The object, __________, has a volume of __________ ml.

5. Hypothesize how you could determine how many drops from the pipette of the dropper bottle equal 1 ml. ____________________

Now perform the operation you suggested. How many drops from the pipette of the dropper bottle equal 1 ml? ____________________

6. Some pipettes are graduated and can be filled to a certain level as a way to measure volume directly. Your instructor will demonstrate this. Are pipettes customarily used to measure large or small volumes? _

2.4 Temperature

There are two temperature scales: the **Fahrenheit (F)** and **Celsius (centigrade, C)** scales (Fig. 2.4). Scientists use the Celsius scale.

Experimental Procedure: Temperature

1. Study the two scales in Figure 2.4, and complete the following information:

   a. Water freezes at either __________ °F or __________ °C.

   b. Water boils at either __________ °F or __________ °C.

2. To convert from the Fahrenheit to the Celsius scale, use the following equation:

$$°C = (°F - 32°)/1.8$$

or

$$°F = (1.8°C) + 32$$

Human body temperature of 98°F is what temperature on the Celsius scale? __________

3. Record any two of the following temperatures in your lab environment. In each case, allow the end bulb of the Celsius thermometer to remain in or on the sample for one minute.

   Room temperature = __________ °C

   Surface of your skin = __________ °C

   Cold tap water in a 50 ml beaker = __________ °C

   Hot tap water in a 50 ml beaker = __________ °C

   Ice water = __________ °C

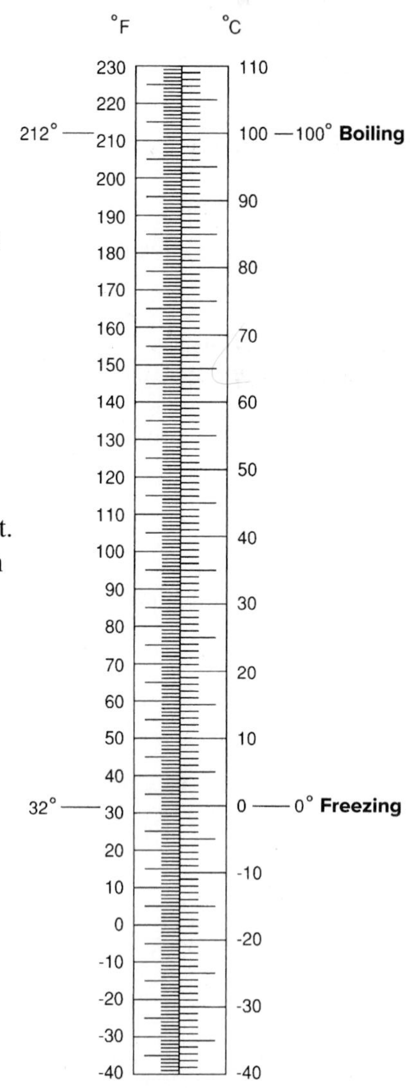

Figure 2.4 Temperature scales.
The Fahrenheit (°F) scale is on the left, and the Celsius (°C) scale is on the right.

Laboratory Review 2

_______________ **1.** What system of measurement is used in science?

_______________ **2.** Which types of measurements were examined in this lab?

_______________ **3.** What is the base unit for length?

_______________ **4.** 19 mm equals how many cm?

_______________ **5.** 880 mm equals how many m?

_______________ **6.** What instruments are used to observe objects smaller than a millimeter?

_______________ **7.** What is the base unit for mass?

_______________ **8.** What instrument is used to measure mass?

_______________ **9.** 2,700 mg equals how many grams?

_______________ **10.** One ml is equal to how many cubic centimeters?

_______________ **11.** 3.4 l equals how many ml?

_______________ **12.** To properly measure 20 ml of water, what must be at the 20 ml mark of the graduated cylinder?

_______________ **13.** Which temperature scale is used in science?

_______________ **14.** 22°C equals how many degrees F?

Thought Questions

15. Describe the advantages of using the metric system you discovered during this lab.

16. Explain how you would measure the volume of a solid object, such as a rock.

17. Why is it advantageous to use a standard measurement system in all sciences?

3

Microscopy

Learning Outcomes

3.1 Microscopy

- Describe similarities and differences between the stereomicroscope (dissecting microscope), the compound light microscope, and the electron microscope.

Prelab Question: Greater magnification enlarges the subject, but what is another equally important attribute of a microscope? Explain.

3.2 Stereomicroscope (Dissecting Microscope)

- Identify the parts and tell how to focus the stereomicroscope.

Prelab Question: If you wanted to examine the external features of a fly, you might decide to use a stereomicroscope. Why?

3.3 Use of the Compound Light Microscope

- Identify and give the function of the basic parts of the compound light microscope.
- List, in proper order, the steps for bringing an object into focus with the compound light microscope.
- Describe how the image is inverted by the compound light microscope.
- Calculate the total magnification and the diameter of field for both low- and high-power lens systems.
- Explain how a slide of colored threads provides information on the depth of field.

Prelab Question: What's the difference between low power and high power when using the compound light microscope?

3.4 Microscopic Observations

- Name and describe the three kinds of cells studied in this exercise.
- State two differences between human epithelial cells and onion epidermal cells.
- Examine a wet mount of *Euglena* and pond water. Contrast the organisms observed in pond water.

Prelab Question: What is there about live *Euglena* that makes them difficult to study microscopically?

Application for Daily Living: **Microscopic Diagnoses**

Introduction

This laboratory examines the features, functions, and use of the compound light microscope and the stereomicroscope (dissecting microscope). Transmission and scanning electron microscopes are explained, and micrographs produced using these microscopes appear throughout this lab manual. The stereomicroscope and the scanning electron microscope view the surface and/or the three-dimensional structure of an object. The compound light microscope and the transmission electron microscope can view only extremely thin sections of a specimen. If a subject was sectioned lengthwise for viewing, the interior of the projections at the top of the cell, called cilia, would appear in the micrograph. A lengthwise cut through any type of specimen is called a **longitudinal section (ls).** On the other hand, if the subject in Figure 3.1 was sectioned crosswise below the area of the cilia, you would see other portions of the interior of the subject. A crosswise cut through any type of specimen is called a **cross section (cs).**

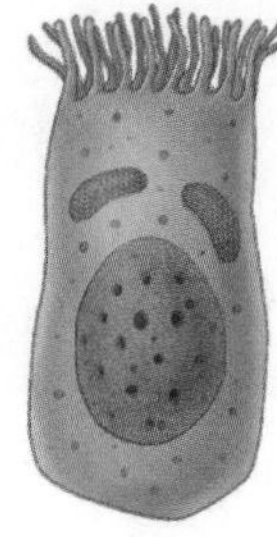

a. The cell

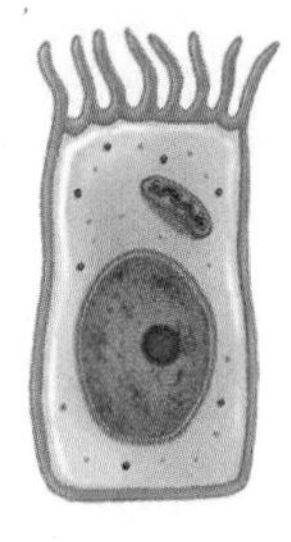

b. Longitudinal section

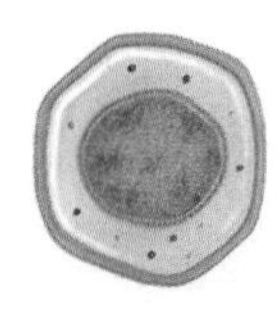

c. Cross section

Figure 3.1 Longitudinal and cross sections. **a.** Transparent view of a cell. **b.** A longitudinal section would show the cilia at the top of the cell. **c.** A cross section shows only the interior where the cut is made.

3.1 Microscopy

Because biological objects can be very small, we often use a microscope to view them. Many kinds of instruments, ranging from the hand lens to the electron microscope, are effective magnifying devices. A short description of two kinds of light microscopes and two kinds of electron microscopes follows.

Light Microscopes

Light microscopes use light rays passing through lenses to magnify the object. The **stereomicroscope (dissecting microscope)** is designed to study entire objects in three dimensions at low magnification. The **compound light microscope** is used for examining small or thinly sliced sections of objects under higher magnification than that of the stereomicroscope. The term **compound** refers to the use of two sets of lenses: the ocular lenses located near the eyes and the objective lenses located near the object. Illumination is from below, and visible light passes through clear portions but does not pass through opaque portions. To improve contrast, the microscopist uses stains or dyes that bind to cellular structures and absorb light. Photomicrographs, also called light micrographs, are images produced by a compound light microscope (Fig. 3.2*a*).

Figure 3.2 Comparative micrographs of a lymphocyte.
A lymphocyte is a type of white blood cell. **a.** A photomicrograph (light micrograph) of a lymphocyte shows less detail than a **(b)** transmission electron micrograph (TEM). **c.** A scanning electron micrograph (SEM) of a lymphocyte shows the cell surface in three dimensions.
(a) © Michael Ross/Science Source; (b) © CNRI/SPL/Science Source; (c) © Steve Gschmeissner/Science Source

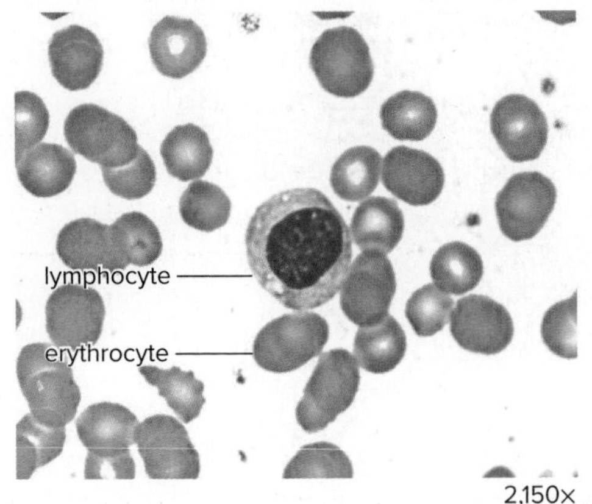

a. Photomicrograph or light micrograph (LM)

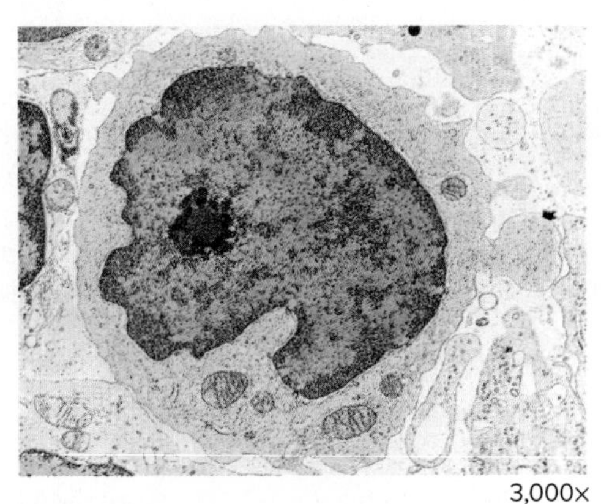

b. Transmission electron micrograph (TEM)

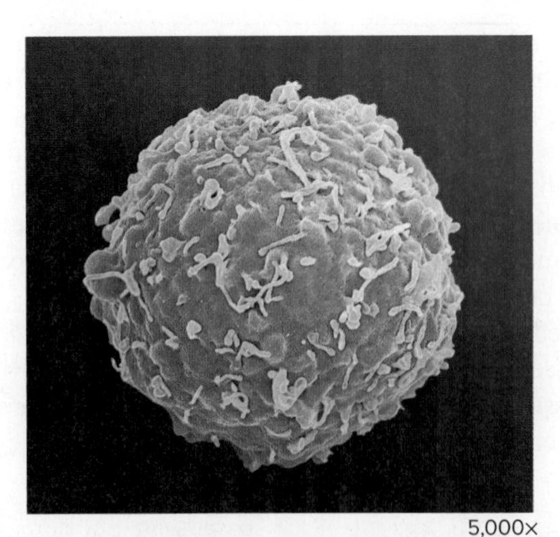

c. Scanning electron micrograph (SEM)

Electron Microscopes

Electron microscopes use beams of electrons to magnify the object. The beams are focused on a photographic plate by means of electromagnets. The **transmission electron microscope** is analogous to the compound light microscope. The object is ultra-thinly sliced and treated with heavy metal salts to improve contrast. Figure 3.2*b* is a micrograph produced by this type of microscope. The **scanning electron microscope** is analogous to the dissecting light microscope. It gives an image of the surface and dimensions of an object, as is apparent from the scanning electron micrograph in Figure 3.2*c*.

The micrographs in Figure 3.2 demonstrate that an object is magnified more with an electron microscope than with a compound light microscope. The difference between these two types of microscopes, however, is not simply a matter of magnification; it is also the electron microscope's ability to show detail. The electron microscope has greater resolving power. **Resolution** is the minimum distance between two objects at which they can still be seen, or resolved, as two separate objects. The use of high-energy electrons rather than light gives electron microscopes a much greater resolving power since two objects that are much closer together can still be distinguished as separate points. Table 3.1 lists several other differences between the compound light microscope and the transmission electron microscope.

Table 3.1 Comparison of the Compound Light Microscope and the Transmission Electron Microscope

Compound Light Microscope	Transmission Electron Microscope
1. Glass lenses	1. Electromagnetic lenses
2. Illumination by visible light	2. Illumination due to beam of electrons
3. Resolution $\cong$ 200 nm	3. Resolution $\cong$ 0.1 nm
4. Magnifies to 2,000×	4. Magnifies to 1,000,000×
5. Costs up to tens of thousands of dollars	5. Costs up to hundreds of thousands of dollars

Conclusions: Microscopy

- Which two types of microscopes view the surface of an object? ______________________
- Which two types of microscopes view objects that have been sliced and treated to improve contrast? ______________________
- Of the microscopes just mentioned, which one resolves the greater amount of detail?

____

3.2 Stereomicroscope (Dissecting Microscope)

The **stereomicroscope (dissecting microscope,** Fig. 3.3) allows you to view objects in three dimensions at low magnifications. It is used to study entire small organisms, any object requiring lower magnification, and opaque objects that can be viewed only by reflected light. It is called a stereomicroscope because it produces a three-dimensional image.

Identifying the Parts

After your instructor has explained how to carry a microscope, obtain a stereomicroscope and a separate illuminator, if necessary, from the storage area. Place it securely on the table. Plug in the power cord, and turn on the illuminator. There is a wide variety of stereomicroscope styles, and your instructor will discuss the specific style(s) available to you. Regardless of style, the following features should be present:

1. **Binocular head:** Holds two eyepiece lenses that move to accommodate for the various distances between different individuals' eyes.
2. **Eyepiece lenses:** The two lenses located on the binocular head. What is the magnification of your eyepieces? ________ Some models have one **independent focusing eyepiece** with a knurled knob to allow independent adjustment of each eye. The nonadjustable eyepiece is called the **fixed eyepiece.**
3. **Focusing knob:** A large, black or gray knob located on the arm; used for changing the focus of both eyepieces together.

Figure 3.3 Binocular dissecting microscope (stereomicroscope).
Label this stereomicroscope with the help of the text material.

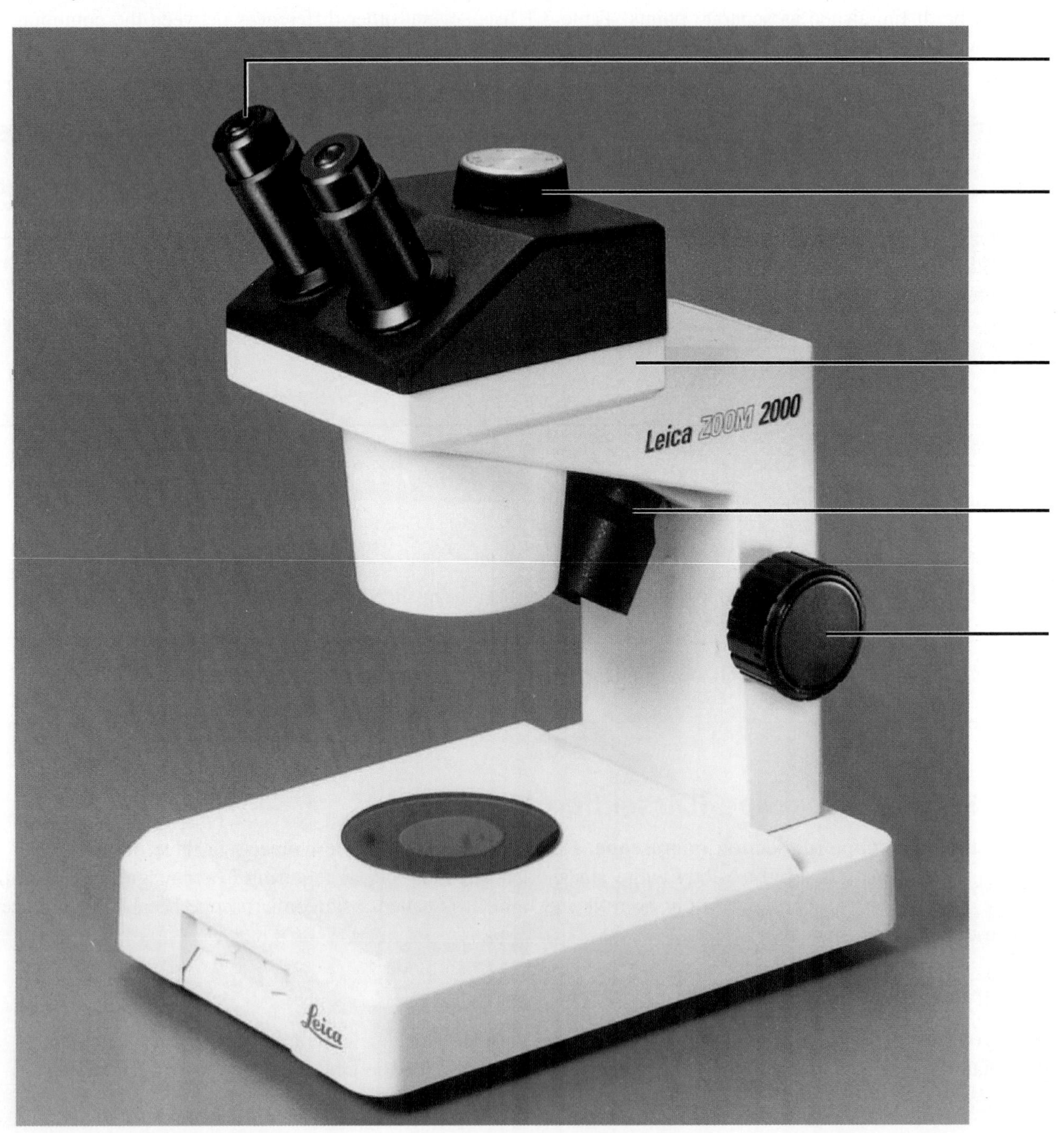

4. **Magnification changing knob:** A knob, often built into the binocular head, used to change magnification in both eyepieces simultaneously. This may be a **zoom** mechanism or a **rotating lens** mechanism of different powers that clicks into place.
5. **Illuminator:** Used to illuminate an object from above; may be built into the microscope or separate.

Locate each of these parts on your stereomicroscope, and label them on Figure 3.3.

Focusing the Stereomicroscope

1. In the center of the stage, place a plastomount that contains small organisms.
2. Adjust the distance between the eyepieces on the binocular head so that they comfortably fit the distance between your eyes. You should be able to see the object with both eyes as one three-dimensional image.
3. Use the focusing knob to bring the object into focus.
4. Does your microscope have an independent focusing eyepiece? __________ If so, use the focusing knob to bring the image in the fixed eyepiece into focus, while keeping the eye at the independent focusing eyepiece closed. Then adjust the independent focusing eyepiece so that the image is clear, while keeping the other eye closed. Is the image inverted? ____
5. Turn the magnification changing knob, and determine the kind of mechanism on your microscope. A zoom mechanism allows continuous viewing while changing the magnification. A rotating lens mechanism blocks the view of the object as the new lenses are rotated. Be sure to click each lens firmly into place. If you do not, the field will be only partially visible. What kind of mechanism is on your microscope? ____
6. Set the magnification changing knob on the lowest magnification. Sketch the object in the following circle as though this represents your entire field of view:

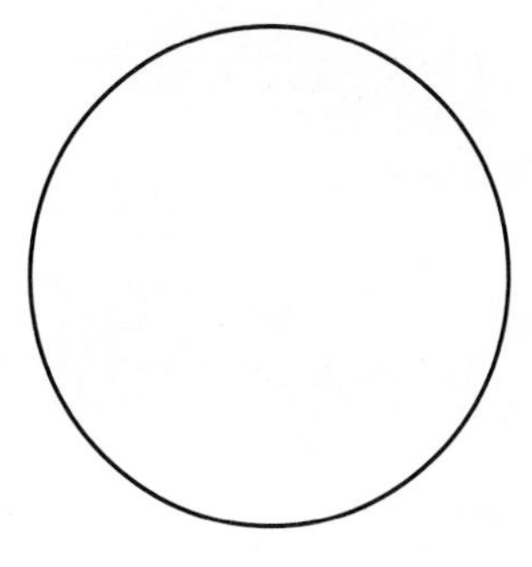

7. Rotate the magnification changing knob to the highest magnification. Draw another circle within the one provided to indicate the reduction of the field of view.
8. Experiment with various objects at various magnifications until you are comfortable with using the stereomicroscope.

3.3 Use of the Compound Light Microscope

As mentioned, the name **compound light microscope** indicates that it uses two sets of lenses and light to view an object. The two sets of lenses are the ocular lenses located near the eyes and the objective lenses located near the object. Illumination is from below, and the light passes through clear portions but does not pass through opaque portions. This microscope is used to examine small or thinly sliced sections of objects under higher magnification than would be possible with the stereomicroscope.

Identifying the Parts

Obtain a compound light microscope from the storage area, and place it securely on the table. *Identify the following parts on your microscope, and label them in Figure 3.4 with the help of the text material.*

Figure 3.4 Compound light microscope.
Compound light microscope with binocular head and mechanical stage. Label this microscope with the help of the text material.

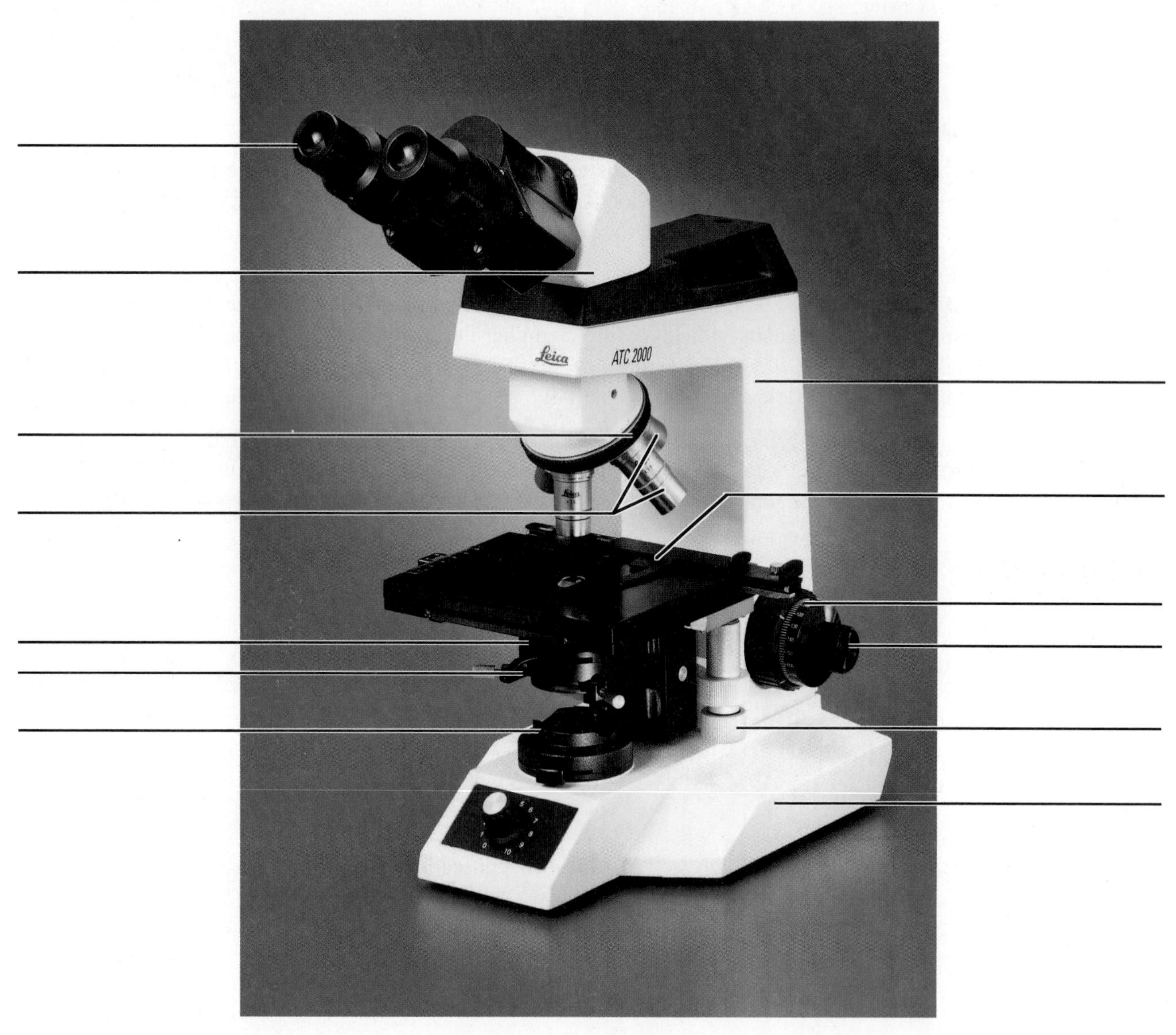

1. **Eyepieces** (ocular lenses): What is the magnifying power of the ocular lenses on your microscope? ________
2. **Viewing head:** Holds the ocular lenses.
3. **Arm:** Supports upper parts and provides carrying handle.
4. **Nosepiece:** Revolving device that holds objectives.
5. **Objectives** (objective lenses):
   a. **Scanning objective:** This is the shortest of the objective lenses and is used to scan the whole slide. The magnification is stamped on the housing of the lens. It is a number followed by an ×. What is the magnifying power of the scanning objective lens on your microscope? _______

b. **Low-power objective:** This lens is longer than the scanning objective lens and is used to view objects in greater detail. What is the magnifying power of the low-power objective lens on your microscope? __________

c. **Power objective:** If your microscope has three objective lenses, this lens will be the longest. It is used to view an object in even greater detail. What is the magnifying power of the high-power objective lens on your microscope? __________

d. **Oil immersion objective** (on microscopes with four objective lenses): Holds a 95× (to 100×) lens and is used in conjunction with immersion oil to view objects with the greatest magnification. Does your microscope have an oil immersion objective? __________ If this lens is available, your instructor will discuss its use when the lens is needed.

6. **Stage:** Platform that holds and supports microscope slides. A mechanical stage is a movable stage that aids in the accurate positioning of the slide. Does your microscope have a mechanical stage? __________
   a. **Stage clips:** Clips that hold a slide in place on the stage.
   b. **Mechanical stage control knobs:** Two knobs that control forward/reverse movement and right/left movement, respectively.

7. **Coarse-adjustment knob:** Knob used to bring object into approximate focus; used only with low-power objective.
8. **Fine-adjustment knob:** Knob used to bring object into final focus.
9. **Condenser:** Lens system below the stage used to focus the beam of light on the object being viewed.
   a. **Diaphragm** or **diaphragm control lever:** Lever that controls the amount of light passing through the condenser.
10. **Light source:** An attached lamp that directs a beam of light up through the object.
11. **Base:** The flat surface of the microscope that rests on the table.

Rules for Microscope Use

Observe the following rules for using a microscope:

1. The lowest power objective (scanning or low) should be in position at both the beginning and the end of microscope use.
2. Use only lens paper for cleaning lenses.
3. Do not tilt the microscope because the eyepieces could fall out, or wet mounts could be ruined.
4. Keep the stage clean and dry to prevent rust and corrosion.
5. Do not remove parts of the microscope.
6. Keep the microscope dust-free by covering it after use.
7. Report any malfunctions.

Focusing the Compound Light Microscope—Lowest Power

1. Turn the nosepiece so that the *lowest* power objective on your microscope is in straight alignment over the stage.
2. Always begin focusing with the *lowest* power objective on your microscope (4× [scanning] or 10× [low power]).
3. With the coarse-adjustment knob, lower the stage (or raise the objectives) until it stops.

4. Place a slide of the letter *e* on the stage, and stabilize it with the clips. (If your microscope has a mechanical stage, pinch the spring of the slide arms on the stage, and insert the slide.) Center the *e* as best you can on the stage or use the two control knobs located below the stage (if your microscope has a mechanical stage) to center the *e*.
5. Again, be sure that the lowest-power objective is in place. Then, as you look from the side, decrease the distance between the stage and the tip of the objective lens until the lens comes to an automatic stop or is no closer than 3 mm above the slide.
6. While looking into the eyepiece, rotate the diaphragm (or diaphragm control lever) to give the maximum amount of light.
7. Using the coarse-adjustment knob, slowly increase the distance between the stage and the objective lens until the object—in this case, the letter *e*—comes into view, or focus.
8. Once the object is seen, you may need to adjust the amount of light. To increase or decrease the contrast, rotate the diaphragm slightly.
9. Use the fine-adjustment knob to sharpen the focus if necessary.
10. Practice having both eyes open when looking through the eyepiece, as this greatly reduces eyestrain.

Inversion

Inversion refers to the fact that a microscopic image is upside down and reversed.

Observation: Inversion

1. Draw the letter *e* as it appears on the slide (with the unaided eye, not looking through the eyepiece). ________
2. Draw the letter *e* as it appears when you look through the eyepiece. ________
3. What differences do you notice? ________
4. Move the slide to the right. Which way does the image appear to move? ________
5. Move the slide toward you. Which way does the image appear to move? ________

Focusing the Compound Light Microscope—Higher Powers

Compound light microscopes are **parfocal;** that is, once the object is in focus with the lowest power, it should also be almost in focus with the higher power.

1. Bring the object into focus under the lowest power by following the instructions in the previous section.
2. Make sure that the letter *e* is centered in the field of the lowest objective.
3. Move to the next higher objective (low power [10×] or high power [40×]) by turning the nosepiece until you hear it click into place. Do not change the focus; parfocal microscope objectives will not "hit" normal slides when changing the focus if the lowest objective is initially in focus. (If you are on low power [10×], proceed to high power [40×] before going on to step 4.)
4. If any adjustment is needed, use only the *fine*-adjustment knob. (*Note:* Always use only the fine-adjustment knob with high power, and do not use the coarse-adjustment knob.)
5. On a drawing of the letter *e* to the right, *draw a circle around the portion of the letter that you are now seeing with high-power magnification.* The letter *e* will not disappear because your microscope is parcentric (the focus remains near the center). ________
6. When you have finished your observations of this slide (or any slide), rotate the nosepiece until the lowest-power objective clicks into place, and then remove the slide.

Total Magnification

Total magnification is calculated by multiplying the magnification of the ocular lens (eyepiece) by the magnification of the objective lens. The magnification of a lens is imprinted on the lens casing.

Observation: Total Magnification

Calculate total magnification figures for your microscope, and record your findings in Table 3.2.

Table 3.2 Total Magnification

Objective	Ocular Lens	Objective Lens	Total Magnification
Scanning power (if present)			
Low power			
High power			
Oil immersion (if present)			

Field of View

A microscope's **field of view** is the circle visible through the lenses. The **diameter of field** is the length of the field from one edge to the other.

Observation: Field of View

Low-Power (10×) Diameter of Field

1. Place a clear plastic ruler across the stage so that the edge of the ruler is visible as a horizontal line along the diameter of the low-power (not scanning) field. Be sure that you are looking at the millimeter side of the ruler.
2. Estimate the number of millimeters, to tenths, that you see along the field: ______________ mm. (*Hint:* Start by placing any millimeter marker at the edge of the field.) Convert the observed number of millimeters to micrometers: __________ µm. This is the **low-power diameter of field (LPD)** for your microscope in micrometers.

High-Power (40×) Diameter of Field

1. To compute the **high-power diameter of field (HPD),** substitute these data into the formula given:
   a. LPD = low-power diameter of field (in micrometers) = ______________
   b. LPM = low-power total magnification (from Table 3.2) = ______________
   c. HPM = high-power total magnification (from Table 3.2) = ______________

Example: If the diameter of field is about 2 mm, then the LPD is 2,000 µm. Using the LPM and HPM values from Table 3.2, the HPD would be 500 µm.

$$HPD = LPD \times \frac{LPM}{HPM}$$

$$HPD = (\quad\quad) \times \frac{(\quad\quad)}{(\quad\quad)} = \underline{\quad\quad\quad}$$

Conclusions: Total Magnification and Field of View

- Does low power or high power have a larger field of view (one that allows you to see more of the object)? ______________________
- Which has a smaller field but magnifies to a greater extent? ______________________
- To locate small objects on a slide, first find them under __________ ; then place them in the center of the field before rotating to __________.

Depth of Field

When viewing an object on a slide under high power, the **depth of field** (Fig. 3.5) is the area—from top to bottom—that comes into focus while slowly focusing up and down with the microscope's fine-adjustment knob.

Observation: Depth of Field

1. Obtain a prepared slide with three or four colored threads mounted together, or prepare a wet-mount slide with three or four crossing threads or hairs of different colors. (Directions for preparing a wet mount are given in section 3.4.)
2. With low power, find a point where the threads or hairs cross. Slowly focus up and down. Notice that when one thread or hair is in focus, the others seem blurred. Remember, as the stage moves upward (or the objectives move downward), objects on top come into focus first. Determine the order of the threads or hairs, and complete Table 3.3.
3. Switch to high power, and notice that the depth of field is more shallow with high power than with low power. Focusing up and down with the fine-adjustment knob when viewing a slide with high power will give you an idea of the specimen's three-dimensional form. For example, viewing a number of sections from bottom to top allows reconstruction of the three-dimensional structure, as demonstrated in Figure 3.5.

Figure 3.5 Depth of field.
A demonstration of how focusing at depths 1, 2, and 3 would produce three different images (views) that could be used to reconstruct the original three-dimensional structure of the object.

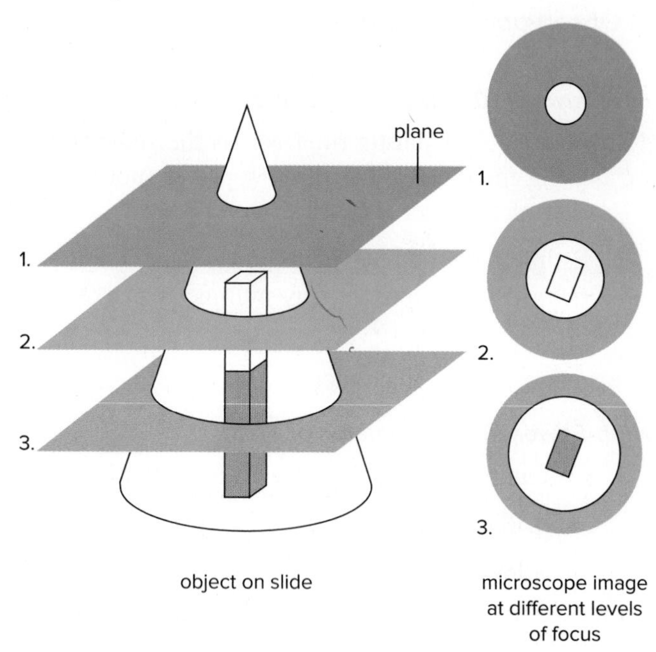

Table 3.3 Order of Threads (or Hairs)

Depth	Thread (or Hair) Color
Top	
Middle	
Bottom	

3.4 Microscopic Observations

When a specimen is prepared for observation, the object should always be viewed as a **wet mount.** A wet mount is prepared by placing a drop of liquid on a slide or, if the material is dry, by placing it directly on the slide and adding a drop of water or stain. The mount is then covered with a coverslip, as illustrated in Figure 3.6. Dry the bottom of your slide before placing it on the stage.

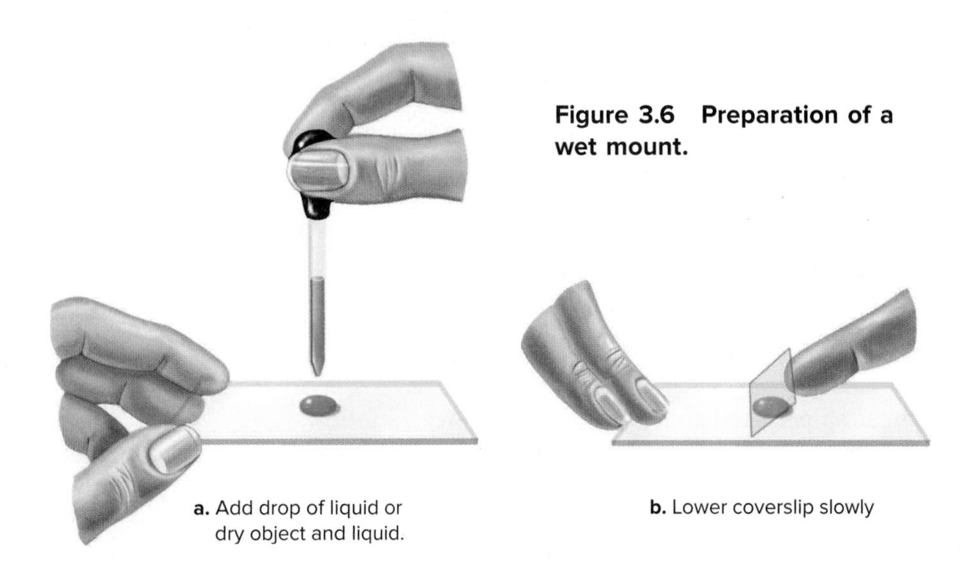

Figure 3.6 Preparation of a wet mount.

Onion Epidermal Cells

Epidermal cells cover the surfaces of plant organs, such as leaves. The bulb of an onion is made up of fleshy leaves.

Observation: Onion Epidermal Cells

1. With a scalpel, strip a small, thin, transparent layer of cells from the inside of a fresh onion leaf.
2. Place it gently on a clean, dry slide, and add a drop of iodine solution (or methylene blue). Cover with a coverslip.
3. Observe under the microscope.
4. Locate the cell wall and the nucleus. *Label Figure 3.7.*
5. Count the number of onion cells that line up end to end in a single line across the diameter of the high-power (40×) field.__________
   Based on what you learned in section 2.4 about measuring diameter of field, what is the high-power diameter of field (HPD) in micrometers? __________ µm

Figure 3.7 Onion epidermal cells. Label the cell wall and the nucleus.

© Ted Kinsman/Science Source

1. ____________________

2. ____________________

100×

Calculate the length of each onion cell:

(HPD ÷ number of cells): __________ μm

6. Wash and reuse this slide for the next exercise (Human Epithelial Cells).

Human Epithelial Cells

Epithelial cells (Fig. 3.8) cover the body's surface and line its cavities.

Observation: Human Epithelial Cells

1. Obtain a prepared slide, or make your own as follows:
   a. Obtain a prepackaged flat toothpick (or sanitize one with alcohol or alcohol swabs).
   b. Gently scrape the inside of your cheek with the toothpick, and place the scrapings on a clean, dry slide. Discard used toothpicks in the biohazard waste container provided.
   c. Add a drop of very weak *methylene blue* or *iodine solution,* and cover with a coverslip.
2. Observe under the microscope.
3. Locate the nucleus (the central, round body), the cytoplasm, and the plasma membrane (outer cell boundary). *Label Figure 3.8.*
4. Because your epithelial slides are biohazardous, they must be disposed of as indicated by your instructor.
5. Note some obvious differences between the onion cells and the human cheek cells, and list them in Table 3.4.

Figure 3.8 Cheek epithelial cells.
Label the nucleus, the cytoplasm, and the plasma membrane.

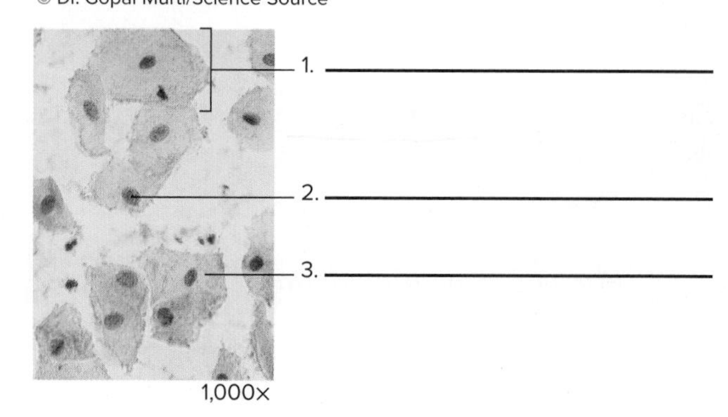

> ⚠ **Methylene blue** Avoid ingestion, inhalation, and contact with skin, eyes, and mucous membranes. If any should spill on your skin, wash the area with mild soap and water. Methylene blue will also stain clothing.

Table 3.4 Differences Between Onion Epidermal and Human Epithelial Cells

Differences	Onion Epidermal Cells	Human Epithelial Cells (Cheek)
Shape		
Orientation		
Boundary		

Euglena

Examination of *Euglena* (a unicellular organism with a flagellum to facilitate movement) will test your ability to observe objects with the microscope, to utilize depth of field, and to control illumination to heighten contrast.

Observation: *Euglena*

1. Make a wet mount of *Euglena* by using a drop of a *Euglena* culture and adding a drop of Protoslo® (methyl cellulose solution) onto a slide. The Protoslo® slows the organism's swimming.
2. Mix thoroughly with a toothpick, and add a coverslip.
3. Scan the slide for *Euglena:* Start at the upper left-hand corner, and move the slide forward and back as you work across the slide from left to right. The *Euglena* may be at the edge of the slide because they show an aversion to Protoslo®. Use Figure 3.9 to help identify the structural details of *Euglena.*

Figure 3.9 ***Euglena.***
Euglena is a unicellular, flagellated organism.

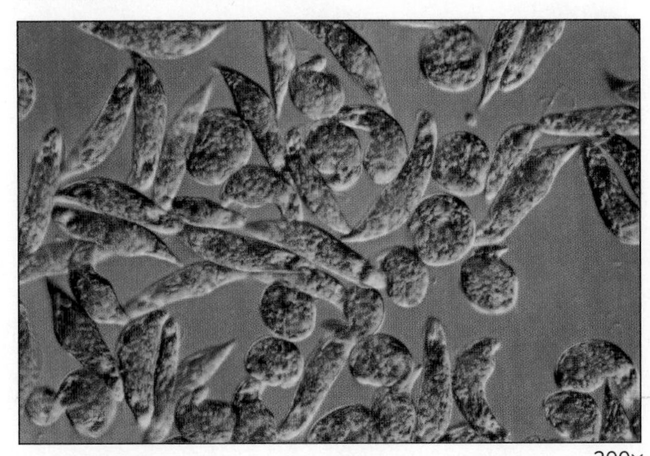

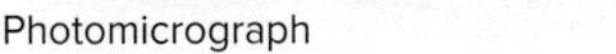

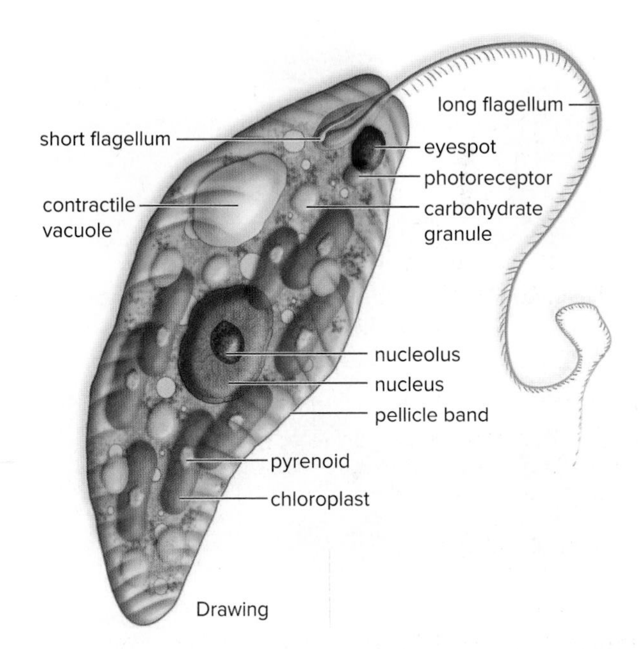

4. Experiment by using scanning, low-power, and high-power objective lenses; by focusing up and down with the fine-adjustment knob; and by adjusting the light so that it is not too bright.
5. Compare your *Euglena* specimens with Figure 3.9. List the labeled features that you can actually see: ______________________________

____

Pond Water

Examination of pond water will also test your ability to observe objects with the microscope, to utilize depth of field, and to control illumination to heighten contrast. Figure 3.10 identifies, for microorganisms found in pond water.

Figure 3.10 Microorganisms found in pond water (not actual size).

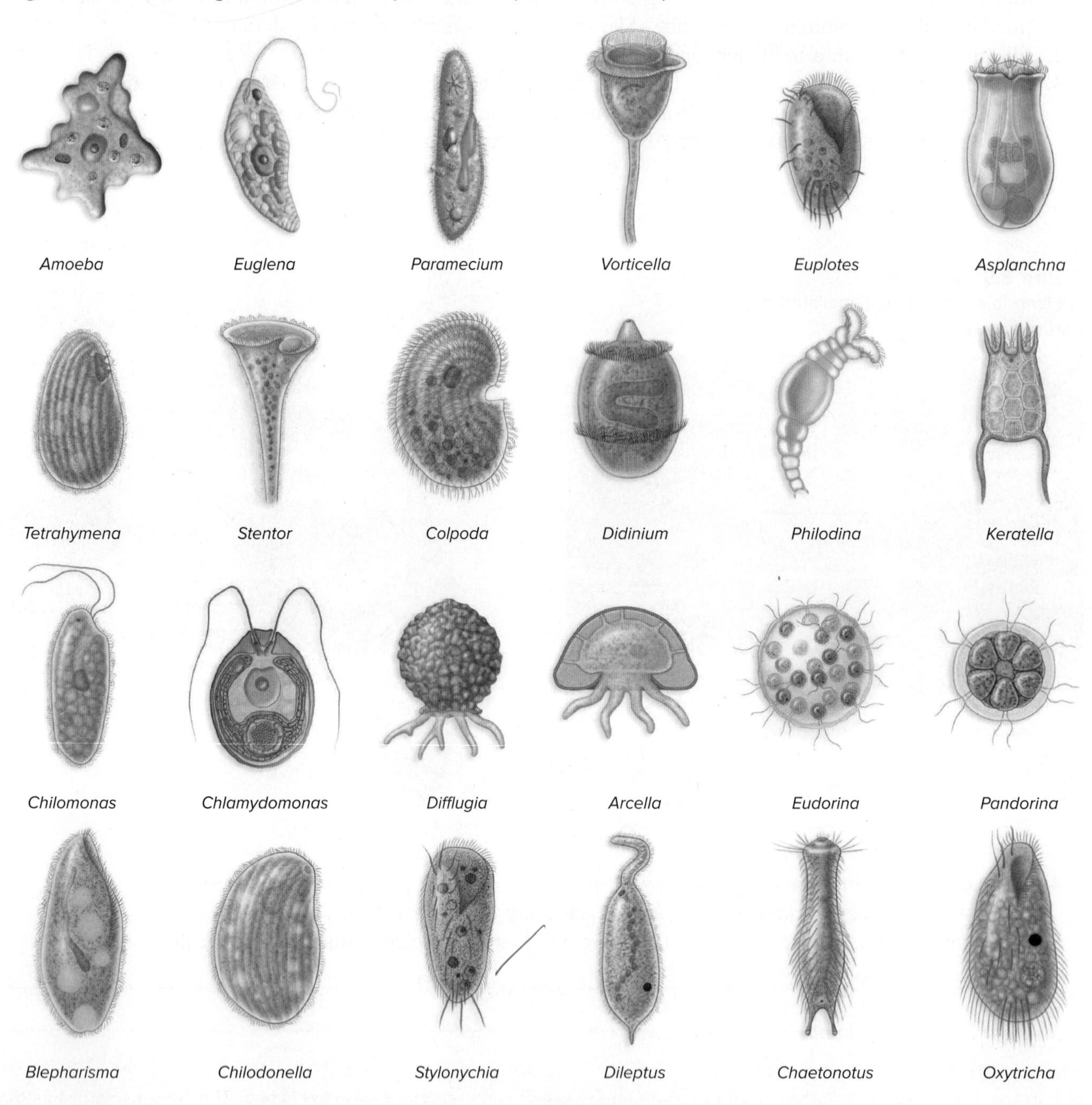

Application for Daily Living

Microscopic Diagnoses

Microscopic examination of discharges, tissues, and the blood plays an important part in a physical exam. For example, your doctor may order a differential blood cell count when he or she believes it will help diagnose your illness. Or, he or she may microscopically examine a discharge if you may have a particular bacterial sexually transmitted disease (STD).

The Pap test, which women have during a routine gynecological exam, is a low-cost, easy microscopic examination of cells taken from the cervix located at the entrance to the womb. Pap tests look for abnormal cells in the lining of the cervix to determine if they are precancerous or cancerous. The more abnormal the appearance of the cells, the greater the concern (Fig. 3.11).

To do a Pap test, a physician merely takes a sample of cells from the cervix, which is then microscopically examined for signs of abnormality. Pap tests are credited with preventing over 90% of deaths from cervical cancer.

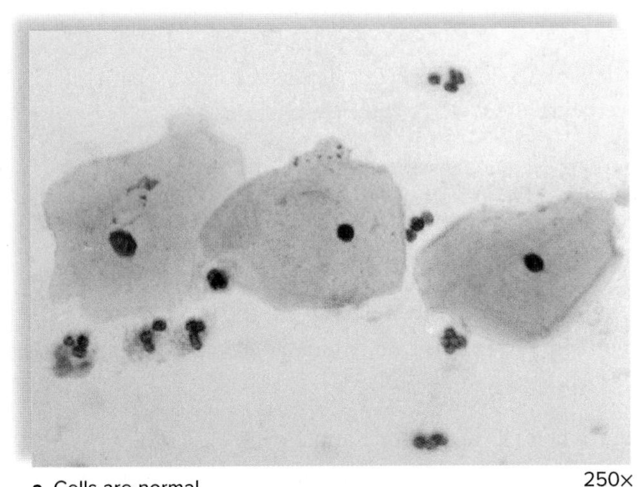

a. Cells are normal. 250×

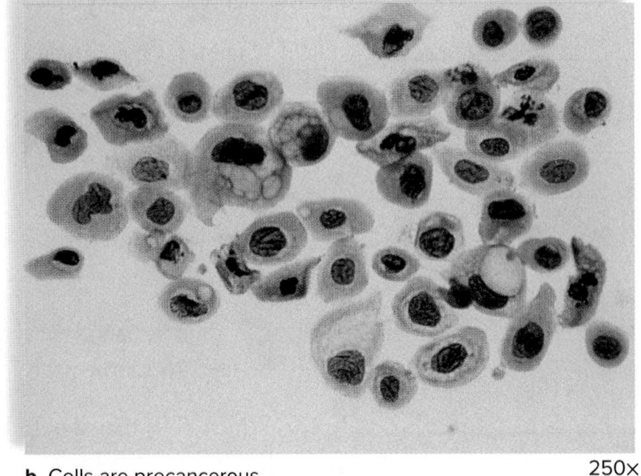

b. Cells are precancerous. 250×

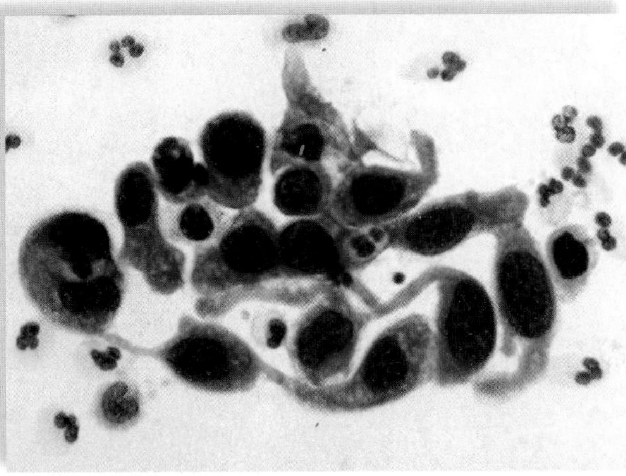

c. Cells are cancerous. 250×

Figure 3.11 Pap test.
A Pap test is a microscopic examination of cells from the cervix at the entrance to the womb. **a.** The cells are normal and cancer is not present. **b.** The cells are precancerous. **c.** The cells are abnormal and cancer is present.

Laboratory Review 3

_______________ **1.** List several types of microscopes.

_______________ **2.** What is the difference between a longitudinal section and a cross section of an object?

_______________ **3.** What are the differences between a stereomicroscope and a compound light microscope?

_______________ **4.** What does the term *compound* mean?

_______________ **5.** What is used to improve contrast when viewing clear portions of cells?

_______________ **6.** What kind of microscope would be used to study a whole or opaque object?

_______________ **7.** What is the name for the lenses located near the eye?

_______________ **8.** Which term describes the minimum distance between two objects required to distinguish them as two separate objects?

_______________ **9.** If the total magnification of a slide is 400× and the ocular lenses are 10×, what is the magnifying power of the objective being used?

_______________ **10.** If the amount of light passing through the condenser needs to be decreased, what microscope part should be adjusted?

_______________ **11.** What word describes a microscope that remains in focus when the objective lenses are changed?

_______________ **12.** If a slide is being viewed with the high-power objective, which adjustment knob should be used to sharpen the focus?

_______________ **13.** Which objective should be in place when the microscope is put away?

_______________ **14.** If threads are layered from top to bottom, brown, green, red, which layer will come into focus first if you are using the microscope properly?

Thought Questions

15. Which type of microscope should be used to view a virus that is 50 nm in size? Justify your choice.

16. Justify your choice of an objective to use when starting your observation of the *Euglena* wet mount; if the *Euglena* are swimming to the left, which way should you move your slide to keep the organism in view?

17. If you use the coarse-adjustment knob to focus on an object with the high-power objective, what problems will you encounter? Explain.

8-21-2018

LABORATORY

4

Chemical Composition of Cells

Learning Outcomes

Introduction

- Describe how cells build and break down polymers.
- Distinguish between positive and negative controls and between positive and negative results.

4.1 Proteins

- State the function of various types of proteins.
- Identify the subunit of a polypeptide, and explain how this subunit connects to others to form a polypeptide.
- Describe a test for the detection of a protein versus a peptide.

Prelab Question: When testing a food for protein, what test might you use and how would you do the test?

4.2 Carbohydrates

- Identify the subunit of starch, and distinguish sugars from starch in terms of subunit number.
- Describe a test for the detection of starch and another for the detection of sugars.
- Explain the varied results of the Benedict's test.

Prelab Question: Why is it a good idea to avoid foods high in sugars?

4.3 Lipids

- Name several types of lipids and state their components.
- Describe a simple test for the detection of fat and tell how an emulsifier works.

Prelab Question: Why is butter solid at room temperature, while an oil is liquid even when placed in the refrigerator?

4.4 Testing Foods and Unknowns

- Explain a procedure for testing the same food for all three components—carbohydrates, protein, and fat.

Prelab Question: Explain why tests for proteins, carbohydrates, and lipids need to be performed separately.

***Application for Daily Living:* Nutrition Labels**

Introduction

Planning Ahead To save time, your instructor may have you start the boiling water bath needed for the experiments to test for proteins and carbohydrates.

All organisms consist of basic units of matter called **atoms.** Molecules form when atoms bond with one another. Inorganic molecules are often associated with nonliving things. Organic biomolecules are associated with organisms. In this laboratory, you will be studying the biomolecules of cells: **proteins, carbohydrates** (monosaccharides, disaccharides, polysaccharides), and **lipids** (i.e., fat).

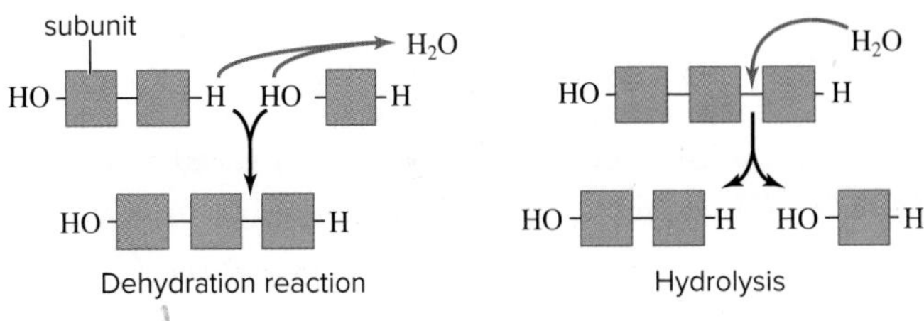

Large biomolecules, sometimes called macromolecules, form during dehydration reactions when smaller molecules bond as water is given off. During *hydrolysis reactions,* bonds are broken as water is added.

A fat contains one glycerol and three fatty acids. Proteins and some carbohydrates (called polysaccharides) are **polymers** because they are made up of smaller molecules called **monomers.** Proteins contain a large number of amino acids (the monomer) joined together by a peptide bond. A polysaccharide, such as starch, contains a large number of glucose molecules joined together. Various chemicals will be used

What is a Control?

The experiments in today's laboratory have both a positive control and a negative control, *which should be saved for comparison purposes until the experiment is complete.* The **positive control** goes through all the steps of the experiment and does contain the substance being tested. Therefore, positive results are expected. The **negative control** goes through all the steps of the experiment except it does not contain the substance being tested. Therefore, negative results are expected.

For example, if a test tube contains glucose (the substance being tested) and Benedict's reagent (blue) is added, a red color develops upon heating. This test tube is the positive control. The red color indicates a positive test for glucose. If a test tube does not contain glucose and Benedict's reagent is added, Benedict's is expected to remain blue. This test tube is the negative control; it tests negative for glucose.

What benefit is a positive control? Positive controls give you a standard by which to tell if the substance being tested is present (or acting properly) in an unknown sample. Negative controls ensure that the experiment is giving reliable results. If a negative control should happen to give a positive result, then the entire experiment may be faulty and unreliable.

in this laboratory to test for the presence of cellular biomolecules. If a color change is observed, the test is said to be *positive* because it indicates that the molecule is present. If the color change is not observed, the test is said to be *negative* because it indicates that the molecule is not present.

4.1 Proteins

Proteins have numerous functions in cells. Antibodies are proteins that combine with pathogens so that the pathogens are destroyed by the body. Transport proteins combine with and move substances from place to place. Hemoglobin transports oxygen throughout the body. Albumin is another transport protein in our blood. Regulatory proteins control cellular metabolism in some way. For example, the hormone insulin regulates the amount of glucose in blood so that cells have a ready supply. Structural proteins include keratin, found in hair, and myosin, found in muscle. **Enzymes** are proteins that speed chemical reactions. A reaction that could take days or weeks to complete can happen within an instant if the correct enzyme is present. Amylase is an enzyme that speeds the breakdown of starch in the mouth and small intestine.

Proteins are made up of **amino acids** (the subunits) joined together. About 20 different common amino acids are found in cells. All amino acids have an acidic group (—COOH) and an amino group (H_2N—). They differ by the ***R* group** (remainder group) attached to a carbon atom, as shown in Figure 4.1. The *R* groups have varying sizes, shapes, and chemical activities.

A chain of two or more amino acids is called a **peptide,** and the bond between the amino acids is called a **peptide bond.** A **polypeptide** is a very long chain of amino acids. A protein can contain one or more polypeptide chains. A single chain forms insulin, while four chains form hemoglobin. A protein has a particular shape, which is important to its function. The shape comes about because the *R* groups of the polypeptide chain(s) can interact with one another in various ways.

Figure 4.1 Formation of a dipeptide.
During a dehydration reaction, a dipeptide forms when an amino acid joins with an amino acid as a water molecule is removed. The bond between amino acids is called a peptide bond. During a hydrolysis reaction, water is added and the peptide bond is broken.

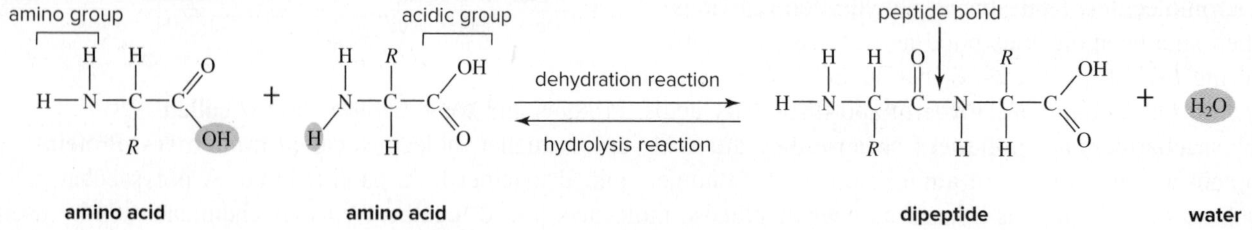

amino group
acidic group
peptide bond
dehydration reaction
hydrolysis reaction
amino acid
amino acid
dipeptide
water

Test for Proteins

Biuret reagent (blue color) contains a strong solution of sodium or potassium hydroxide (NaOH or KOH) and a small amount of dilute copper sulfate ($CuSO_4$) solution. The reagent changes color in the presence of proteins or peptides because the peptide bonds of the protein or peptide chemically combine with the copper ions in biuret reagent (Table 4.1).

Table 4.1 Biuret Test for Protein and Peptides

	Protein	Peptides
Biuret reagent (blue)	Purple	Pinkish-purple

Biuret test for protein and peptides

© David S. Moyer

Experimental Procedure: Test for Proteins

1. Label four clean test tubes (1 to 4).
2. Using the designated graduated transfer pipets, add 1 ml of the experimental solutions listed in Table 4.2 to the test tubes according to their numbers.
3. Then add five drops of biuret reagent to the tubes, swirling to mix.
4. The reaction is almost immediate. Record your observations in Table 4.2.

> ⚠ **Biuret reagent** Biuret reagent is highly corrosive. Exercise care in using this chemical. If any should spill on your skin, wash the area with mild soap and water. Follow your instructor's directions for its disposal.

Table 4.2 Biuret Test for Protein

Tube	Contents	Final Color	Conclusions (+ or –)
1	Distilled water	clear teal	–
2	Albumin	Foggy	–
3	Pepsin	translucent	–
4	Starch	Foggy Blue	–

Conclusions: Proteins

- From your test results, conclude if a protein is present (+) or absent (–). Enter your conclusions in Table 4.2.
- Pepsin is an enzyme. Enzymes are composed of what type of organic molecule? Amylase
- According to your results, is starch a protein? No
- Which of the four tubes is the negative control sample? 3 Why? No protein peptides
- Why do experimental procedures include control samples? to be saved for comparison purposes until the experiment is complete
- If your results are not as expected, inform your instructor, who will advise you how to proceed.

4.2 Carbohydrates

Carbohydrates include sugars and molecules that are chains of sugars. **Glucose,** which has only one sugar unit, is a monosaccharide; **maltose,** which has two sugar units, is a disaccharide (Fig. 4.2). Glycogen, starch, and cellulose are polysaccharides, made up of chains of glucose units (Fig. 4.3).

Glucose is used by all organisms as an energy source. Energy is released when glucose is broken down to carbon dioxide and water. This energy is used by the organism to do work. Animals store glucose as glycogen and plants store glucose as starch. Plant cell walls are composed of cellulose.

Figure 4.2 Formation of a disaccharide.
During a dehydration reaction, a disaccharide, such as maltose, forms when a glucose joins with a glucose as a water molecule is removed. During a hydrolysis reaction, the components of water are added, and the bond is broken.

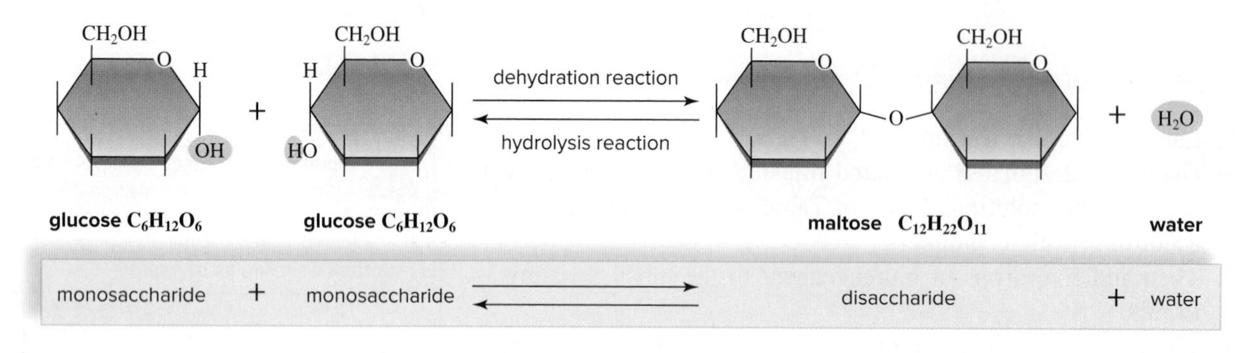

Figure 4.3 Starch.
Starch is a polysaccharide composed of many glucose units. **a.** Photomicrograph of starch granules in cells of a potato. **b.** Structure of starch. Starch consists of amylose that is nonbranched and amylopectin that is branched.

(a) © Jeremy Burgess/SPL/Science Source

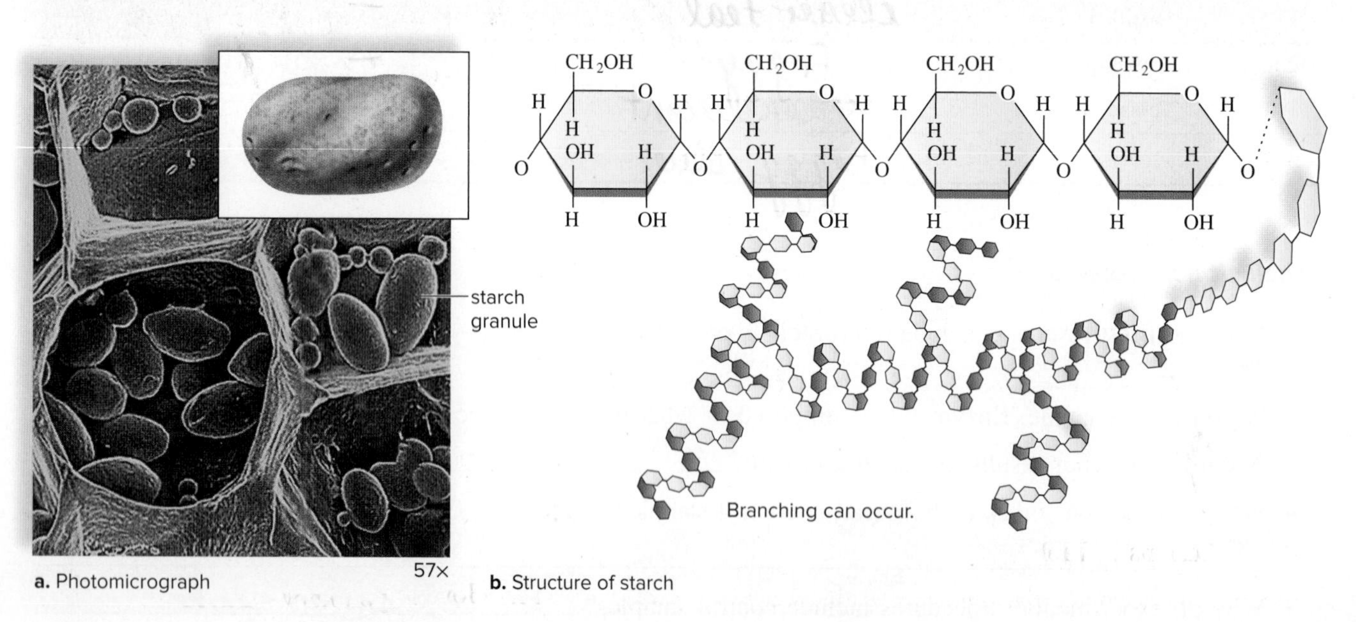

a. Photomicrograph **b.** Structure of starch

Test for Starch

In the presence of starch, iodine solution (yellowish-brown) reacts chemically with starch to form a blue-black color (Table 4.3).

Table 4.3 Iodine Test for Starch

	Starch
Iodine solution	Blue-black

Experimental Procedure: Test for Starch

1. Label five clean test tubes (1 to 5).
2. Using the designated graduated transfer pipets, add 1 ml of the experimental solutions listed in Table 4.4 to the test tubes according to their numbers.
3. Then add five drops of iodine solution to the tubes at the same time.
4. Note the final color changes and record your observations in Table 4.4.

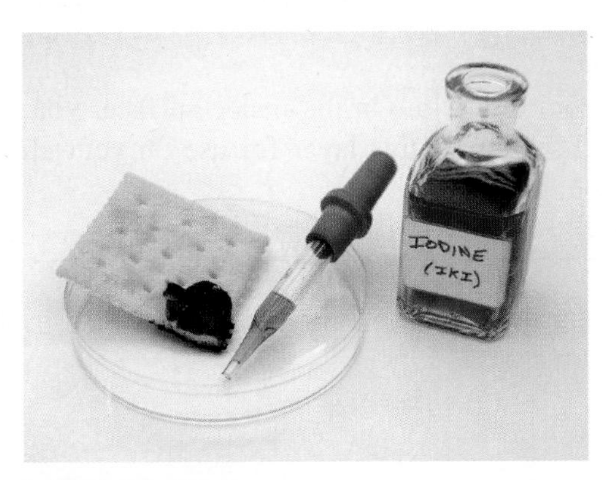

Iodine test for starch

Table 4.4 Iodine (IKI) Test for Starch

Tube	Contents	Color	Conclusions
1	Water	urine yellow	No starch
2	Starch suspension	Blue-Black	starch
3	Onion juice		No —
4	Potato juice	charcol grey	No starch
5	Glucose solution	urine yellow	—

Conclusions: Starch

- From your test results, draw conclusions about what organic compound is present in each tube. Write these conclusions in Table 4.4.
- Does the potato or the onion store glucose as starch? Neither How do you know? Neither is yellowish-brown
- If your results are not as expected, offer an explanation. Then inform your instructor, who will advise you how to proceed.

Experimental Procedure: Microscopic Study

Potato

1. With a scalpel, slice a very thin piece of potato. Place it on a microscope slide, add a drop of water and a coverslip, and observe under low power with your compound light microscope. Compare your slide with the photomicrograph of starch granules (see Fig. 4.3*a*). Find the cell wall (large, geometric compartments) and the starch grains (numerous clear, oval-shaped objects).
2. Without removing the coverslip, place two drops of iodine solution onto the microscope slide so that the iodine touches the coverslip. Draw the iodine under the coverslip by placing a small piece of paper towel in contact with the water on the **opposite** side of the coverslip.
3. Microscopically examine the potato again on the side closest to where the iodine solution was applied.

   What is the color of the small, oval bodies? ______________________________

   What is the chemical composition of these oval bodies? ______________________________

Onion

1. Peel a single layer of onion from the bulb. On the inside surface, you will find a thin, transparent layer of onion skin. Peel off a small section of this layer for use on your slide.
2. Add a large drop of iodine solution.
3. Does onion contain starch? ______________________________
4. Are these results consistent with those you recorded for onion juice in Table 4.4? ______________

Test for Sugars

> ⚠ **Benedict's reagent** Benedict's reagent is highly corrosive. Exercise care in using this chemical. If any should spill on your skin, wash the area with mild soap and water. Follow your instructor's directions for disposal of this chemical.

Monosaccharides and some disaccharides will react with **Benedict's reagent** after being heated in a boiling water bath. In this reaction, copper ion (Cu^{2+}) in the Benedict's reagent reacts with part of the sugar molecule, causing a distinctive color change. The color change can range from green to red, and increasing concentrations of sugar will give a continuum of colored products (Table 4.5).

Table 4.5 Benedict's Test for Sugars (Some Typical Reactions)

Chemical	Chemical Category	Benedict's Reagent (After Heating)
Water	Inorganic	Blue (no change)
Glucose	Monosaccharide (carbohydrate)	Varies with concentration: very low—green low—yellow moderate—yellow-orange high—orange very high—orange-red
Maltose	Disaccharide (carbohydrate)	Varies with concentration—see "Glucose"
Starch	Polysaccharide (carbohydrate)	Blue (no change)

Benedict's test for sugar

Experimental Procedure: Test for Sugars

1. Prepare a boiling water bath and label five clean test tubes (1 to 5).
2. Using the designated graduated transfer pipets, add 1 ml of the experimental solutions listed in Table 4.6 to the test tubes according to their numbers.
3. Then add five drops of Benedict's reagent to all the tubes at this time.
4. Place the tubes into the boiling water bath at the same time.
5. When, after a few minutes, you see a change of colors, remove all of the tubes from the water bath and record your observations in Table 4.6.
6. Save your tubes for comparison purposes when you do section 4.4.

Table 4.6 Benedict's Test for Sugars

Tube	Contents	Color (After Heating)	Conclusions
1	Water	ocean blue	No Sugar
2	Glucose solution	Orange Red	Very High
3	Starch suspension	sky Blue	No Sugar
4	Onion juice	yellow-orange	high
5	Potato juice	clear green	very low

Conclusions: Sugars

- From your test results, conclude what kind of chemical is present. Enter your conclusions in Table 4.6.
- Which tube served as a negative and which as a positive control? Negative - water and starch Suspension. Positive - glucose Solution,

- Compare Table 4.4 with Table 4.6. Sugars are an immediate energy source in cells. In plant cells, glucose (a primary energy molecule) is often stored in the form of starch. Is glucose stored as starch in the potato? No Is glucose stored as starch in the onion? Yes Does this explain your results in Table 4.6? ________ Why? colours and conclusions add up

4.3 Lipids

Lipids are compounds that are insoluble in water and soluble in solvents, such as alcohol and ether. Lipids include fats, oils, phospholipids, steroids, and cholesterol. Typically, **fat,** such as in the adipose tissue of animals, and **oils,** such as the vegetable oils from plants, are composed of three molecules of fatty acids bonded to one molecule of glycerol (Fig. 4.4). **Phospholipids** have the same structure as fats, except that in place of the third fatty acid there is a phosphate group (a grouping that contains phosphate). **Steroids** are derived from **cholesterol.** These molecules have skeletons of four fused rings of carbon atoms, but they differ by functional groups (attached side chains). Fat, as we know, is long-term stored energy in the human body. Phospholipids are found in the plasma membrane of a cell. Cholesterol, a molecule transported in the blood, has been implicated in causing cardiovascular disease. Regardless, steroids are very important compounds in the body. For example, the sex hormones, like testosterone and estrogen, are steroids.

Figure 4.4 Formation of a fat.
During a dehydration reaction, a fat molecule forms when glycerol joins with three fatty acids as three water molecules are removed. During a hydrolysis reaction, water is added, and the bonds are broken between glycerol and the three fatty acids.

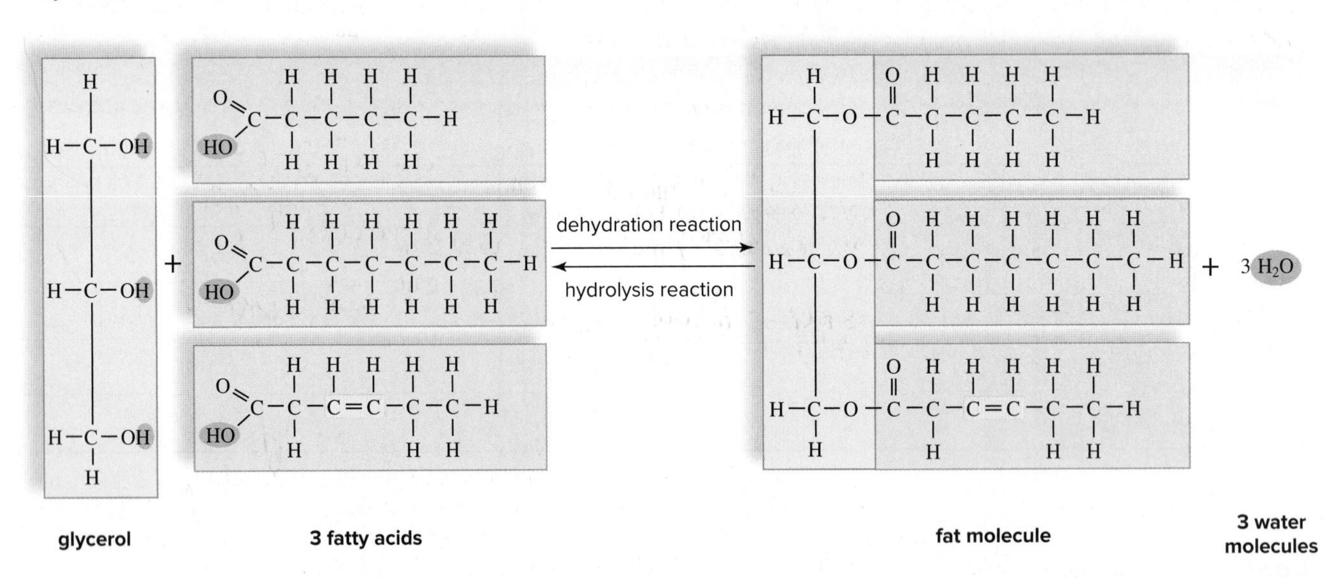

Test for Fat

Fats and oils do not evaporate from brown paper or loose-leaf paper; instead, they leave an oily spot.

Experimental Procedure: Paper Test for Fat

1. Place a small drop of water on a square of brown paper or loose-leaf paper. Describe the immediate effect. ______________________________
2. Place a small drop of vegetable oil on a square of the paper. Describe the immediate effect. ______________________________
3. Wait at least 15 minutes for the paper to dry. Evaluate which substance penetrates the paper and which is subject to evaporation. Record your observations and conclusions in Table 4.7. Save the paper for comparison use with section 4.4.

Table 4.7 Paper Test for Fat

Sample	Observations	Conclusions
Water spot		
Oil spot		

Emulsification of Oil

Some molecules are **polar,** meaning that they have charged groups or atoms, and some are **nonpolar,** meaning that they have no charged groups or atoms. A water molecule is polar, and therefore, water is a good solvent for other polar molecules. When the charged ends of water molecules interact with the charged groups of polar molecules, these polar molecules disperse in water.

Water is not a good solvent for nonpolar molecules, such as fats. A fat has no polar groups to interact with water molecules. An **emulsifier,** however, can cause a fat to disperse in water. An emulsifier contains molecules with both polar and nonpolar ends. When the nonpolar ends interact with the fat and the polar ends interact with the water molecules, the fat disperses in water, and an **emulsion** results (Fig. 4.5).

Bile salts (emulsifiers found in bile produced by the liver) are used in the digestive tract. Today milk, such as 1% milk, has been homogenized so that fat droplets do not congregate and rise to the top of the container. Homogenization requires the addition of natural emulsifiers such as phospholipids—the phosphate part of the molecule is polar and the lipid portion is nonpolar.

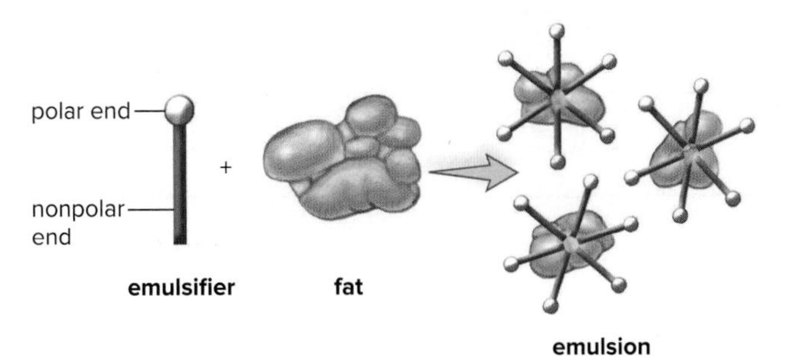

Figure 4.5 Emulsification. An emulsifier contains molecules with both a polar and a nonpolar end. The nonpolar ends are attracted to the nonpolar fat, and the polar ends are attracted to the water. This causes droplets of fat molecules to disperse.

Experimental Procedure: Emulsification of Lipids

Label three clean test tubes (1 to 3), and use the appropriate graduated pipet to add solutions to the test tubes as follows:

Tube 1

1. Add 3 ml of water and 1 ml of vegetable oil. Shake.
2. Observe for the initial dispersal of oil, followed by rapid separation into two layers. Is vegetable oil soluble in water? ______________________________
3. Let the tube settle for 5 minutes. Label a microscope slide as 1.
4. Use a dropper to remove a sample of the solution that is just below the layer of oil. Place the drop on the slide, add a coverslip, and examine with the low power of your compound light microscope.
5. Record your observations in Table 4.8.

Tube 2

1. Add 2 ml of water, 3 ml of vegetable oil, and 1 ml of the available emulsifier (Tween or bile salts). Shake.
2. Describe how the distribution of oil in tube 2 compares with the distribution in tube 1.

   ____

   ____
3. Let the tube settle for 5 minutes. Label a microscope slide as 2.
4. Use a different dropper to remove a sample of the solution that is just below the layer of oil. Place the drop on the slide, add a coverslip, and examine with the low power of your compound light microscope.
5. Record your observations in Table 4.8.

Tube 3

1. Add 1 ml of milk and 2 ml of water. Shake well.
2. Use a different dropper to remove a sample of the solution that is just below the layer of oil. Place the drop on the slide, add a coverslip, and examine with the low power of your compound light microscope.
3. Record your observations in Table 4.8.

Table 4.8 Emulsification

Tube	Contents	Observations	Conclusions
1	Oil Water		
2	Oil Water Emulsifier		
3	Milk Water		

Conclusions: Emulsification

- From your observations, conclude why the contents of each tube appear as they do under the microscope. Record your conclusions in Table 4.8.
- Explain the correlation between your macroscopic observations (how the tubes look to your unaided eye) and your microscopic observations. ______________________________

  ____

4.4 Testing Foods and Unknowns

It is common for us to associate the term *organic* with the foods we eat, including carbohydrate foods (Fig. 4.6), protein foods (Fig. 4.7), and lipid foods (Fig. 4.8). Though we may recognize foods as being organic, often we are not aware of what specific types of compounds are found in what we eat. In the following Experimental Procedure, you will use the same tests you used previously to determine the composition of everyday foods and unknowns.

Figure 4.6 Carbohydrate foods.

© McGraw-Hill Education/John Thoeming, photographer

Figure 4.7 Protein foods.

© McGraw-Hill Education/John Thoeming, photographer

Figure 4.8 Lipid foods.

© McGraw-Hill Education/John Thoeming, photographer

Experimental Procedure: Testing Foods and Unknowns

Your instructor will provide you with several everyday foods including unknowns, and your task is to

1. Recall how you will test substances for protein, carbohydrates, and fat.

___

___

___

___

2. Have your instructor okay your procedures, and then conduct the necessary tests.
3. Record your results as positive (+) or negative (–) in Table 4.9.

Table 4.9 Testing Foods and Unknowns				
Sample Name	Protein (Biuret)	Starch (Iodine)	Sugar (Benedict's)	Fat (Brown or loose-leaf paper)
Unknown A				
Unknown B				

Conclusions: Testing Foods and Unknowns

- What foods tested positive for only one of the organic compounds? ____________________
- What does more than one positive test tell you about these foods?

Application for Daily Living

Nutrition Labels

For good health, the diet should be balanced and should not contain an overabundance of either protein, carbohydrate, or fat. Everyone should become familiar with reading nutrition labels. A portion of one is shown in Figure 4.9.

It is very important to note the serving size and servings per container. For comparison purposes, you need to compare the same serving sizes. The % daily values are based on an intake of 2,000 calories per day. A calorie is an indication of the amount of energy provided, and the number of calories taken in should match the number required per day, unless you wish to gain weight.

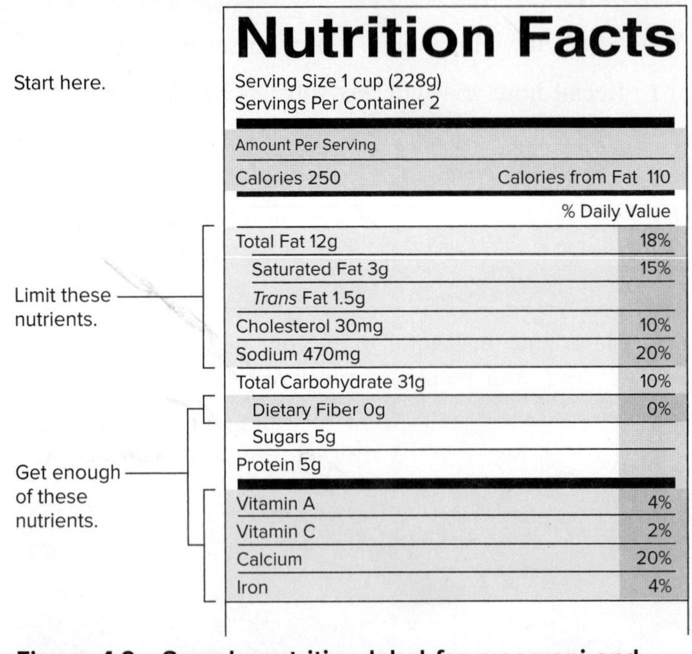

Figure 4.9 Sample nutrition label for macaroni and cheese.
As of January 2006, foods must include the quantity of trans fat. There is no % Daily Value for trans fats, and they should be avoided. See page 38.

Most people are extremely interested in the amount of fat provided by a food. One serving of this food provides 18% of the total fat required for the day. If you find this excessive, you could look for another macaroni and cheese that supplies less total fat and less saturated fat. If you decide that you prefer this product, you might wish to balance it with a food that contains very little fat and one that provides a carbohydrate that contains fiber. This food contains no fiber. Complex carbohydrates containing fiber, such as those found in whole grains, are recommended and not those that contain simply starch.

Making wise decisions about the foods we buy can lead to a longer, healthier life.

Laboratory Review 4

______________ 1. What kind of reaction adds water to break large biomolecules into subunits?

______________ 2. What kind of protein is lactase, the biomolecule that speeds up the breakdown of lactose?

______________ 3. What kind of bond forms during a dehydration synthesis reaction involving two amino acids?

______________ 4. If maltose undergoes hydrolysis, what subunits result?

______________ 5. As what molecule do plants store glucose?

______________ 6. If starch is present, what color will the iodine turn?

______________ 7. If a solution contains a large amount of glucose, what color will the Benedict's reagent become?

______________ 8. What kinds of biomolecules are insoluble in water but soluble in alcohol or ether?

______________ 9. If two fatty acids, glycerol, and a phosphate group undergo a dehydration reaction, what biomolecule forms?

______________ 10. What is the name used for plant fats?

______________ 11. What must be present to successfully break down fats during digestion?

______________ 12. If you test an unknown substance with the biuret reagent and it turns purple, what is present?

______________ 13. If you test a sample of potato with the biuret reagent, what do you expect the results to be?

______________ 14. If you want to test a sample to see if glucose is present, what reagent should you use?

Thought Questions

15. If amylase is added to a sample of potato and you test it with Benedict's reagent 10 minutes later, what do you expect the results of the test to be? Justify your reply.

16. After enjoying a meal of fish and chips, you notice your shirt has an oily stain on it. Why won't the oily stain come out when you dab at it with just water?

17. If the sample of water you test with biuret reagent turns purple, what should you conclude about the results of your experiment? Explain.

LABORATORY

5

Cell Structure and Function

Learning Outcomes

5.1 Anatomy of a Human Cell

- Using a model or drawing, identify the parts of a human cell and state a function for each part.

Prelab Question: Why is it beneficial for different parts of the cell to have specific functions?

5.2 Diffusion of Solutes

- Describe the process of diffusion as a physical phenomenon independent of the plasma membrane.
- Predict which solutes can cross the plasma membrane by diffusion and which cannot cross the plasma membrane by diffusion.

Prelab Question: Why is it beneficial for cellular substances to cross the plasma membrane by diffusion?

5.3 Osmosis: Diffusion of Water Across the Plasma Membrane

- Define osmosis and explain the movement of water across a membrane.
- Define isotonic, hypertonic, and hypotonic solutions and give examples in terms of NaCl concentrations.
- Predict the effect that solutions of different tonicities have on red blood cells.

Prelab Question: Why is 0.9% the usual tonicity of intravenous solutions?

5.4 Enzyme Activity

- Explain how enzymes have the ability to speed chemical reactions in cells.
- List some factors that can affect the speed of enzymatic reactions.

Prelab Question: Why is a warm body temperature advantageous to metabolism?

***Application for Daily Living:* Dehydration and Water Intoxication**

Introduction

Planning Ahead To save time, your instructor may have you start a boiling water bath and the potato strip experiment at the beginning of the laboratory.

We are accustomed to observing the outward appearance of human beings and rarely have an opportunity to become aware that humans are composed of small entities called **cells**. Though we often think that the heart, liver, or intestines enable the human body to function, it is actually cells that do the work of these organs. In a previous lab, you may have, you observed human cheek cells (see Figure 3.8) using a compound light microscope. The much more powerful **electron microscope** led to the discovery that cells are actually very complicated because they contain many **organelles** that carry out enzymatic functions.

The model of an animal cell available in the laboratory is based on electron micrographs. In today's laboratory, we review the structure and function of **organelles**, subcompartments of a cell, before actually observing how its outer surface, the **plasma membrane**, serves as a selective regulator of what enters and exits cells. We will discover that the passage of water into a cell depends on the difference in concentration of solutes (particles) between the **cytoplasm** (contents of a cell) and the surrounding medium or solution.

Enzymes are proteins that carry out metabolic reactions within organelles, and we will have an opportunity to marvel at the rapidity of an enzymatic reaction, even at room temperature.

5.1 Anatomy of a Human Cell

Figure 5.1 shows that an animal cell is partitioned into a number of compartments. Just like a house works more efficiently when each room has a specialized function, so does a cell with different compartments for varied functions.

Figure 5.1 Human (animal) cell.

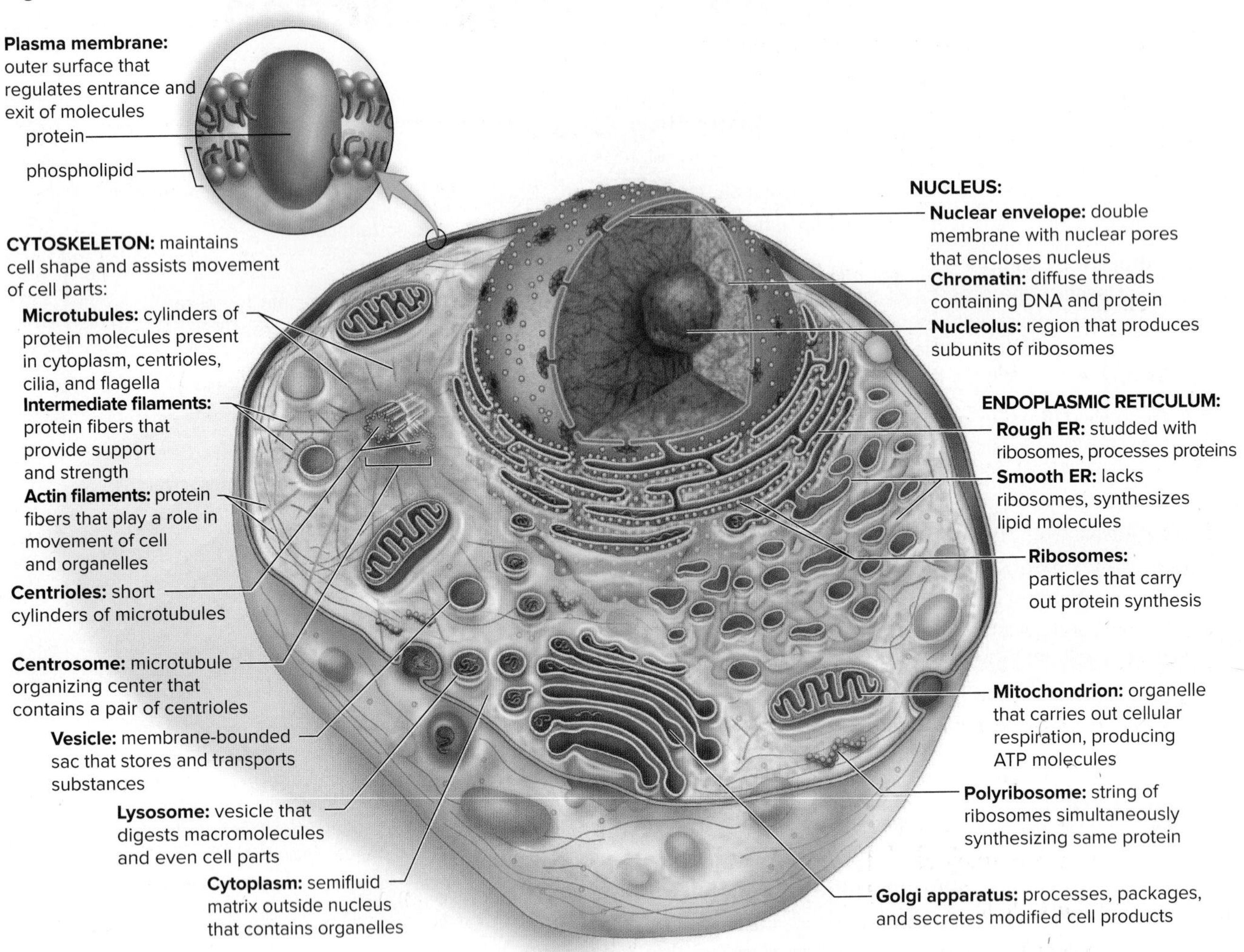

Learning the Organelles of the Cell

Identify these parts of a cell:

Composition and Function	Structure
1. Stack of membranous saccules; functions in processing, packaging, and distribution of molecules	vesicle
2. Membranous sacs; storage and transport of substances	vesicle
3. Have a double membrane; responsible for cellular respiration and production of ATP molecules	
4. Particles that carry out protein synthesis	Ribosomes
5. Outer surface that regulates entrance and exit of molecules	Plasma membrane
6. Region in nucleus that produces subunits of ribosomes	
7. Short cylinders, present in centrosomes, of unknown function	
8. Central body, having diffuse threads of DNA and protein	
9. Vesicle that digests macromolecules and even cell parts	
10. Composed of microtubules, actin filaments, and intermediate filaments; responsible for the shape of the cell and movement of its parts	
11. Membranous saccules and canals having no ribosomes; synthesize lipid molecules	

Learning That the Organelles Work Together

1. Imagine that this cell produces digestive enzymes that are sent to the digestive tract:
   - **a.** Which part of the endoplasmic reticulum would produce these enzymes? ____________
   - **b.** How would they be transported to another part of the cell? ____________
   - **c.** Which organelle would process and package these enzymes for export? ____________
2. Imagine that this cell produces a sex hormone (a lipid molecule):
   - **a.** Which part of the endoplasmic reticulum would produce these lipid molecules? ____________
   - **b.** How would they be transported to another part of the cell? ____________
   - **c.** Which organelle would process and package this hormone for export? ____________
3. The nucleus produces the subunits of ribosomes.
   - **a.** Where in the nucleus are the subunits produced? ____________
   - **b.** What part of the nuclear envelope allows them to get out of the nucleus? ____________
   - **c.** Where do the subunits go and what happens to them? ____________

     ____
4. How a cell breaks down engulfed substances.
   - **a.** *Label Figure 5.1* where a vesicle is forming in order to take in a substance that will be digested.
   - **b.** This vesicle will fuse with a ____________ that contains digestive enzymes.

5.2 Diffusion of Solutes

Diffusion is the random movement of molecules from the area of higher concentration to the area of lower concentration until they are equally distributed. A **solution** consists of a a liquid solvent (most often water) and dissolved particles called solutes.

Environmental Factors and Diffusion of Molecules

If you spray a deodorant in one corner of the room, it will soon spread to fill the room because diffusion has occurred. (Notice, therefore, that diffusion can occur independently of a plasma membrane.) Environmental factors such as temperature and the resistance of the medium can affect the speed of diffusion. Air offers little resistance to the random motion of molecules, followed by a liquid, and then by any type of a solid.

Observation: Environmental Factors and Diffusion

You will calculate the speed of diffusion (1) through a semisolid gel and (2) through a liquid. Hypothesize whether you expect diffusion to occur faster through a semisolid or through a liquid, and give a reason for your hypothesis. ______________________________

____

Diffusion Through a Semisolid

1. Observe a petri dish containing 1.5% gelatin (or agar) to which a crystal of potassium permanganate ($KMnO_4$) was added previously (time zero) in the center depression.
2. Obtain time zero from your instructor and record time zero and the final time (now) in Table 5.1. Calculate the length of time in hours and minutes; then convert the entire time to hours. ______________ hr
3. Using a ruler placed over the petri dish, measure (in mm) the movement of color from the center of the depression outward in one direction. ______________ mm
4. Calculate the speed of diffusion. ______________ mm/hr (Divide the number of millimeters by the number of hours.)
5. Record all data in Table 5.1.

> **Potassium permanganate $KMnO_4$**
> $KMnO_4$ is highly poisonous and is a strong oxidizer. Avoid contact with skin and eyes and wash combustible materials. If spillage occurs, wash all surfaces thoroughly. $KMnO_4$ will also stain clothing.

Diffusion Through a Liquid

1. Add water to a glass petri dish.
2. Place the petri dish over a thin, flat ruler.
3. With tweezers, add a crystal of potassium permanganate ($KMnO_4$) directly over a millimeter measurement line. Note time zero in Table 5.1.
4. After 10 minutes, note the distance the color has moved (Fig. 5.2). Record the final time, length of time, and distance moved in Table 5.1.
5. Calculate the speed of diffusion by multiplying the length of time and the distance moved by 6.

______________ mm/hr. Record in Table 5.1.

Figure 5.2 Process of diffusion.
Diffusion is apparent when dye molecules have equally dispersed.

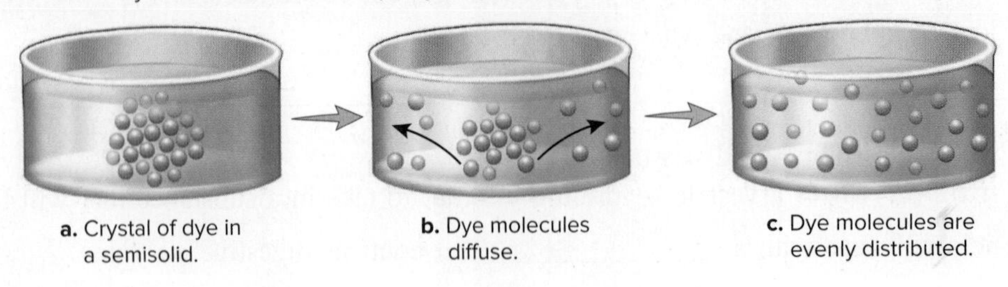

Table 5.1 Speed of Diffusion					
Medium	**Time Zero**	**Final Time**	**Length of Time (hr)**	**Distance Moved (mm)**	**Speed of Diffusion (mm/hr)**
Semisolid	4:58pm			2 min	
Liquid					

Conclusions: Diffusion

- Why did the dye molecules move rather than stay where they were originally? ______________________

____

- In which experiment was diffusion the fastest? ______________________
- What accounts for the difference in speed? ______________________

____

The Plasma Membrane and Diffusion of Molecules

The plasma membrane regulates the passage of molecules into and out of cells. It is said to be **selectively permeable,** because only small, noncharged molecules can diffuse across the plasma membrane without assistance (Fig. 5.3*a*). Carriers, proteins embedded in plasma membrane, can assist the passage of molecules across a membrane. Each carrier is specific to a particular molecule. During facilitated transport, a carrier (or a channel protein) assists a molecule diffusing across the membrane (Fig. 5.3*b*). Diffusion and facilitated diffusion require no energy because molecules move from areas of high concentration to areas of low concentration—that is, down a concentration gradient. Sometimes, substances need to be moved against the concentration gradient—from low to high. This occurs through active transport, where a carrier protein expends energy to move molecules across the plasma membrane.

Figure 5.3 Passage of molecules across a plasma membrane.
a. During diffusion, molecules move from the higher to the lower concentration. **b.** During facilitated transport, carrier proteins transport molecules from the higher to the lower concentration. **c.** During active transport, molecules move from the lower to the higher concentration; a protein carrier and energy are required.

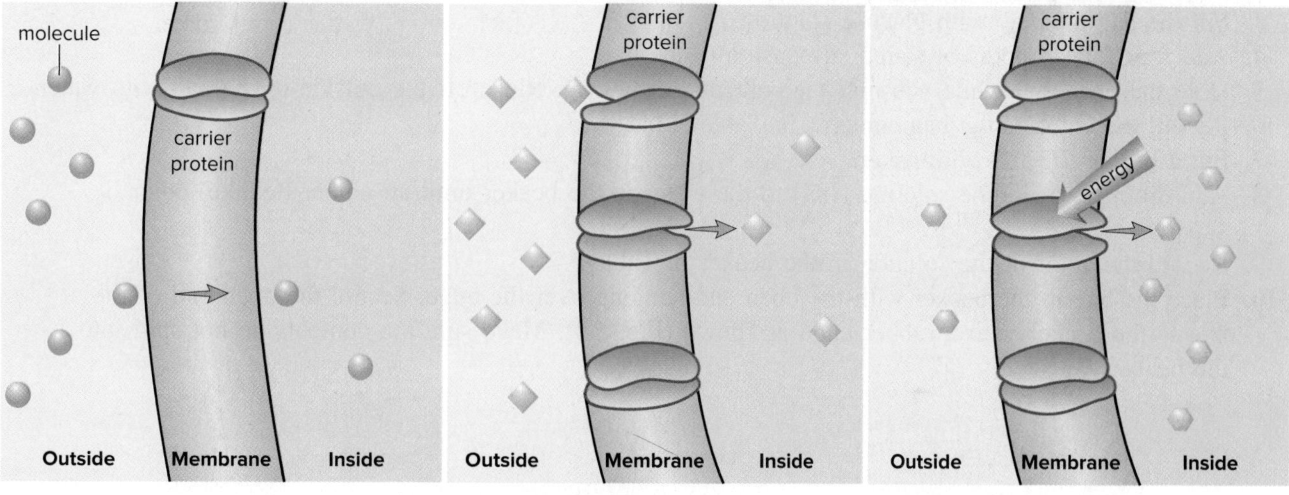

Experimental Procedure: The Plasma Membrane and Diffusion of Solutes

In this experiment, an artificial membrane is used to simulate the plasma membrane; the artificial membrane is also semipermeable. Only certain molecules can cross the membrane. The artificial membrane contains no carriers to assist the movement of molecules across the membrane.

Notice in Figure 5.4 that at the start of this experiment, glucose (small molecule) and starch (large molecule) will be inside the membranous bag and iodine (small molecule) will be outside the bag. Hypothesize which molecules will cross the membrane and in which direction they will move.

to inside the bag ______________________________

to outside the bag ______________________________

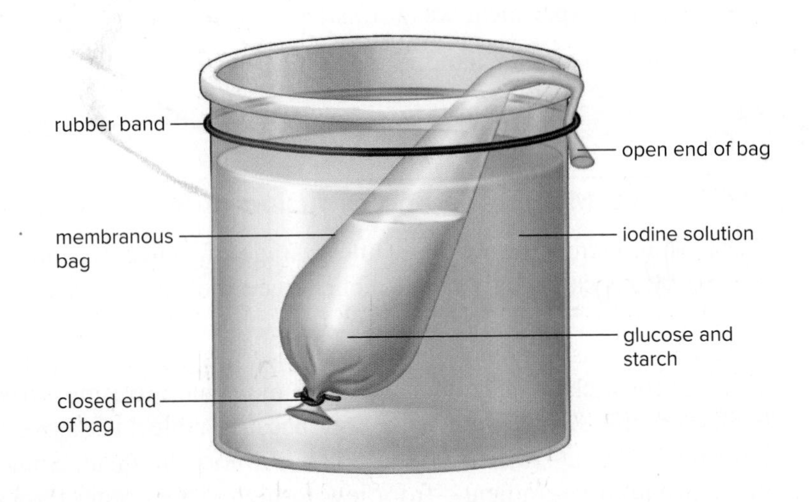

Figure 5.4 Diffusion experiment.

Diffusion Through a Selectively Permeable Membrane

At the start of this experiment:

1. Cut a piece of membranous tubing approximately 40 cm (approximately 16 in) long. Soak the tubing in water until it is soft and pliable.
2. Close one end of the tubing with two knots.
3. Fill the bag halfway with glucose solution.
4. Add four full droppers of starch suspension to the bag.
5. Hold the open end while you mix the contents of the bag. Rinse off the outside of the bag with water.
6. Record the color of the bag contents in Table 5.2.
7. Fill a beaker 2/3 full with water.
8. Add droppers of iodine solution (IKI) to the water in the beaker until an amber (tealike) color is apparent.
9. Record the color of the solution in the beaker in Table 5.2.
10. Place the bag in the beaker with the open end hanging over the edge. Secure the open end of the bag to the beaker with a rubber band as shown (Fig. 5.4). Make sure the contents do not spill into the beaker.

After about 5 minutes, at the end of the experiment:

Benedict's reagent Benedict's reagent is highly corrosive. Exercise care in using this chemical. If any should spill on your skin, wash the area with mild soap and water. Follow your instructor's directions for disposal of this chemical.

11. You will note a color change. Record the present color of the bag contents and the beaker contents in Table 5.2.
12. Obtain a small test tube. Using a graduated transfer pipet, draw 1 ml from the bottom of the beaker (near the bag) and place it in the test tube. Using a designated transfer pipet, add 3 ml of Benedict's reagent. Heat in a boiling water bath for 5 to 10 minutes, observe any color change, and record your results as positive or negative in Table 5.2. (Optional use of glucose test strip: Dip glucose test strip into beaker. Compare stick with chart provided by instructor.)
13. Remove the dialysis bag from the beaker. Dispose of it and the used Benedict's reagent solution in the manner directed by your instructor.

Table 5.2 Solute Diffusion Through a Membrane

	At Start of Experiment		At End of Experiment		
	Contents	Color	Color	Benedict's Test (+) or (–)	Conclusion
Bag	Glucose Starch			________	
Beaker	Water Iodine				

Conclusions: Plasma Membrane and Diffusion of Solutes

- Based on the color change noted in the bag, conclude what molecule diffused across the membrane from the beaker to inside the bag, and record your conclusion in Table 5.2.
- From the results of the Benedict's test on the beaker contents, conclude what molecule diffused across the membrane from the bag to the beaker, and record your conclusion in Table 5.2.
- Which molecule did not diffuse across the membrane from the bag to the beaker? ____________________

  Explain. ____

5.3 Osmosis: Diffusion of Water Across the Plasma Membrane

Osmosis is the diffusion of water molecules across a selectively permeable membrane. Like other molecules, water follows its concentration gradient and moves from a region of higher concentration to a region of lower concentration. To cross the plasma membrane, water passes through channel proteins called **aquaporins**. Therefore, osmosis (the diffusion of water) is a form of facilitated transport.

Tonicity is the relative concentration of solute (e.g., salt molecules) and solvent (water molecules) outside the cell compared to inside the cell.[1] An **isotonic solution** has the same concentration of solute (and therefore of water) as the cell. When cells are placed in an isotonic solution, there is no net movement of water. A **hypertonic solution** has a higher solute (therefore, lower water) concentration than the cell. When cells are placed in a hypertonic solution, water moves out of the cell into the solution. A **hypotonic solution** has a lower solute (therefore, higher water) concentration than the cell. When cells are placed in a hypotonic solution, water moves from the solution into the cell.

The next two Experimental Procedures explore tonicity using potato strips and red blood cells.

[1]Percent solutions are grams of solute per 100 ml of solvent. Therefore, a 10% solution is 10 g of sugar with water added to make up 100 ml of solution.

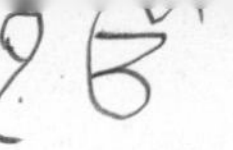

Experimental Procedure: Tonicity and Potato Strips

This procedure runs for 1 hour. Prior setup can maximize your time efficiency.

1. Cut two strips of potato, each about 7 cm long and 1 1/2 cm wide.
2. Label two test tubes 1 and 2. Place one potato strip in each tube.
3. Fill tube 1 with water to cover the potato strip.
4. Fill tube 2 with 10% sodium chloride (NaCl) to cover the potato strip. NaCl is table salt.
5. After 1 hour, observe each strip for limpness (water loss) or stiffness (water gain). Which tube has the limp potato strip? ____________ Why did water diffuse out of the potato strip? ____________ Which tube has the stiff potato strip? ____________ Why did water diffuse into the potato strip?

____

____

Red Blood Cells (Animal Cells)

A solution of 0.9% NaCl is isotonic to red blood cells. In such a solution, red blood cells maintain their normal appearance (Fig. 5.5*a*). A solution greater than 0.9% NaCl is hypertonic to red blood cells. In such a solution, the cells shrivel up, a process called **crenation** (Fig. 5.5*b*). A solution of less than 0.9% NaCl is hypotonic to red blood cells. In such a solution, the cells swell to bursting, a process called **hemolysis** (Fig. 5.5*c*).

Complete Table 5.3 by following these instructions. In the second column, state whether the solution is isotonic, hypertonic, or hypotonic to red blood cells. In the third column, hypothesize the effect on the shape of the cell after being in this solution. In the fourth column, explain why you hypothesized this outcome. Base your explanation on the movement of water.

Table 5.3 Effect of Tonicity on Red Blood Cells

Concentration (NaCl)	Tonicity	Effect on Cells	Explanation
0.9% 3	isotonic	Normal appearance.	
Higher than 0.9% 2	Hypertonic	shrivels due to H_2O	
Lower than 0.9% 1	Hypotonic	Burst due to gain of H_2O (Bright Red)	

Figure 5.5 Tonicity and red blood cells.

(a) © David M. Phillips/Science Source, (b) © David M. Phillips/Science Source, (c) © David M. Phillips/Science Source

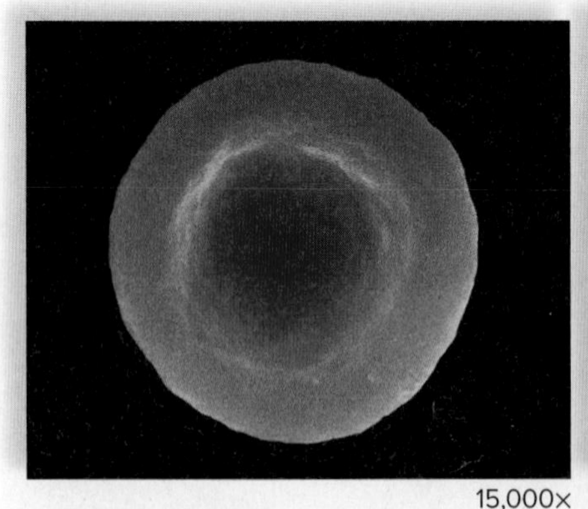

15,000×

a. Isotonic solution. Red blood cell has normal appearance due to no net gain or loss of water.

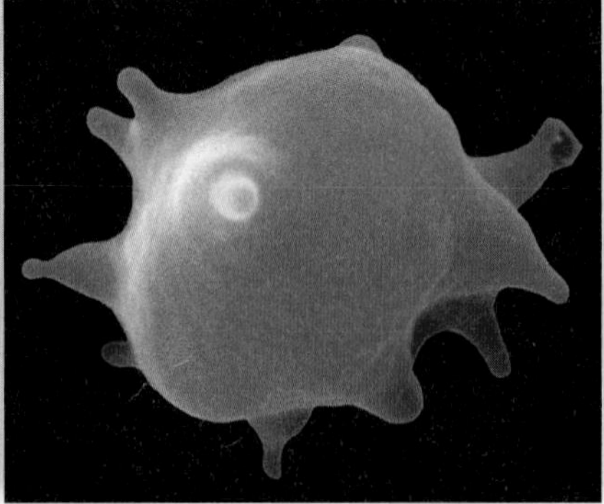

15,000×

b. Hypertonic solution. Red blood cell shrivels due to loss of water.

15,000×

c. Hypotonic solution. Red blood cell fills to bursting due to gain of water.

Experimental Procedure: Tonicity and Red Blood Cells

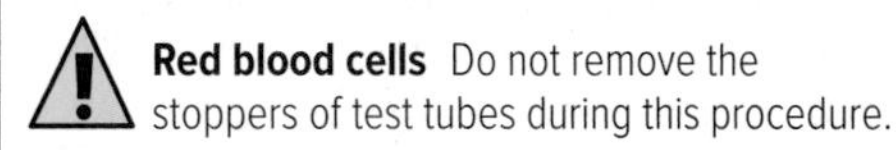

Red blood cells Do not remove the stoppers of test tubes during this procedure.

Three numbered and stoppered test tubes are on display. Each test tube contains NaCl and a few drops of blood.

1. Shake each tube as shown in Figure 5.6*a*. Then place the tube in front of your lab manual as shown in Figure 5.6*b*. Determine whether you can see the print on the page and record your decision under Print Visibility in Table 5.4.
2. Dependent on how difficult it is to see the print, which tube is hypotonic (less than 0.9% NaCl), hypertonic (10% NaCl), or isotonic (0.9% NaCl)? Record your deduction under Tonicity in Table 5.4.
3. In the last column of Table 5.4, explain how you arrived at this deduction.

Table 5.4 Print Visibility and Tonicity

Tube	Print Visibility	Tonicity	Explanation
1			
2			
3			

Figure 5.6 Proof of hemolysis.
a. Shake the tube as shown here. **b.** Once the red blood cells burst, you can read print placed behind a tube of diluted blood.

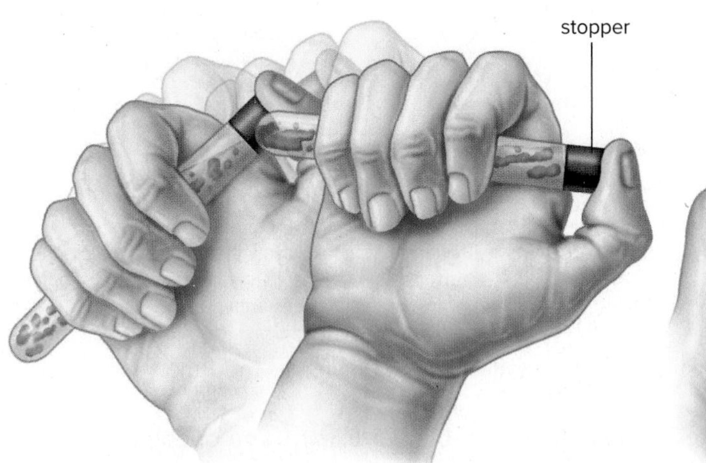

a. Gently invert tube several times.

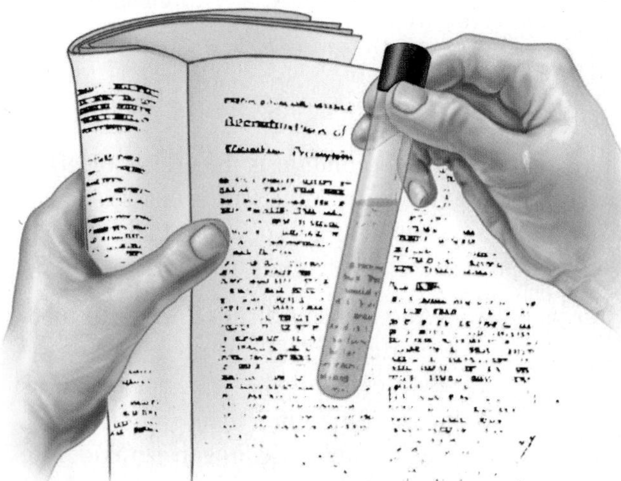

b. Determine whether print is readable through tube.

5.4 Enzyme Activity

Enzymes are organic catalysts that speed metabolic reactions, either degradation or synthesis (Fig. 5.7). Each enzyme has a three-dimensional shape that accommodates its substrate(s), the reactant(s) in the enzyme's reaction. This shape, therefore, determines the specificity of the enzyme and is important to the action of the enzyme. Although the shape of the enzyme and its substrate are compatible, the favored model for enzyme action suggests that the enzyme initially interacts with its substrate, changes shape slightly to improve the interaction, and then proceeds with a more efficient reaction. Certain environmental effects ensure that enzymes can function speedily. A warm temperature, sufficient enzyme and substrate concentrations, and the correct pH (whether the solution should be acidic, basic, or neutral) are all important. Each enzyme has a pH at which the speed of the reaction is optimum. Any pH higher or lower than the optimum affects the shape of the enzyme, leading to reduced activity.

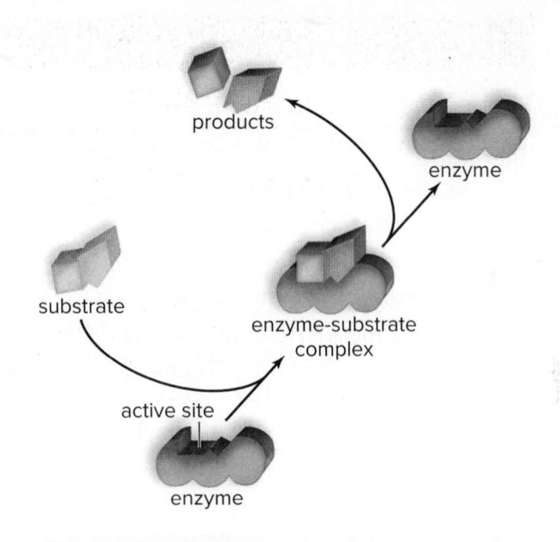

Figure 5.7 Enzymatic action. During degradation, the substrate is broken down into smaller products.

Experiment with the Enzyme Catalase

In the Experimental Procedure that follows, you will be working with the enzyme catalase. **Catalase** is present in cells, where it speeds the breakdown of the toxic chemical hydrogen peroxide (H_2O_2) to water and oxygen:

$$\underset{\text{hydrogen peroxide}}{2H_2O_2} \xrightarrow{\text{catalase}} \underset{\text{water}}{2H_2O} + \underset{\text{oxygen}}{O_2}$$

In this example, the enzyme is catalase; the substrate that fits within the active site of the enzyme is hydrogen peroxide; and the products are water and oxygen. Catalase performs a useful function in organisms because hydrogen peroxide is harmful to cells. Hydrogen peroxide is a powerful oxidizer that can attack and denature cellular molecules such as DNA. Knowing its harmful nature, humans use hydrogen peroxide as a commercial antiseptic to kill germs. In reduced concentration, hydrogen peroxide is a whitening agent used to bleach hair and teeth. It is also used industrially to clean most anything from tubs to sewage.

Experimental Procedure: Catalase Activity

This Experimental Procedure tests the effects of pH on the activity of catalase. Potato will be the source of catalase. As the reaction occurs, easily observable bubbling will develop. Label two clean test tubes and use the appropriate graduated transfer pipets to follow these directions.

> ⚠ **Protective eyewear** Protective eyewear should be worn for this procedure. HCl is a strong caustic acid. Exercise care in using this chemical, and follow your instructor's directions for disposal of this tube. If any acid should spill on your skin, rinse immediately with water.

Tube 1

1. Add 2 ml of distilled water to tube 1. (This is **neutral pH.**)
2. Macerate a cube (1 cc) of potato using a mortar and pestle. Transfer to the tube.
3. Wait 3 minutes.
4. Add 4 ml of hydrogen peroxide. Record bubbling in Table 5.5 using 0 (no bubbling) or + signs (e.g., +, ++, +++, most bubbling).

Tube 2

1. Carefully add 2 ml of hydrochloric acid (5M HCl) to this tube.
2. Macerate a cube (1 cc) of potato using a mortar and pestle. Transfer to the tube.
3. Wait 3 minutes.
4. Add 4 ml hydrogen peroxide. Record the degree of bubbling in Table 5.5.

Table 5.5 Effect of pH on Catalase Activity

Tube	Contents	Bubbling*	Explanation
1	Distilled water Potato, macerated Hydrogen peroxide		
2	Hydrochloric acid Potato, macerated Hydrogen peroxide		

*Ignore large bubbles and look for small bubbles.

Conclusions: Catalase Activity

- Give explanations for your results in Table 5.5.
- From your test results, decide whether enzymes can be negatively affected by environmental conditions and explain your answer. ____________________

____

Application for Daily Living

Dehydration and Water Intoxication

Most people have heard of dehydration, but they may not realize that dehydration occurs because blood has become hypertonic to cells and, in response, the cells lose water. Dehydration can be due to excessive sweating, perhaps during exercise, or it can be a side effect of many illnesses that cause prolonged vomiting or diarrhea. The signs of moderate dehydration are a dry mouth, sunken eyes, and skin that will not bounce back after light pinching. Most people have never heard of water intoxication, which occurs when blood becomes hypotonic to cells. In response to hypotonicity, cells gain water. Water intoxication can lead to pulmonary edema (the lungs gain water) and swelling in the brain. In extreme cases, it is fatal. It can be due to an intake of too much pure water during vigorous exercise, such as a marathon race. The cure is introducing an intravenous solution containing high amounts of sodium in a hospital setting. To prevent both dehydration and water intoxication, athletes should replace lost fluids slowly. Pure water is a good choice if the exercise period is short. Low-sodium solutions, such as sports drinks, are a good choice for longer-duration events like marathons.

Human Laboratory Review 5

_______________ **1.** What is the function of rough endoplasmic reticulum?

_______________ **2.** Which organelle carries on intracellular digestion?

_______________ **3.** What is the function of the nucleus?

_______________ **4.** Which organelle is responsible for protein synthesis?

_______________ **5.** What regulates the movement of molecules into and out of the cell?

_______________ **6.** What term is used to describe the movement of molecules from an area of higher concentration to an area of lower concentration?

_______________ **7.** Which types of molecules can pass through the plasma membrane?

_______________ **8.** What is the term for the movement of water across a selectively permeable membrane?

_______________ **9.** Is 10% NaCl isotonic, hypertonic, or hypotonic to red blood cells?

_______________ **10.** What appearance will red blood cells have when they are placed in 0.0009% NaCl?

_______________ **11.** In which direction does water move when cells are placed in a hypertonic solution?

_______________ **12.** The active site of an enzyme brings together the _______________ of a reaction.

_______________ **13.** In general, what do unfavorable environmental conditions do to the speed of an enzymatic reaction?

_______________ **14.** An unfavorable pH causes an enzyme to lose its normal _______________.

Thought Questions

15. If a dialysis bag filled with water is placed in a starch solution, what do you predict will happen to the weight of the bag over time? Why?

16. Ocean water is hypertonic to the internal environment of the body. Predict what would happen to your cells if you consumed large quantities of ocean (salt) water.

15. Why does the human body strive to maintain a near-neutral pH?

LABORATORY

6

Body Tissues

Learning Outcomes

6.1 Epithelial Tissue

- Identify slides and models of various types of epithelium.
- Explain where particular types of epithelium are located in the body, and describe their function.

Prelab Question: Where in the digestive tract would you find epithelial tissue?

6.2 Connective Tissue

- Identify slides and models of various types of connective tissue.
- Explain where particular connective tissues are located in the body, and describe their function.

Prelab Question: Name six types of connective tissue and give a location for each type in the body.

6.3 Muscular Tissue

- Identify slides and models of three types of muscular tissue.
- Explain where each type of muscular tissue is located in the body, and describe their function.

Prelab Question: Name each type of muscle tissue, and name a specific organ containing each type.

6.4 Nervous Tissue

- Identify a slide and model of a neuron.
- Explain where nervous tissue is located in the body, and describe its function.

Prelab Question: Which part of a neuron do you expect to see concentrated in the brain?

6.5 Organ Level of Organization

- Identify a slide of the intestinal wall and the layers in the wall. Describe the function of each tissue.
- Identify a slide of skin and the two regions of skin. Describe the function for each region of skin.

Prelab Question: Choose an organ and describe which types of tissues it contains.

***Application for Daily Living:* Tissue Engineering**

Introduction

Humans, as well as all other organisms, are made up of **cells.** Groups of cells that have the same structural characteristics and perform the same functions are called **tissues.** Figure 6.1 shows the four categories of tissues in the human body. An **organ** is composed of different types of tissues, and various organs form **organ systems.** Humans thus have the following levels of biological organization: cells ⟶ tissues ⟶ organs ⟶ organ systems.

In this lab, we will examine tissues in the human body. Figure 6.1 shows the four categories of tissues including epithelial, connective, muscular, and nervous. While some organs of the body are known for having a particular type of tissue—for example, the brain contains nervous tissue and a muscle contains muscular tissue—other organs have several types of tissues.

Figure 6.1 The major tissues in the human body.
The many kinds of tissues in the human body are grouped into four types: epithelial tissue, muscular tissue, nervous tissue, and connective tissue.

(simple squamous, pseudostratified columnar, simple cuboidal, simple columnar, cardiac): © Ed Reschke; (smooth): © Ed Reschke/Photolibrary/Getty Images; (skeletal, nervous): © Ed Reschke; (blood, adipose): © McGraw-Hill Education/Al Telser, photographer; (bone): © McGraw-Hill Education/Dennis Strete, photographer; (cartilage): © Ed Reschke; (dense): © Ed Reschke/Photolibrary/Getty Images

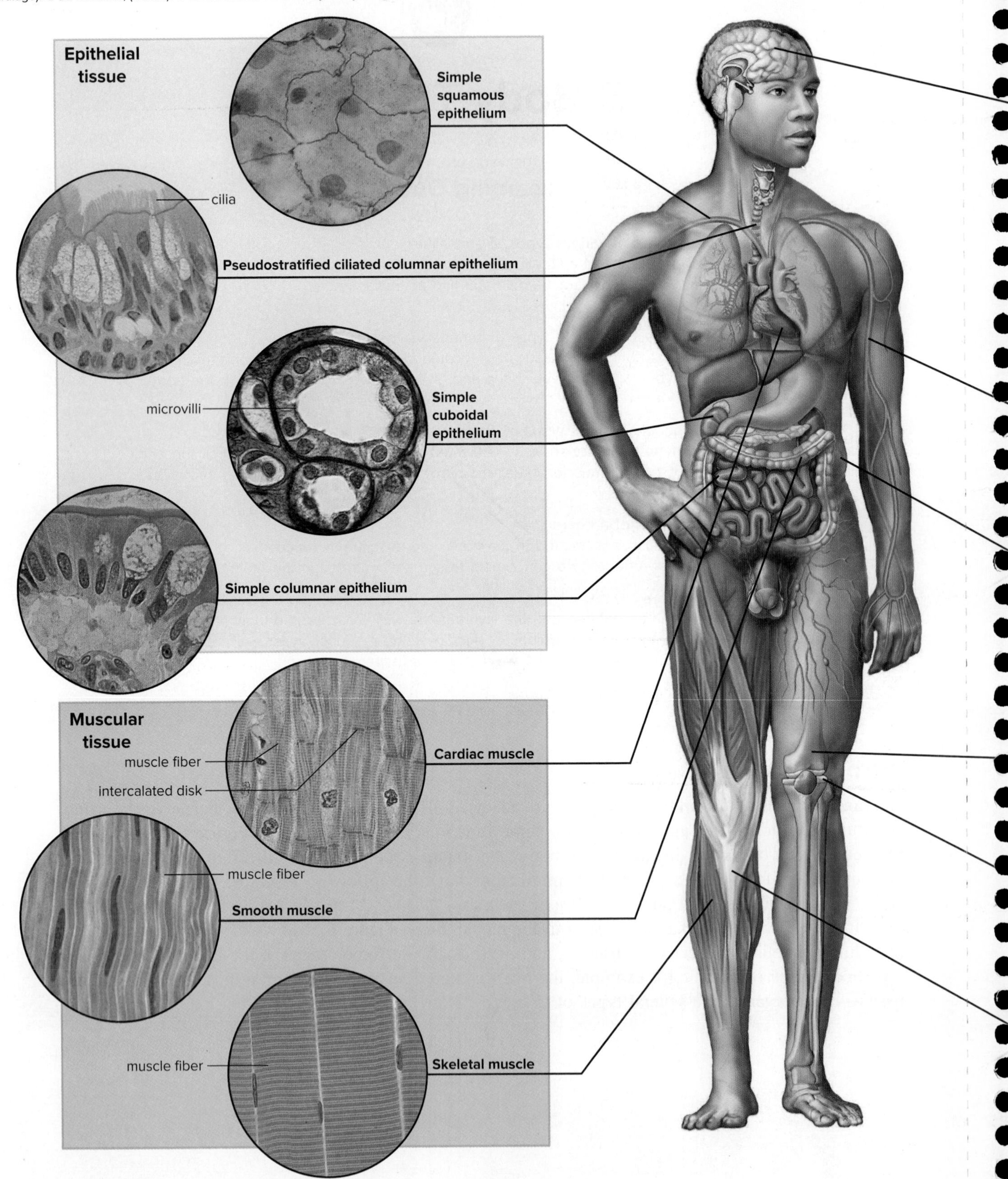

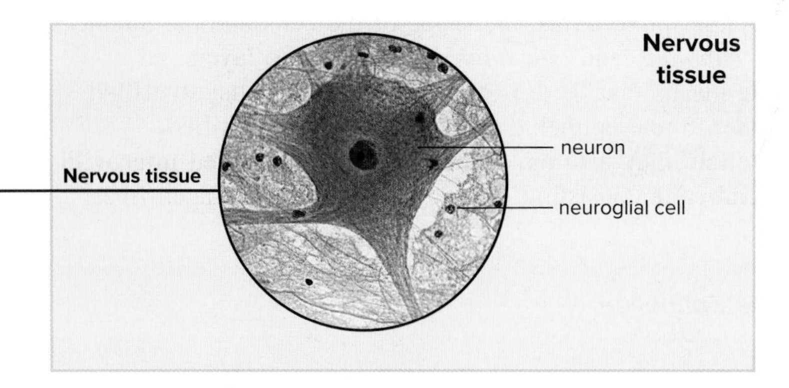
Nervous tissue
Nervous tissue
neuron
neuroglial cell

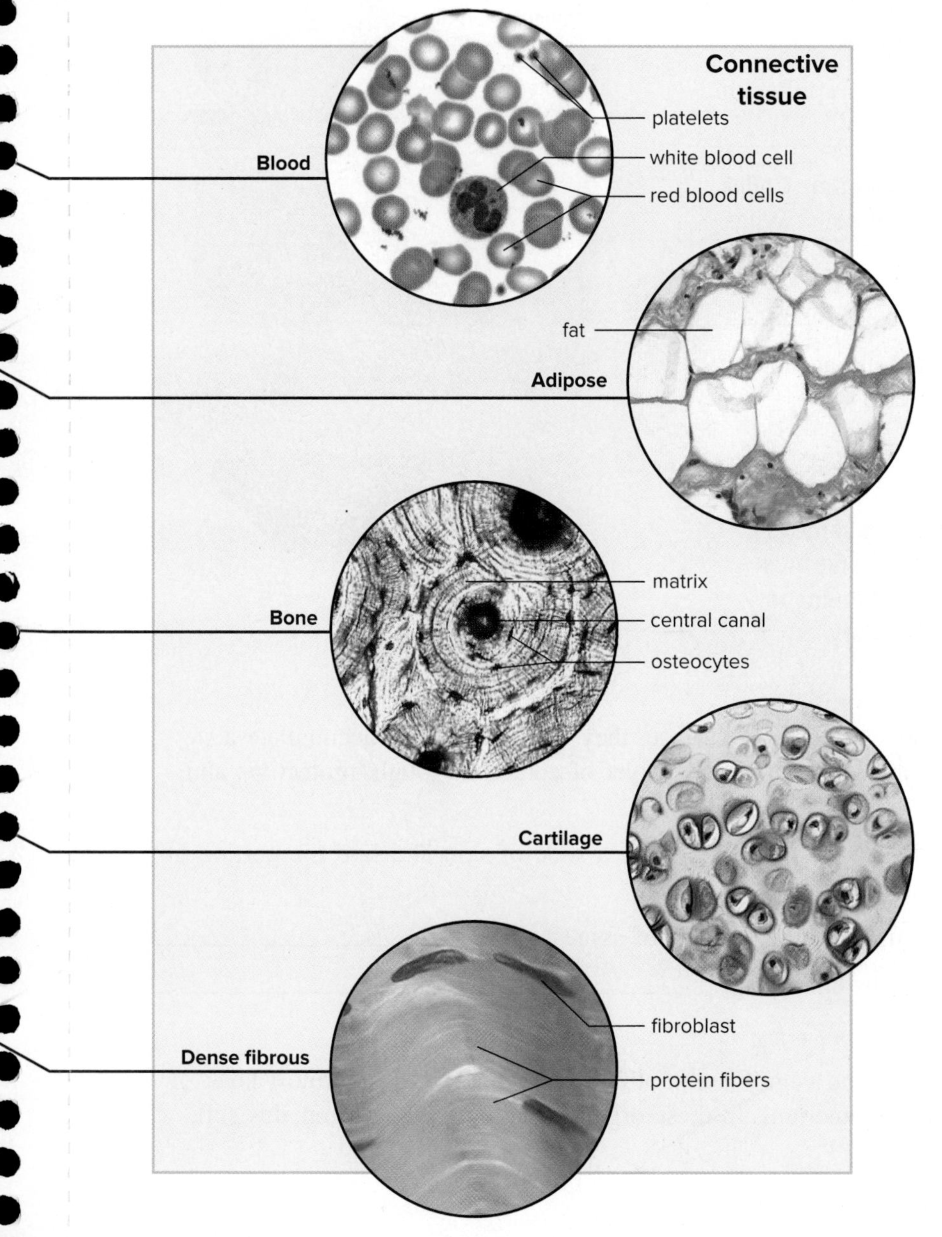
Connective tissue
Blood
platelets
white blood cell
red blood cells
fat
Adipose
Bone
matrix
central canal
osteocytes
Cartilage
Dense fibrous
fibroblast
protein fibers

6.1 Epithelial Tissue

Epithelial tissue (epithelium) forms a continuous layer, or sheet, over the entire body surface and most of the body's inner cavities. Externally, it forms a covering that protects the animal from infection, injury, and drying out. Some epithelial tissues produce and release secretions. Others absorb nutrients.

The name of an epithelial tissue includes two descriptive terms: the shape of the cells and the number of layers. The three possible shapes are *squamous, cuboidal,* and *columnar.* With regard to layers, an epithelial tissue may be simple or stratified. **Simple** means that there is only one layer of cells; **stratified** means that cell layers are placed on top of each other. Some epithelial tissues are **pseudostratified,** meaning that they only appear to be layered. Epithelium may also have cellular extensions called **microvilli** or hairlike extensions called **cilia.** A **basement membrane** consisting of glycoproteins and collagen fibers joins an epithelium to underlying connective tissue.

Observation: Simple and Stratified Squamous Epithelium

Simple Squamous Epithelium

Simple squamous epithelium is a single layer of thin, flat, many-sided cells, each with a central nucleus. It lines internal cavities, the heart, and all the blood vessels. It also lines parts of the urinary, respiratory, and male reproductive tracts.

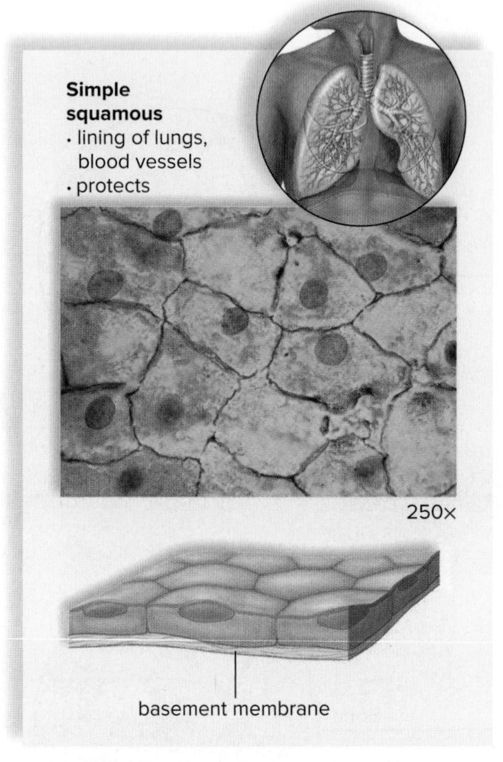

© Ed Reschke

1. Study a model or diagram of simple squamous epithelium.

   What does squamous mean? ____________________

2. Examine a prepared slide of squamous epithelium. Under low power, note the close packing of the flat cells. What shapes are the cells? ____________________

   ____

3. Under high power, examine an individual cell, and identify the plasma membrane, cytoplasm, and nucleus.
4. Add a sketch of this tissue to Table 6.1.

Stratified Squamous Epithelium

As would be expected from its name, stratified squamous epithelium consists of many layers of cells. The innermost layer produces cells that are first cuboidal or columnar in shape, but as the cells push toward the surface, they become flattened.

The outer region of the skin, called the epidermis, is stratified squamous epithelium. As the cells move toward the surface, they flatten, begin to accumulate a protein called **keratin,** and eventually die. Keratin makes the outer layer of epidermis tough, protective, and able to repel water.

1. Either now or when you are studying skin in section 6.2, examine a slide of skin and find the portion of the slide that is stratified squamous epithelium.
2. Approximately how many layers of cells make up this portion of skin? ____________________
3. Which layers of cells best represent squamous epithelium? ____________________
4. Add a sketch of this tissue to Table 6.1.

The linings of the mouth, throat, anal canal, and vagina are stratified epithelium. The outermost layer of cells surrounding the cavity is simple squamous epithelium. In these organs, this layer of cells remains soft, moist, and alive.

Observation: Simple Cuboidal Epithelium

Simple cuboidal epithelium is a single layer of cube-shaped cells, each with a central nucleus. It is found in tubules of the kidney and in the ducts of many glands, where it has a protective function. It also occurs in the secretory portions of some glands—that is, where the tissue produces and releases secretions.

1. Study a model or diagram of simple cuboidal epithelium.
2. Examine a prepared slide of simple cuboidal epithelium. Move the slide until you locate cube-shaped cells that line a lumen (cavity). Are these cells ciliated? ______________
3. Add a sketch of this tissue to Table 6.1.

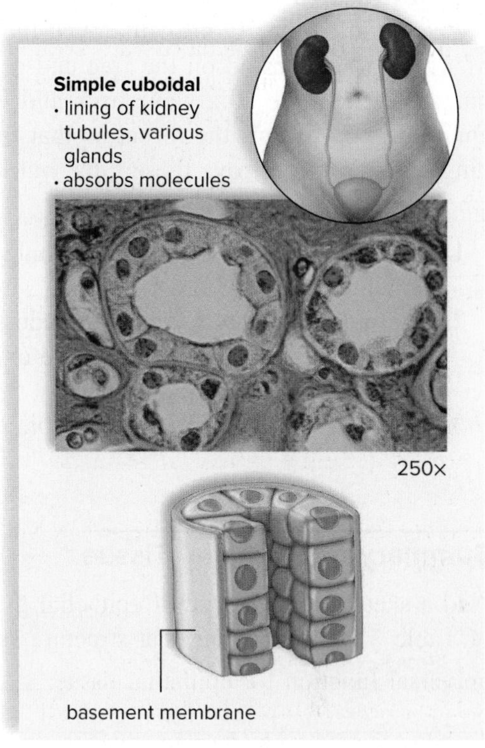

© Ed Reschke

Observation: Simple Columnar Epithelium

Simple columnar epithelium is a single layer of tall, cylindrical cells, each with a nucleus near the base. This tissue, which lines the digestive tract from the stomach to the anus, protects, secretes, and allows absorption of nutrients.

1. Study a model or diagram of simple columnar epithelium.
2. Examine a prepared slide of simple columnar epithelium. Find tall and narrow cells that line a lumen. Under high power, focus on an individual cell. Identify the plasma membrane, the cytoplasm, and the nucleus. Epithelial tissues are attached to underlying tissues by a basement membrane composed of extracellular material containing protein fibers.
3. The tissue you are observing contains mucus-secreting cells. Search among the columnar cells until you find a **goblet cell,** so named because of its goblet-shaped, clear interior. This region contains mucus, which may be stained a light blue. In the living animal, the mucus is discharged into the gut cavity and protects the lining from digestive enzymes.
4. Add a sketch of this tissue to Table 6.1.

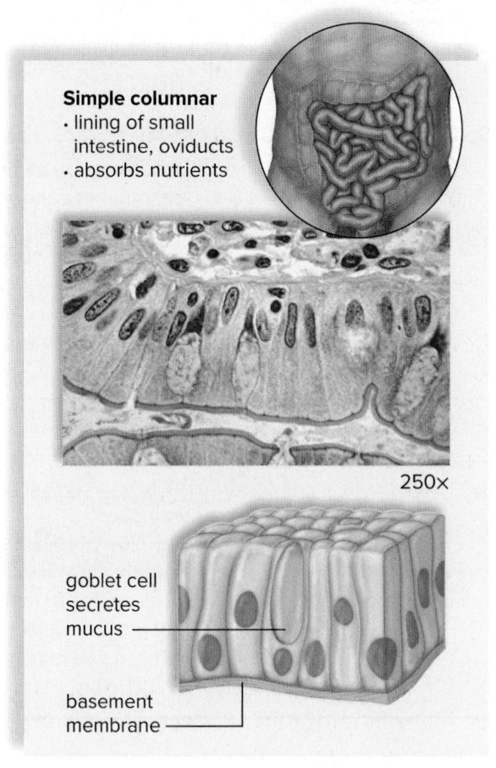

© Ed Reschke

Observation: Pseudostratified Ciliated Columnar Epithelium

Pseudostratified ciliated columnar epithelium appears to be layered, while actually all cells touch the basement membrane. Many cilia are located on the free end of each cell. In the human trachea, the cilia wave back and forth, moving mucus and debris up toward the throat so that they cannot enter the lungs. Smoking destroys these cilia, but they will grow back if smoking is discontinued.

1. Study a model or diagram of pseudostratified ciliated columnar epithelium.
2. Examine a prepared slide of pseudostratified ciliated columnar epithelium. Concentrate on the part of the slide that resembles the model. Identify the cilia.
3. Add a sketch of this tissue to Table 6.1.

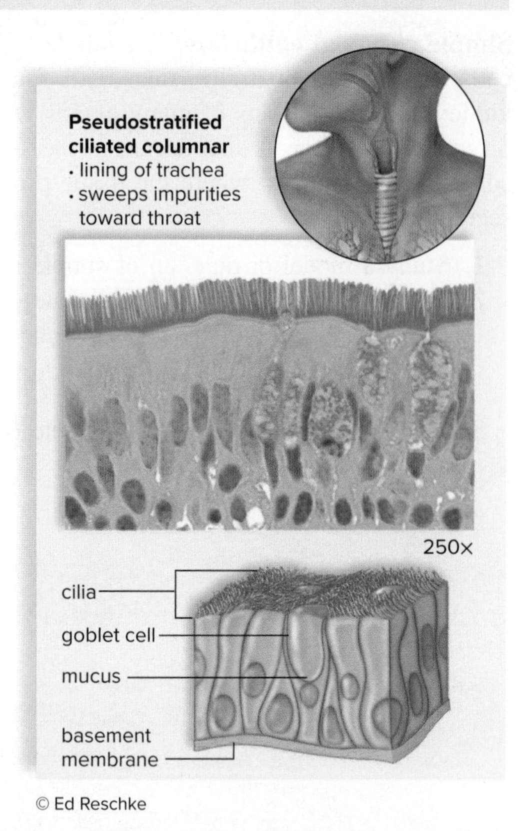

© Ed Reschke

Summary of Epithelial Tissue

Add a sketch of each type of epithelial tissue in the third column of Table 6.1. Recognizing that structure suits function, state a universal function for epithelial tissue. ______________________

____

Table 6.1 Epithelial Tissue

Sketch	Structure	Sketch	Location
Simple squamous	Tightly packed thin flat cells		Walls of capillaries, lining of blood vessels, air sacs of lungs, lining of internal cavities
Stratified squamous	Innermost layers are cuboidal or columnar; outermost layers are flattened dead cells.		Skin, linings of mouth, throat, anal canal, vagina
Simple cuboidal	Tightly packed cuboidal cells that often have microvilli at one end		Surface of ovaries, linings of ducts and glands, lining of kidney tubules
Simple columnar	Columnlike—tall, cylindrical nucleus at base; may contain goblet cells.		Lining of uterus, tubes of digestive tract
Pseudostratified ciliated columnar	Tightly packed columnar cells appear to be layered but are not. Cells are ciliated and may contain goblet cells.		Linings of respiratory passages

6.2 Connective Tissue

Connective tissue joins different parts of the body together. There are four general classes of connective tissue: connective tissue proper, bone, cartilage, and blood. All types of connective tissue consist of cells surrounded by a matrix that usually contains fibers. Elastic fibers are composed of a protein called elastin. Collagenous fibers contain the protein collagen.

Observation: Connective Tissue

There are several different types of connective tissue. The accompanying illustrations will help you understand the name of each tissue, where it occurs, and its functions. We will study loose fibrous connective tissue, dense fibrous connective tissue, adipose tissue, bone, cartilage, and blood. **Loose fibrous connective tissue,** so named because it has space between components, supports epithelium and also many internal organs, such as muscles, blood vessels, and nerves. Its loose construction allows organs to freely move. **Dense fibrous connective tissue** contains many collagenous fibers packed closely together, as in tendons, which connect muscles to bones, and in ligaments, which connect bones to other bones at joints.

1. Examine a slide of loose fibrous connective tissue, and compare it to the figure below (*left*). What is the function of loose fibrous connective tissue? ______________________________
2. Examine a slide of dense fibrous connective tissue, and compare it to the figure below (*right*). What two kinds of structures in the body contain dense fibrous connective tissue? ____________________

    ____
3. Add sketches of these tissues to Table 6.2.

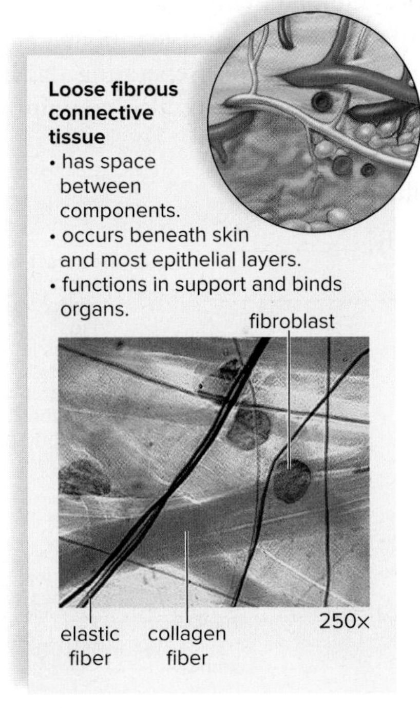

© Ed Reschke

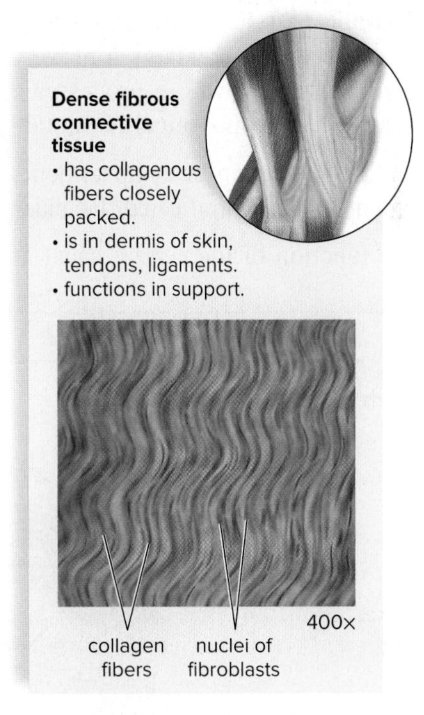

© McGraw-Hill Education/Dennis Strete, photographer

Observation: Adipose Tissue

In **adipose tissue,** the cells have a large, central, fat-filled vacuole that causes the nucleus and cytoplasm to be at the perimeter of the cell. Adipose tissue occurs beneath the skin, where it insulates the body, and around internal organs, such as the kidneys and heart. It cushions and helps protect these organs.

1. Examine a prepared slide of adipose tissue. Why is the nucleus pushed to one side? ______________________________

   ____

2. State a location for adipose tissue in the body. ______________________________

   ____

   ____

   What are two functions of adipose tissue at this location? ______________________________

   ____

   ____

3. Add a sketch of this tissue to Table 6.2.

Adipose tissue
- cells are filled with fat.
- occurs beneath skin, around heart and other organs.
- functions in insulation, stores fat.

250×

nucleus

Observation: Compact Bone

Compact bone is found in the bones that make up the skeleton. It consists of **osteons** (Haversian system) with a **central canal,** and concentric rings of spaces called **lacunae,** connected by tiny crevices called **canaliculi.** The central canal contains a nerve and blood vessels, which service bone. The lacunae contain bone cells called **osteocytes,** whose processes extend into the canaliculi. Separating the lacunae is a matrix that is hard because it contains minerals, notably calcium salts. The matrix also contains collagenous fibers.

1. Study a model or diagram of compact bone. Then look at a prepared slide and identify the central canal, lacunae, and canaliculi.
2. What is the function of the central canal and canaliculi? ______________________________

   ____

   ____

3. Add a sketch of this tissue to Table 6.2.

Compact bone
- has cells in concentric rings.
- occurs in bones of skeleton.
- functions in support and protection.

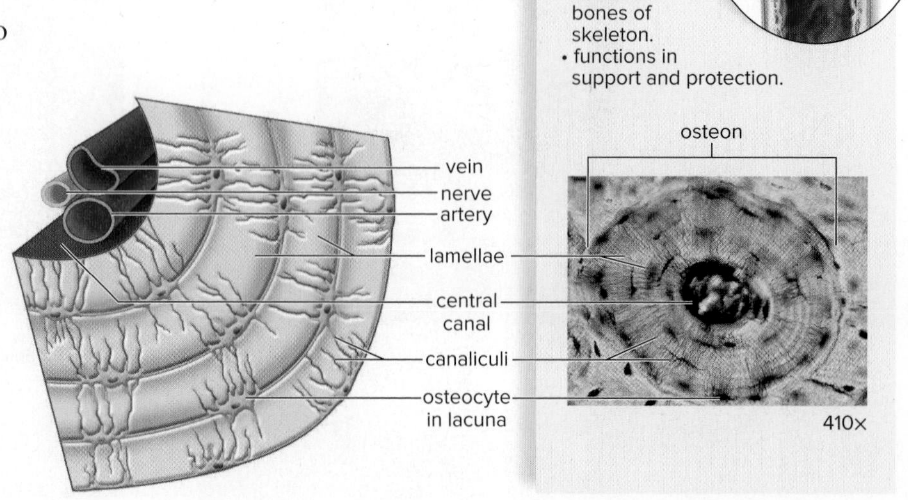

Observation: Hyaline Cartilage

In **hyaline cartilage,** cells called **chondrocytes** are found in twos or threes in lacunae. The lacunae are separated by a flexible matrix containing weak collagenous fibers.

1. Study the diagram and photomicrograph of hyaline cartilage in the figure at the right. Then study a prepared slide of hyaline cartilage, and identify the matrix, lacunae, and chondrocytes.
2. Compare compact bone and hyaline cartilage. Which of these types of connective tissue is more organized? ______________________

   Why? ______________________
3. Which of these two types of connective tissue lends more support to body parts? Why? ______________________

   ____
4. Add a sketch of this tissue to Table 6.2.

Hyaline cartilage
- has cells in lacunae.
- occurs in nose and walls of respiratory passages; at ends of bones, including ribs.
- functions in support and protection.

chondrocyte within lacunae

matrix

250×

Observation: Blood

Blood is a connective tissue in which the matrix is an intercellular fluid called **plasma. Red blood cells** (erythrocytes) have a biconcave appearance and lack a nucleus. These cells carry oxygen combined with the respiratory pigment hemoglobin. **White blood cells** (leukocytes) have a nucleus and are typically larger than the more numerous red blood cells. These cells fight infection.

1. Study a prepared slide of human blood. With the help of Figure 6.2, identify the red blood cells and the white blood cells, which appear faint because of the stain.
2. Try to identify a neutrophil, which has a multilobed nucleus, and a lymphocyte, which is the smallest of the white blood cells, with a spherical or slightly indented nucleus.
3. Add a sketch of this tissue to Table 6.2.

Figure 6.2 Blood cells.
Red blood cells are more numerous than white blood cells. White blood cells can be separated into five distinct types. If you have blood work done that includes a complete blood count (CBC), the doctor is getting a count of each of these types of white blood cells. (a-e: Magnification 1,050×) (a-c) © Ed Reschke; (d) © Biophoto Associates/Science Source; (e) © Ed Reschke/Getty Images

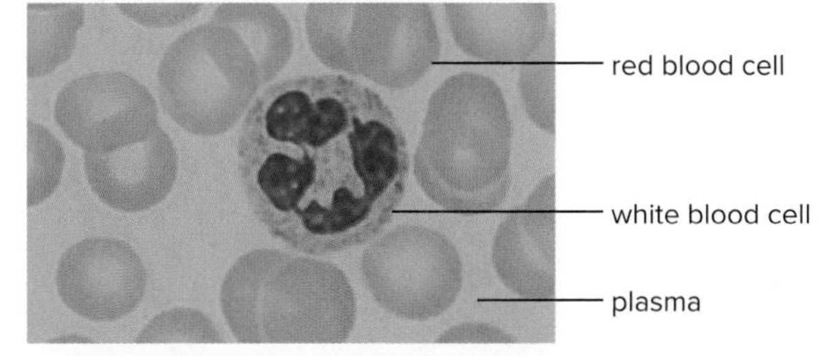

a. Neutrophil

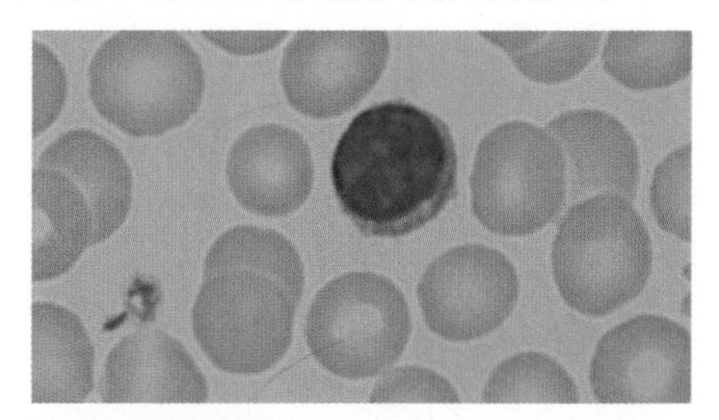

b. Lymphocyte

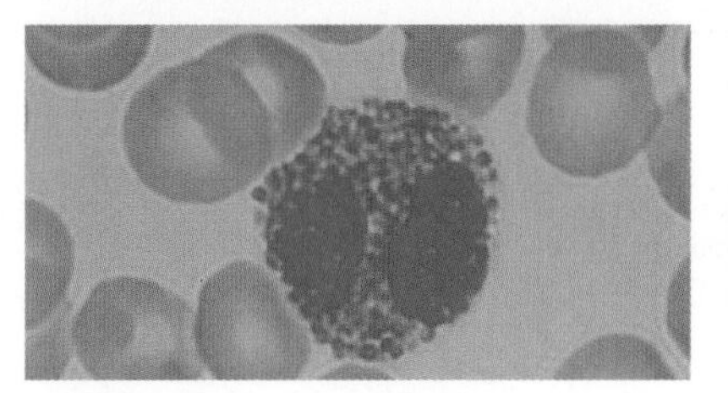

c. Eosinophil

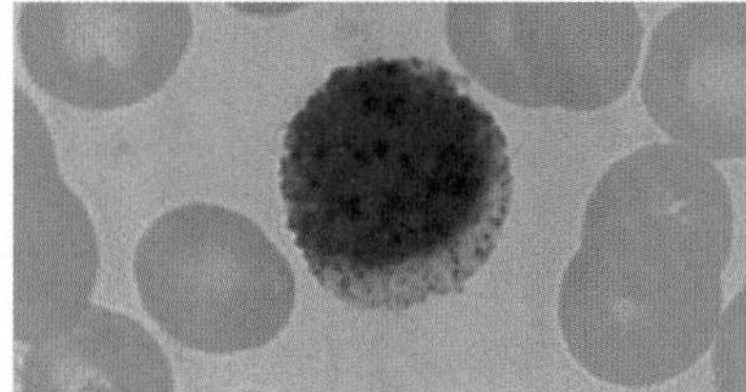

d. Basophil

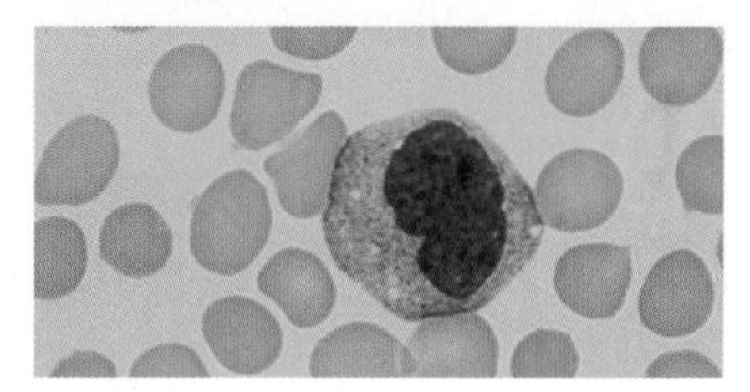

e. Monocyte

Summary of Connective Tissue

1. Add a sketch of each type of connective tissue in the third column of Table 6.2. Recognizing that structure suits function, state a universal function for connective tissue. ______________________

____

Table 6.2 Connective Tissue

Type	Structure	Sketch	Location
Loose fibrous connective tissue	Fibers are widely separated.		Between the muscles; beneath the skin; beneath most epithelial layers
Dense fibrous connective tissue	Fibers are closely packed.		Tendons, ligaments
Adipose	Large cells with fat-filled vacuoles; nuclei are pushed to one side.		Beneath the skin; around the kidney and heart; in the breast
Compact bone	Concentric circles of cells separated by a hard matrix		Bones of skeleton
Hyaline cartilage	Cells in lacunae within a flexible matrix		Nose; ends of bones; rings in walls of respiratory passages; between ribs and sternum
Blood	Red and white cells floating in plasma		Blood vessels

2. Working with others in a group, decide how the structure of each connective tissue suits its function.

Loose fibrous connective tissue ______________________

Dense fibrous connective tissue ______________________

Adipose tissue ______________________

Compact bone ______________________

Hyaline cartilage ______________________

Blood ______________________

6.3 Muscular Tissue

Muscular (contractile) tissue is composed of cells called muscle fibers. Muscular tissue has the ability to contract, and contraction usually results in movement. The body contains skeletal, cardiac, and smooth muscle.

Observation: Skeletal Muscle

Skeletal muscle occurs in the muscles attached to the bones of the skeleton. The contraction of skeletal muscle is said to be **voluntary** because it is under conscious control. Skeletal muscle is striated; it contains light and dark bands. The striations are caused by the arrangement of contractile filaments (actin and myosin filaments) in muscle cells often called fibers. Each fiber contains many nuclei, all peripherally located.

1. Study a model or diagram of skeletal muscle, and note that striations are present. You should see several muscle fibers, each marked with striations.
2. Examine a prepared slide of skeletal muscle. The striations may be difficult to make out, but bringing the slide in and out of focus may also help.
3. Each skeletal muscle cell is not only multinucleated, but also cylindrical and elongated. Counting partial cells, how many muscle cells are in the adjoining micrograph? ______________

____

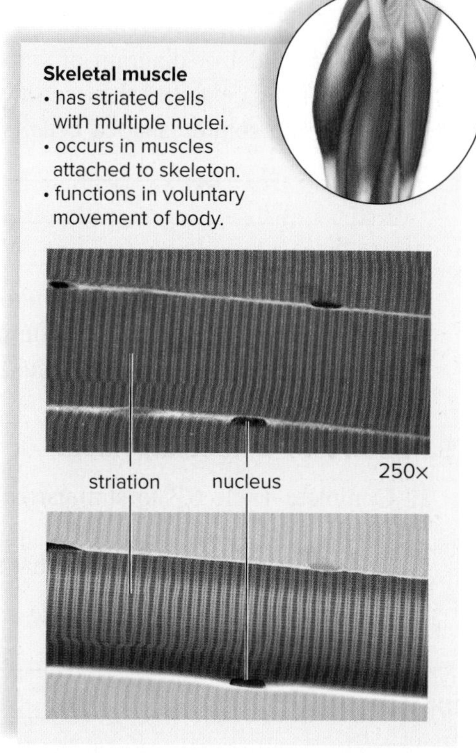

Observation: Cardiac Muscle

Cardiac muscle is found only in the heart. It is called **involuntary** because its contraction does not require conscious effort. Cardiac muscle is striated in the same way as skeletal muscle. However, the fibers are branched and bound together at **intercalated disks,** where their folded plasma membranes touch. This arrangement aids communication between fibers.

1. Study a model or diagram of cardiac muscle, and note that striations are present.
2. Examine a prepared slide of cardiac muscle. Find an intercalated disk. What is the function of cardiac muscle? ____

____

____

3. Aside from being striated, cardiac muscle cells are rectangular and branched. What is the benefit of this arrangement? ______

____

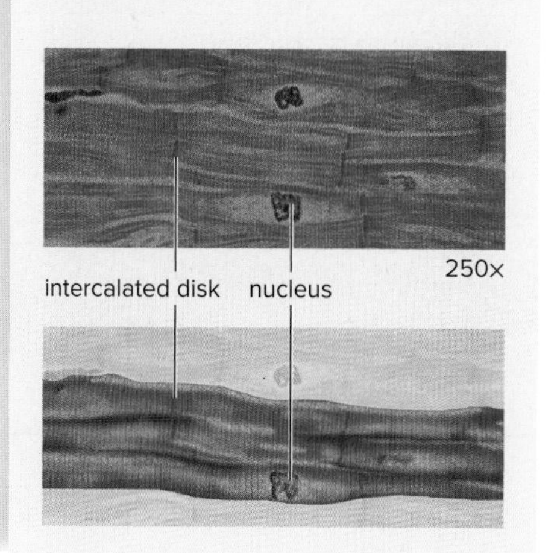

Observation: Smooth Muscle

Smooth muscle is sometimes called **visceral muscle** because it makes up the walls of the internal organs, such as the intestines and the blood vessels. Smooth muscle is involuntary because its contraction does not require conscious effort.

1. Study a model or diagram of smooth muscle, and note the shape of the cells and the centrally placed nucleus. Smooth muscle has spindle-shaped cells. What does *spindle-shaped* mean? ______________________________

____

____

2. Examine a prepared slide of smooth muscle. Distinguishing the boundaries between the different cells may require you to take the slide in and out of focus.

Smooth muscle
- has spindle-shaped cells, each with a single nucleus.
- cells have no striations.
- functions in movement of substances in lumens of body.
- is involuntary.
- is found in blood vessel walls and walls of the digestive tract.

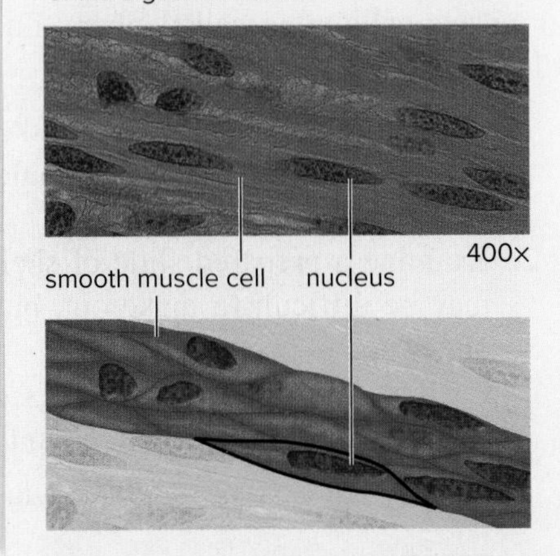

Summary of Muscular Tissue

1. Complete Table 6.3 to summarize your study of muscular tissue.
2. How does it benefit an animal that skeletal muscle is voluntary while cardiac and smooth muscle are involuntary? ______________________________

____

____

Table 6.3 Muscular Tissue

Type	Striations (Yes or No)	Branching (Yes or No)	Conscious Control (Yes or No)
Skeletal			
Cardiac			
Smooth			

6.4 Nervous Tissue

Nervous tissue is found in the brain, spinal cord, and nerves. Nervous tissue receives and integrates incoming stimuli before conducting nerve impulses, which control the glands and muscles of the body. Nervous tissue is composed of two types of cells: **neurons** that transmit messages and **neuroglia** that support and nourish the neurons. Motor neurons, which take messages from the spinal cord to the muscles, are often used to exemplify typical neurons. Motor neurons have several **dendrites,** processes that take signals to a **cell body,** where the nucleus is located, and an **axon** that takes nerve impulses away from the cell body.

Observation: Nervous Tissue

1. Study a model or diagram of a neuron, and then examine a prepared slide. Most likely, you will not be able to see neuroglia because they are much smaller than neurons.
2. Identify a cell body, the nucleus, a dendrite, and the axon in Figure 6.3*a* and *label the micrograph. Also label the neuroglia surrounding the neuron.*
3. Explain the appearance and function of the parts of a motor neuron:
   a. Dendrites ____
   ____
   b. Cell body ____
   ____
   c. Axon ____
   ____

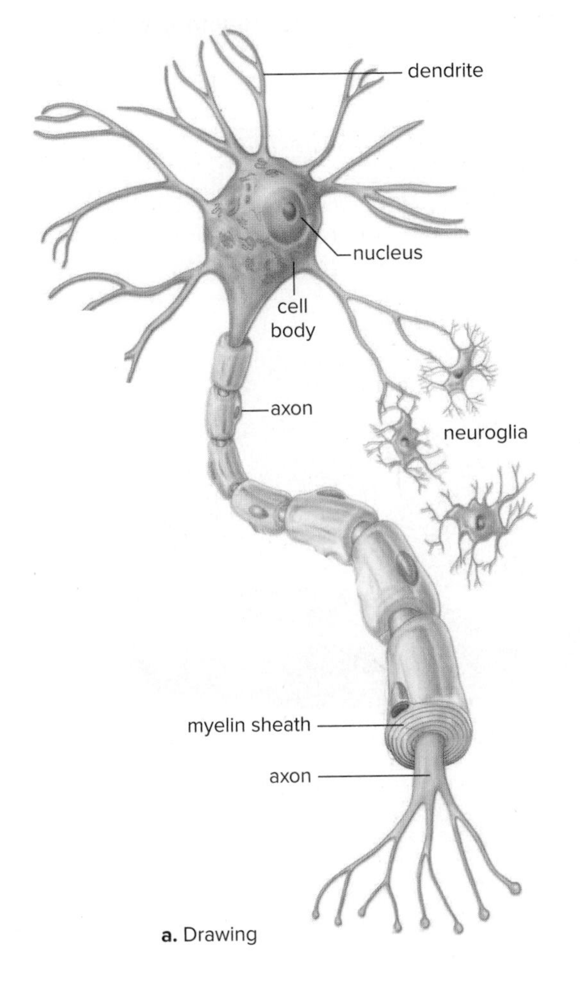

a. Drawing

Figure 6.3 Motor neuron anatomy.

© Ed Reschke

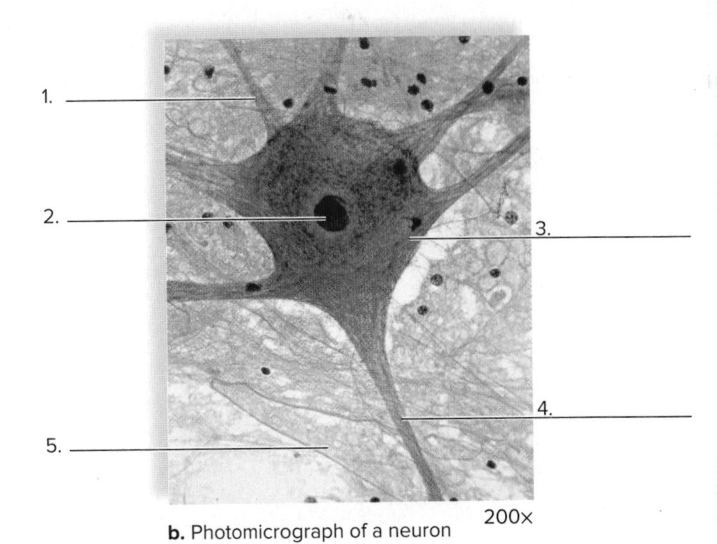

b. Photomicrograph of a neuron

6.5 Organ Level of Organization

Organs are structures composed of two or more types of tissue that work together to perform particular functions. You may tend to think that a particular organ contains only one type of tissue. For example, muscular tissue is usually associated with muscles and nervous tissue with the brain. However, muscles and the brain also contain other types of tissue—for example, loose connective tissue and blood. Here we will study the compositions of two organs—the intestine and the skin.

Intestine

The **intestine,** a part of the digestive system, processes food and absorbs nutrient molecules.

Observation: Intestinal Wall

Study a slide of a cross section of intestinal wall. With the help of Figure 6.4, identify the following layers:

1. **Mucosa** (mucous membrane layer): This layer, which lines the central lumen (cavity), is made up of columnar epithelium overlying the submucosa. This epithelium is glandular—that is, it secretes mucus from goblet cells and digestive enzymes from the rest of the epithelium. The membrane is arranged in deep folds (fingerlike projections) called **villi,** which increase the small intestine's absorptive surface.
2. **Submucosa** (submucosal layer): This loose fibrous connective tissue layer contains nerve fibers, blood vessels, and lymphatic vessels. The products of digestion are absorbed into these blood and lymphatic vessels.
3. **Muscularis** (smooth muscle layer): Circular muscular tissue and then longitudinal muscular tissue are found in this layer. Rhythmic contraction of these muscles causes **peristalsis,** a wavelike motion that moves food along the intestine.
4. **Serosa** (serous membrane layer): In this layer, a thin sheet of loose fibrous connective tissue underlies a thin, outermost sheet of squamous epithelium. This membrane is part of the **peritoneum,** which lines the entire abdominal cavity.

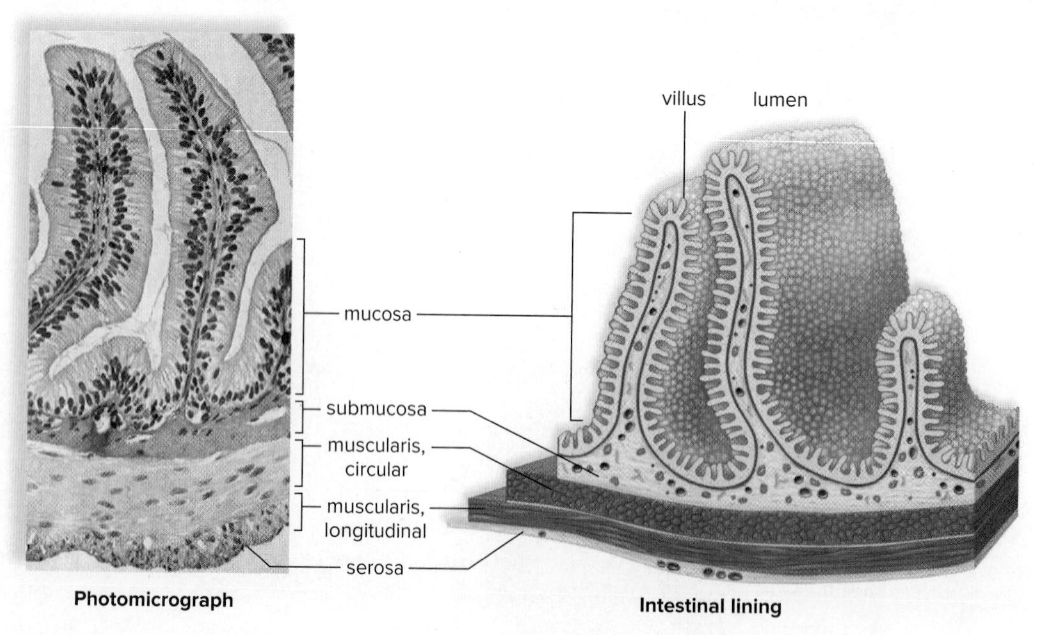

Figure 6.4 Tissues of the intestinal wall.
© Ed Reschke

In Table 6.4, list the types of tissue found in the layers of the intestinal wall.

Table 6.4 Tissues of the Intestinal Wall

	Mucosa	Submucosa	Muscularis	Serosa
Tissue(s)				

Skin

The skin covers the entire exterior of the human body. Skin functions include protection, water retention, sensory reception, body temperature regulation, and vitamin D synthesis.

Observation: Skin

Study a model or diagram and also a prepared slide of the skin. Identify these two skin regions and the subcutaneous layer.

1. **Epidermis:** This region is composed of stratified squamous epithelial cells. The outer cells of the epidermis are nonliving and create a waterproof covering that prevents excessive water loss. These cells are always being replaced because an inner layer of the epidermis is composed of living cells that constantly produce new cells.
2. **Dermis:** This region is a connective tissue containing blood vessels, nerves, sense organs, and the expanded portions of oil (sebaceous) and sweat glands and hair follicles.

   List the structures you can identify on your slide: ____

   ____

   ____

3. **Subcutaneous layer:** This is a layer of loose connective tissue and adipose tissue that lies beneath the skin proper and serves to insulate and protect inner body parts.

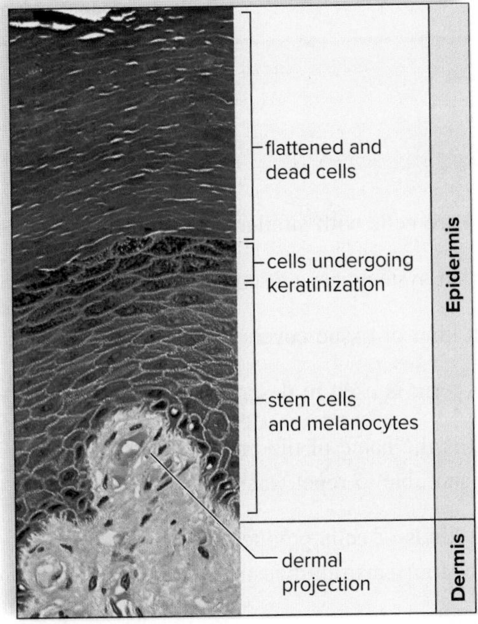

Photomicrograph of skin

Application for Daily Living

Tissue Engineering

Just a few years ago, scientists believed that transplant tissues and organs had to come directly from humans. Now, however, tissue engineering is demonstrating that it is possible to make replacement tissues and organs in the lab. For example, skin products have now been approved for use in humans. One is composed of dermal cells growing on a degradable polymer, which can be used to temporarily cover the wounds of burn patients while their own skin regenerates (Fig. 6.5). The other uses only live human skin cells to treat diabetic leg and foot ulcers. Similarly, the damaged cartilage of a knee can be replaced with a tissue produced after chondrocytes are harvested from a patient. Soon to come is a host of other products, including replacement corneas, heart valves, bladder valves, and breast tissue.

After nine years, researchers were able to produce a working urinary bladder in the laboratory. This bladder can be implanted in humans whose bladders have been damaged by accident or disease. Another group of scientists have been able to grow arterial blood vessels in the lab using a pig's small intestine as the mold. Tissue engineers recently announced they had produced a human heart in the lab.

Figure 6.5 Acid burn victim.
This young woman, whose face was damaged by acid, received skin, produced in the laboratory, as a transplant.

Laboratory Review 6

_______________ **1.** What are cells with similar structural features and a common function called?

_______________ **2.** If organ systems are present in an animal, what else must be present?

_______________ **3.** What kind of tissue covers the entire body surface and most inner cavities?

_______________ **4.** What term is used to describe cell tissues that only appear to be layered?

_______________ **5.** What is the name of the protein that makes the outer layer of epidermis tough, protective, and able to repel water?

_______________ **6.** What kinds of cells produce mucus that protects the simple columnar epithelium lining in the stomach from digestive enzymes?

_______________ **7.** What type of connective tissue has a hard matrix that contains calcium salts?

_______________ **8.** What type of connective tissue has a fluid matrix called plasma?

_______________ 9. Which type of muscle is controlled voluntarily?

_______________ 10. What feature aids communication between cardiac muscle fibers?

_______________ 11. Which type of muscle is found in the bladder?

_______________ 12. What is the name of the cells in nervous tissue that transmit messages?

_______________ 13. When two or more tissues work together to perform a specific function, what is the structure formed by the multiple tissues called?

_______________ 14. In what region of the skin are blood vessels, nerves, sense organs, glands, and hair follicles located?

Thought Questions

15. Why is a muscle, like the biceps brachii in your arm, considered an organ and not just a tissue? What kind of muscle tissue is part of the biceps brachii?

16. Why is an increased consumption of dairy products recommended after someone breaks a bone?

17. Based on your understanding of the various functions of epithelium, explain why long-time smokers tend to cough a lot.

LABORATORY

7

Organization of the Body

Learning Outcomes

7.1 External Anatomy

- Compare the limbs of a pig to the limbs of a human.
- Identify the sex of a fetal pig.

Prelab Question: Aside from the mouth, name two orifices of the human body.

7.2 Oral Cavity and Pharynx

- Find and identify the teeth, tongue, and hard and soft palates.
- Identify and state a function for the epiglottis, glottis, and esophagus.
- Name the two pathways that cross in the pharynx.

Prelab Question: Explain why food does not enter the trachea.

7.3 Thoracic and Abdominal Incisions

- Identify the thoracic cavity and the abdominal cavity.
- Find and identify the diaphragm.

Prelab Question: Why is it important to point scissors up and not down?

7.4 Neck Region

- Find, identify, and state a function for the thymus gland, the larynx, and the thyroid gland.

Prelab Question: How does the thymus gland differ between pigs and humans?

7.5 Thoracic Cavity

- Identify the three compartments and the organs of the thoracic cavity.

Prelab Question: Which is more posterior, the lungs or liver?

7.6 Abdominal Cavity

- Find, identify, and state a function for the liver, stomach, spleen, small intestine, gallbladder, pancreas, and large intestine. Describe where these organs are positioned in relation to one another.

Prelab Question: In which intestine does most nutrient absorption occur?

7.7 Human Anatomy

- Using a human torso model, find, identify, and state a function for the organs studied in this laboratory.
- Associate each organ with a particular system of the body.

Prelab Question: Describe the path of food from the mouth to the anus.

Application for Daily Living: **The Systems Work Together**

Introduction

In this laboratory, you will dissect a fetal pig. Alternately your instructor may choose to have you observe a pig that has already been dissected. Both pigs and humans are mammals; therefore, you will be studying mammalian anatomy. The period of pregnancy, or gestation, in pigs is approximately 17 weeks (compared with an average of 40 weeks in humans). The piglets used in class will usually be within 1 to 2 weeks of birth.

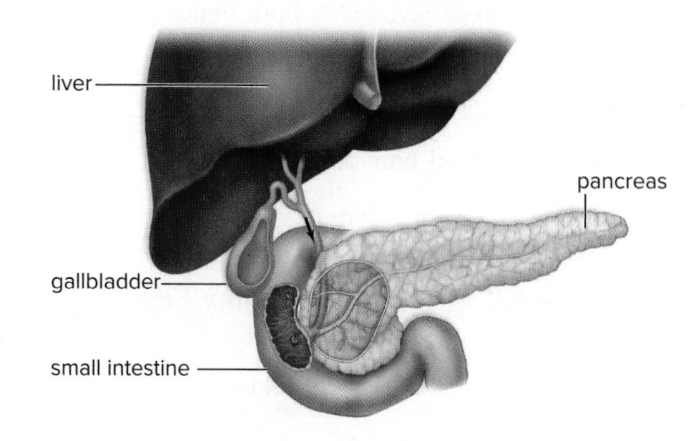

The pigs may have a slash in the right neck region, indicating the site of blood drainage. A red latex solution may have been injected into the **arterial system,** and a blue latex solution may have been injected into the **venous system** of the pigs. If so, when a vessel appears red, it is an artery, and when a vessel appears blue, it is a vein.

As a result of this laboratory, you should gain an appreciation of which organs work together. For example, the liver and the pancreas help to digest fat in the small intestine.

7.1 External Anatomy

Mammals are characterized by the presence of mammary glands and hair. Mammals also occur in two distinct sexes, males and females, often distinguishable by their external **genitals,** the reproductive organs.

Both pigs and humans are placental mammals, which means that development occurs within the uterus of the mother. An **umbilical cord** stretches externally between the fetal animal and the **placenta,** where carbon dioxide and organic wastes are exchanged for oxygen and organic nutrients.

Pigs and humans are tetrapods—that is, they have four limbs. Pigs walk on all four of their limbs; in fact, they walk on their toes, and their toenails have evolved into hooves. In contrast, humans walk only on the feet of their legs.

Observation: External Anatomy

Body Regions and Limbs

> **Latex gloves:** Wear protective latex gloves when handling preserved animal organs. Use protective eyewear and exercise caution when using sharp instruments during this experiment. Wash hands thoroughly upon completion of this experiment.

1. Place your animal in a dissecting pan, and observe the following body regions: the rather large head; the short, thick neck; the cylindrical trunk with two pairs of appendages (forelimbs and hindlimbs); and the short tail (Fig. 7.1*a*). The tail is an extension of the vertebral column.
2. Examine the four limbs, and feel for the joints of the digits, wrist, elbow, shoulder, hip, knee, and ankle.
3. Determine which parts of the forelimb correspond to your arm, elbow, forearm, wrist, and hand.
4. Do the same for the hindlimb, comparing it with your leg.
5. The pig walks on its toenails, which would be like a ballet dancer on "tiptoe." Notice how your heel touches the ground when you walk. Where is the heel of the pig? ______________________

Umbilical Cord

1. Locate the umbilical cord arising from the ventral (toward the belly) portion of the abdomen.
2. Note the cut ends of the umbilical blood vessels. If they are not easily seen, cut the umbilical cord near the end and observe this new surface.
3. What is the function of the umbilical cord? ______________________

____

Nipples and Hair

1. Locate the small **nipples,** the external openings of the **mammary glands.** The nipples are *not* an indication of sex, since both males and females possess them. How many nipples does a pig have? ________ When is it advantageous for a pig to have so many nipples? ______________________

____

2. Can you find hair on the pig? __________ Where? ______________________

____

Directional Terms for Dissecting Fetal Pig

Anterior: toward the head end	Ventral: toward the belly
Posterior: toward the hind end	Dorsal: toward the back

Figure 7.1 External anatomy of the fetal pig.
a. Body regions and limbs. **b, c.** The sexes can be distinguished by the external genitals.

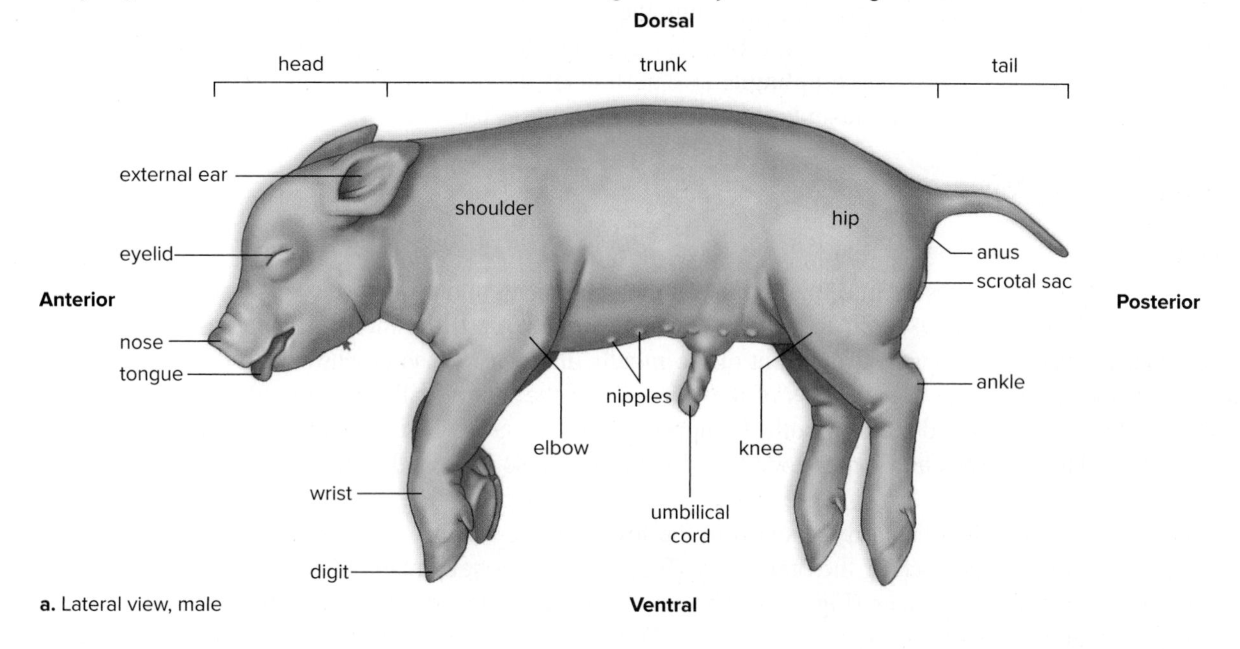

a. Lateral view, male

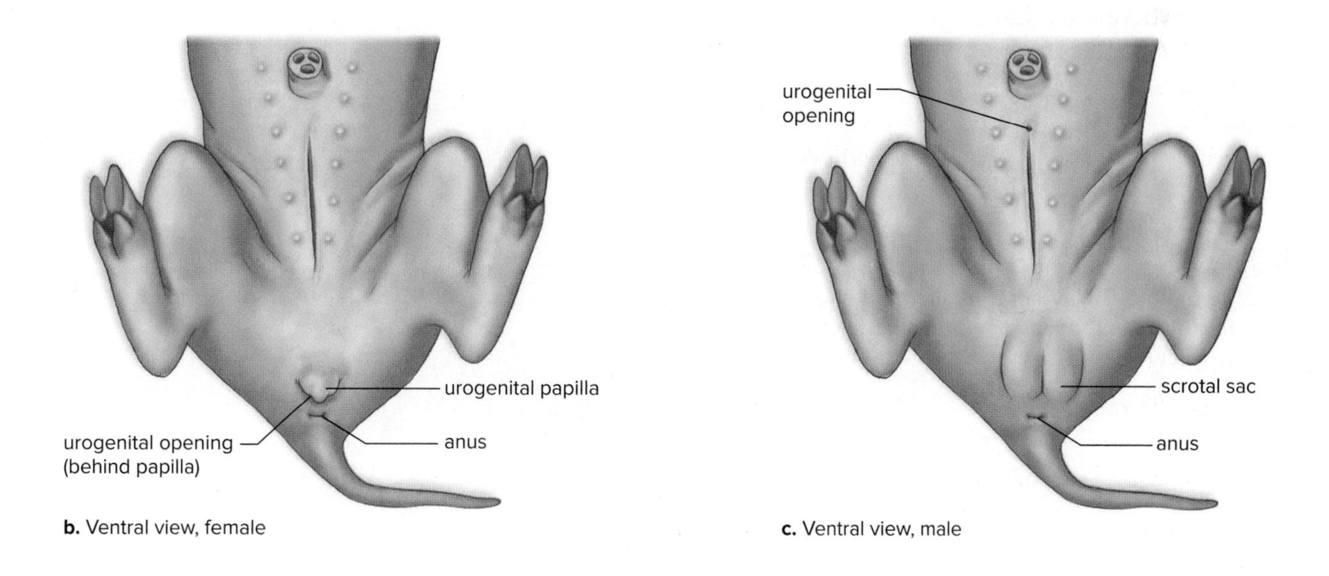

b. Ventral view, female

c. Ventral view, male

Anus and External Genitals

1. Locate the **anus** under the tail. Name the organ system that ends in the opening called the anus. ___________
2. In females, locate the **urogenital opening,** just anterior to the anus, and a small, fleshy **urogenital papilla** projecting from the urogenital opening (Fig. 7.1*b*).
3. In males, locate the urogenital opening just posterior to the umbilical cord (Fig. 7.1*c*). The duct leading to it runs forward from between the legs in a long, thick tube, the **penis,** which can be felt under the skin. In males, the urinary system and the genital system are always joined.
4. You are responsible for identifying pigs of both sexes. What sex is your pig? ___________________
Be sure to look at a pig of the opposite sex that another group of students is dissecting.

7.2 Oral Cavity and Pharynx

The **oral cavity** is the space in the mouth that contains the tongue and the teeth. The **pharynx** is dorsal to the oral cavity and has three openings: The **glottis** is an opening through which air passes on its way to the **trachea** (the windpipe) and lungs. The **esophagus** is a portion of the digestive tract that leads through the neck and thorax to the stomach. The **nasopharynx** leads to the nasal passages.

Observation: Oral Cavity and Pharynx

Oral Cavity

1. Insert a sturdy pair of scissors into one corner of the specimen's mouth, and cut posteriorly (toward the hind end) for approximately 4 cm. Repeat on the opposite side until the mouth is open as in Figure 7.2.
2. Place your thumb on the tongue at the front of the mouth, and gently push downward on the lower jaw. This will tear some of the tissue in the angles of the jaws so that the mouth will remain partly open (Fig. 7.2).
3. Note small, underdeveloped teeth in both the upper and lower jaws. Care should be taken because teeth can be very sharp. Other embryonic, nonerupted teeth may also be found within the gums. The teeth are used to chew food.
4. Examine the tongue, which is partly attached to the lower jaw region but extends posteriorly and is attached to a bony structure at the back of the oral cavity (Fig. 7.2). The tongue manipulates food for swallowing.
5. Locate the hard and soft palates (Fig. 7.2). The **hard palate** is the ridged roof of the mouth that separates the oral cavity from the nasal passages. The **soft palate** is a smooth region posterior to the hard palate. An extension of the soft palate—the **uvula**—hangs down into the throat in humans. (A pig does not have a uvula.)

Figure 7.2 Oral cavity of the fetal pig. The roof of the oral cavity contains the hard and soft palates, and the tongue lies above the floor of the oral cavity.

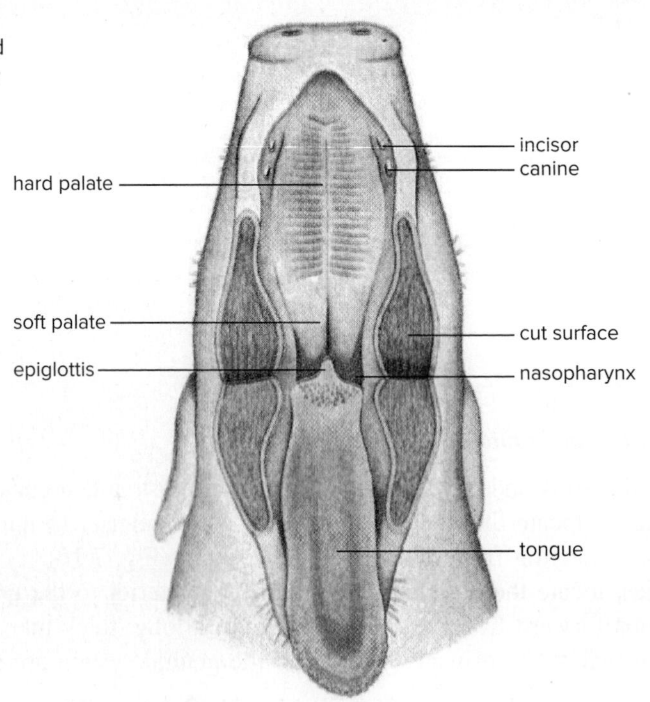

Pharynx

1. Push down on the tongue until you open the jaws far enough to see a slightly pointed flap of tissue pointing dorsally (toward the back) (Fig. 7.2). This flap is the **epiglottis,** which covers the glottis. The **glottis** leads to the trachea (Fig. 7.3*a*).
2. Posterior and dorsal to the glottis, find the opening into the **esophagus,** a tube that takes food to the stomach. Note the proximity of the glottis and the opening to the esophagus. Each time the pig—or a human—swallows, the epiglottis instantly closes to keep food and fluids from going into the lungs via the trachea.
3. Insert a blunt probe into the glottis, and note that it enters the trachea. Remove the probe, insert it into the esophagus, and note the position of the esophagus beneath (dorsal to) the trachea.
4. Make a midline cut in the soft palate from the epiglottis to the hard palate. Then make two lateral cuts at the edge of the hard palate.
5. Posterior to the soft palate, locate the openings to the nasal passages.
6. Explain why it is correct to say that the air and food passages cross in the pharynx. ________________

____

____

____

Figure 7.3 Air and food passages in the fetal pig.
The air and food passages cross in the pharynx. **a.** Drawing. **b.** Dissection of specimen.
(b) © Ken Taylor/Wildlife Images

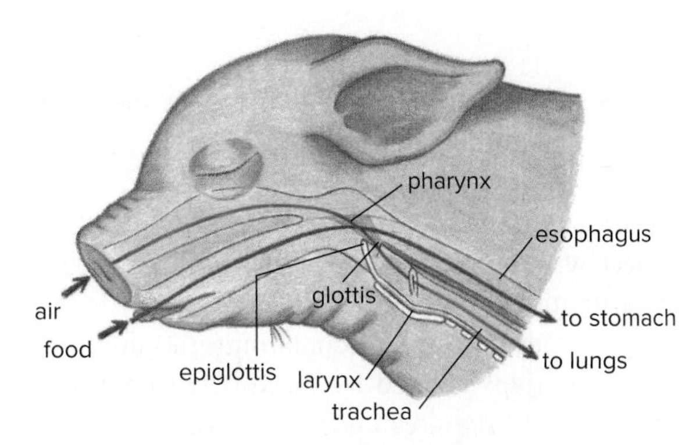

a.

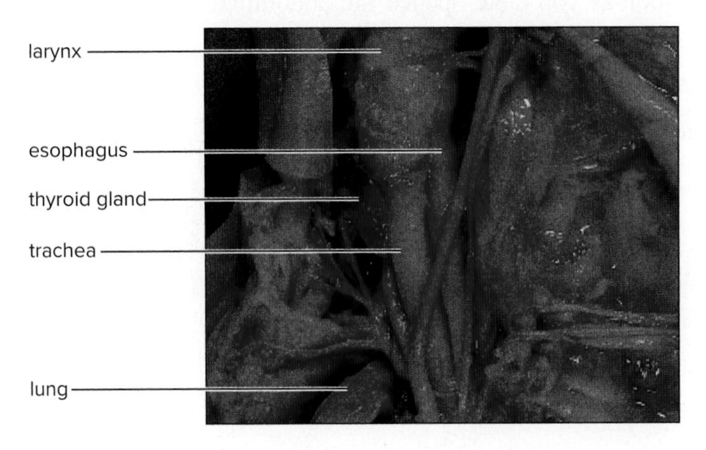

b.

7.3 Thoracic and Abdominal Incisions

First, prepare your pig according to the following directions, and then make thoracic and abdominal incisions so that you will be able to study the internal anatomy of your pig.

Preparation of Pig for Dissection

1. Place the fetal pig on its back in the dissecting pan.
2. Tie a cord around one forelimb, and then bring the cord around underneath the pan to fasten back the other forelimb.
3. Spread the hindlimbs in the same way.
4. With scissors always pointing up (never down), make the following incisions to expose the thoracic and abdominal cavities. The incisions are numbered on Figure 7.4 to correspond with the following steps.

Thoracic Incisions

1. Starting at the diaphragm, a structure that separates the thoracic cavity from the abdominal cavity, cut anteriorly until you reach the hairs in the throat region.
2. Make two lateral cuts, one on each side of the midline incision anterior to the forelimbs, taking extra care not to damage the blood vessels around the heart.
3. Make two lateral cuts, one on each side of the midline just posterior to the forelimbs and anterior to the diaphragm, following the ends of the ribs. Pull back the flaps created by these cuts (do not remove them) to expose the **thoracic cavity.** List the organs you find in the thoracic cavity. ________________

   ____

Abdominal Incisions

4. With scissors pointing up, cut posteriorly from the diaphragm to the umbilical cord.
5. Make a flap containing the umbilical cord by cutting a semicircle around the cord and by cutting posteriorly to the left and right of the cord.
6. Make two cuts, one on each side of the midline incision posterior to the diaphragm. Examine the diaphragm, attached to the chest wall by radially arranged muscles. The central region of the diaphragm, called the **central tendon,** is a membranous area.
7. Make two more cuts, one on each side of the flap containing the umbilical cord and just anterior to the hindlimbs. Pull back the side flaps created by these cuts to expose the **abdominal cavity.**
8. Lifting the flap with the umbilical cord requires cutting the **umbilical vein.** Before cutting the umbilical vein, tie a thread on each side of where you will cut to mark the vein for future reference.
9. Rinse out your pig as soon as you have opened the abdominal cavity. If you have a problem with excess fluid, obtain a disposable plastic pipet to suction off the liquid.
10. Name the two cavities separated by the diaphragm. ________________________________
11. List the organs located in the abdominal cavity. ________________________________

    ____

Figure 7.4 Ventral view of the fetal pig indicating incisions.
These incisions are to be made preparatory to dissecting the internal organs. They are numbered here in the order they should be done.

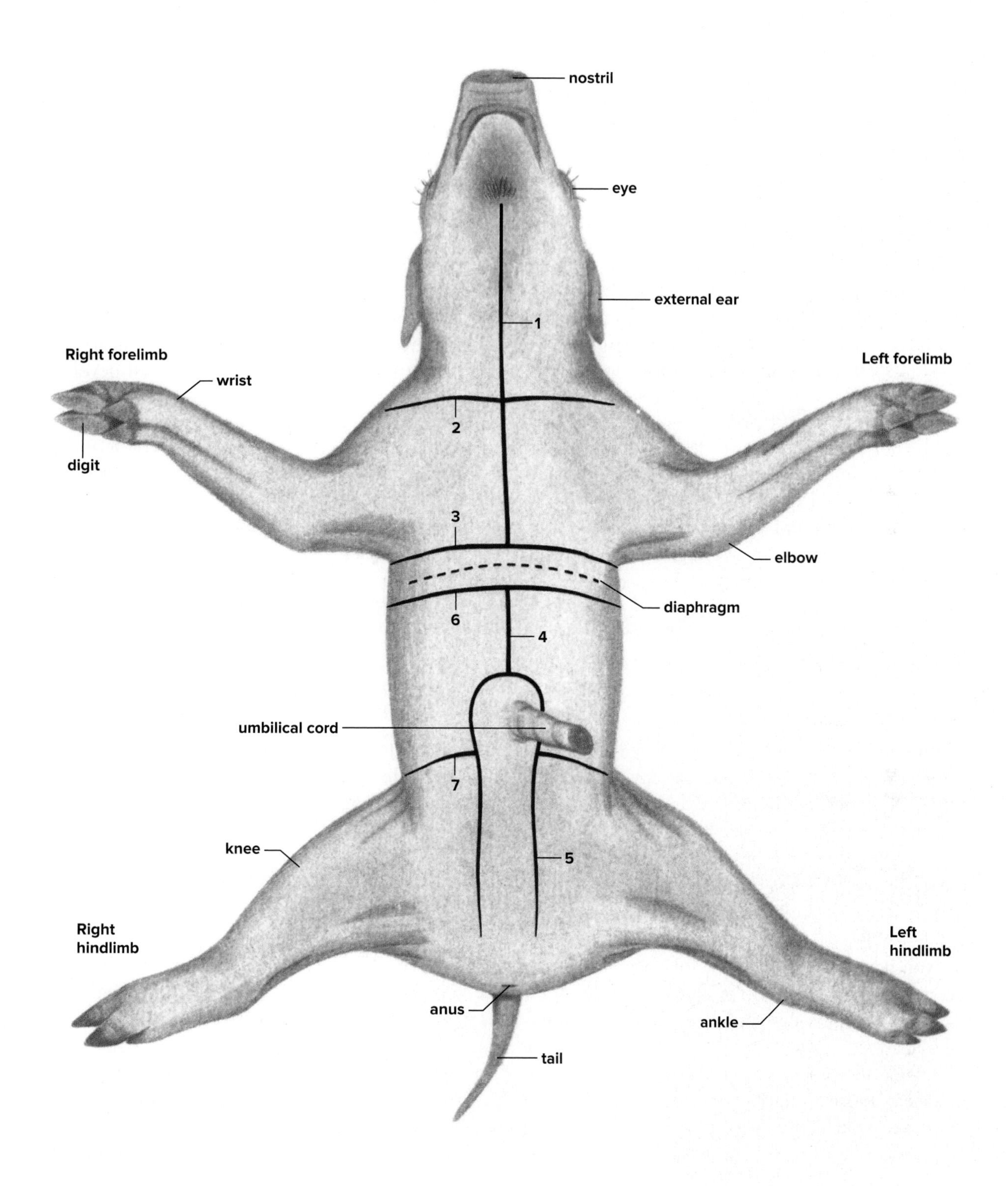

7.4 Neck Region

You will locate several organs in the neck region. Use Figures 7.3*b* and 7.5 as a guide, but *keep all the flaps* in order to close the thoracic and abdominal cavities at the end of the laboratory session.

The **thymus gland** is a part of the lymphatic system. Certain white blood cells called T (for thymus) lymphocytes mature in the thymus gland and help us fight disease. The **larynx,** or voice box, sits atop (anterior to) the **trachea,** or windpipe. The esophagus is a portion of the digestive tract that leads to the stomach. The **thyroid gland** secretes hormones that travel in the blood and act upon other body cells. These hormones (e.g., thyroxine) regulate the rate at which metabolism occurs in cells.

Observation: Neck Region

Thymus Gland

1. Move the skin apart in the neck region just below the hairs mentioned earlier. If necessary, cut the body wall laterally to make flaps. You will most likely be viewing exposed muscles.
2. *Cut through and clear away muscle* to expose the thymus gland, a diffuse gland that lies among the muscles. Later you will notice that the glandular thymus flanks the thyroid and overlies the heart (Fig. 7.5). The thymus is particularly large in fetal pigs, since their immune systems are still developing.

Larynx, Trachea, and Esophagus

1. Probe down into the deeper layers of the neck. Medially (toward the center), beneath several strips of muscle, find the hard-walled larynx and the trachea, which are parts of the respiratory passage to be examined later. Dorsal to the trachea, find the esophagus.
2. Open the mouth and insert a probe into the glottis and esophagus from the pharynx to better understand the orientation of these two organs.

Thyroid Gland

Locate the thyroid gland just posterior to the larynx, lying ventral to (on top of) the trachea.

7.5 Thoracic Cavity

As previously mentioned, the body cavity of mammals, including humans, is divided by the diaphragm into the thoracic cavity and the abdominal cavity. The heart and lungs are in the thoracic cavity (Figs. 7.5 and 7.6). The **heart** is a pump for the cardiovascular system, and the **lungs** are organs of the respiratory system where gas exchange occurs.

Observation: Thoracic Cavity

Heart and Lungs

1. If you have not yet done so, fold back the chest wall flaps. To do this, you will need to tear the thin membranes that divide the thoracic cavity into three compartments: the **left pleural cavity** containing the left lung, the **right pleural cavity** containing the right lung, and the **pericardial cavity** containing the heart.
2. Examine the lungs. Locate the four lobes of the right lung and the three lobes of the left lung. The trachea, dorsal to the heart, divides into the **bronchi,** which enter the lungs. Later, when the heart is removed, you will be able to see the trachea and bronchi.
3. Sequence the organs of the respiratory tract to trace the path of air from the nasal passages to the lungs.

____

____

Figure 7.5 Internal anatomy of the fetal pig.
The major organs are featured in this drawing. In the fetal pig, a red color tells you a vessel is an artery, and a blue color tells you it is a vein. (It does not tell you whether this vessel carries O_2-rich or O_2-poor blood.) Contrary to this drawing, *keep all the flaps on your pig* so you can close the thoracic and abdominal cavities at the end of the laboratory session.

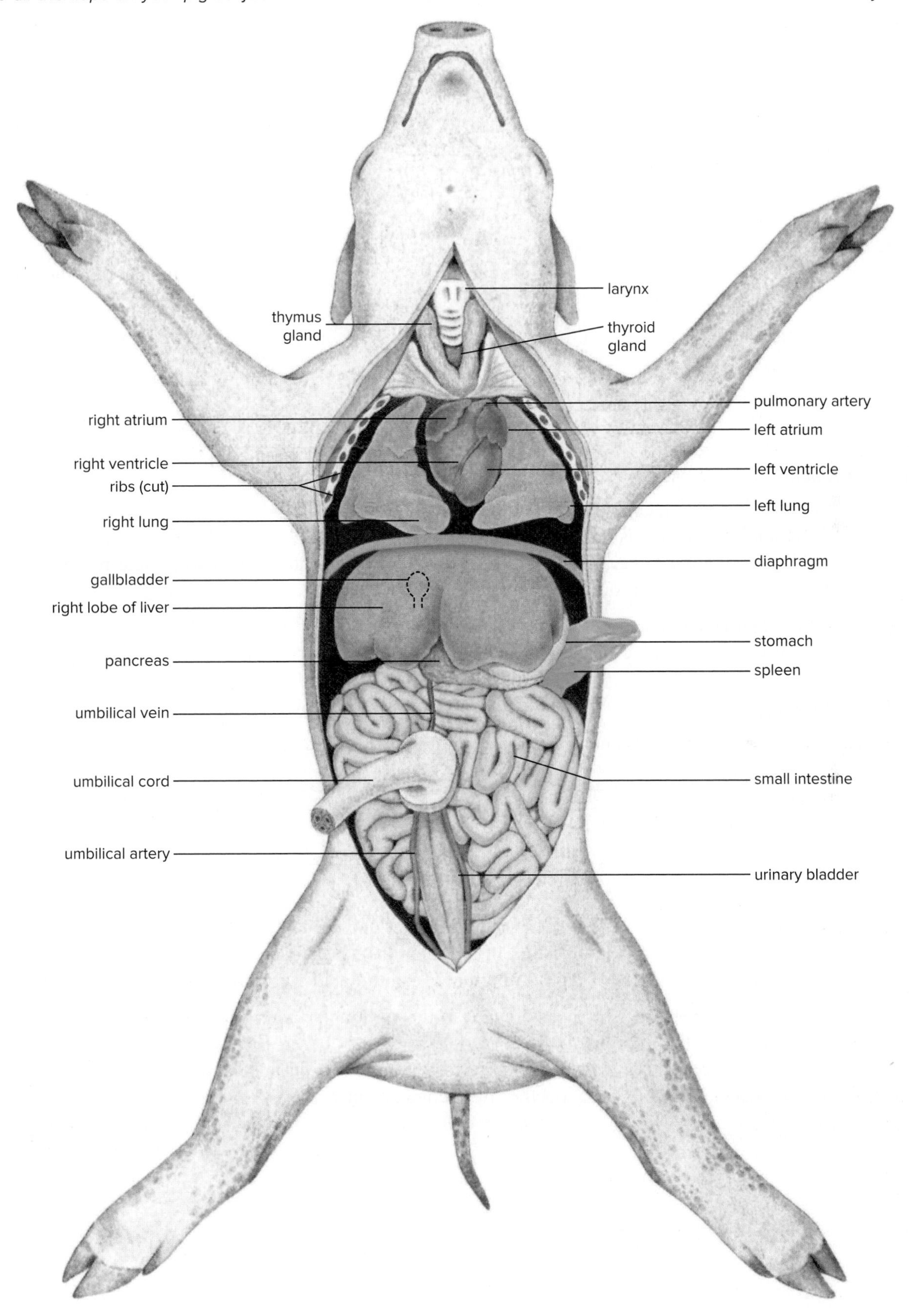

7.6 Abdominal Cavity

The abdominal wall and organs are lined by a membrane called **peritoneum,** consisting of epithelium supported by connective tissue. Double-layered sheets of peritoneum, called **mesenteries,** project from the body wall and support the organs.

Observation: Abdominal Cavity

If your pig is partially filled with dark, brownish material, take your animal to the sink and rinse it out. This material is clotted blood. Consult your instructor before removing any red or blue latex masses, since they may enclose organs you will need to study.

Liver

The **liver,** the largest organ in the abdomen (Fig. 7.6), performs numerous vital functions, including (1) disposing of worn-out red blood cells, (2) producing bile, (3) storing glycogen, (4) maintaining the blood glucose level, and (5) producing blood proteins.

1. Locate the liver, a large, brown organ. Its anterior surface is smoothly convex and fits snugly into the concavity of the diaphragm.
2. Name several functions of the liver. ____

   ____

Stomach and Spleen

The organs of the digestive tract include the stomach, small intestine, and large intestine. The **stomach** (see Fig. 7.5) stores food and has numerous gastric glands. These glands secrete a juice that digests protein. The **spleen** (see Fig. 7.5) is a lymphoid organ in the lymphatic system that contains both white and red blood cells. It purifies blood and disposes of worn-out red blood cells.

1. Push aside and identify the stomach, a large sac dorsal to the liver on the left side.
2. Locate the point near the midline of the body where the **esophagus** penetrates the diaphragm and joins the stomach.
3. Find the spleen, a long, flat, reddish organ attached to the stomach by mesentery.
4. The stomach is a part of what system? ____

   What is its function? ____
5. The spleen is a part of what system? ____

   What is its function? ____

Small Intestine

The **small intestine** is the part of the digestive tract that receives secretions from the pancreas and gallbladder. Besides being an area for the digestion of all components of food, carbohydrate, protein, and fat, the small intestine absorbs the products of digestion: glucose, amino acids, glycerol, and fatty acids.

1. Look posteriorly where the stomach makes a curve to the right and narrows to join the anterior end of the small intestine called the **duodenum.**
2. From the duodenum, the small intestine runs posteriorly for a short distance and is then thrown into an irregular mass of bends and coils held together by a common mesentery.
3. The small intestine is a part of what system? ____

   What is its function? ____

Figure 7.6 Internal anatomy of the fetal pig.
Most of the major organs are shown in this photograph. The stomach has been removed. The spleen, gallbladder, and pancreas are not visible. *Do not* remove any organs or flaps from your pig.

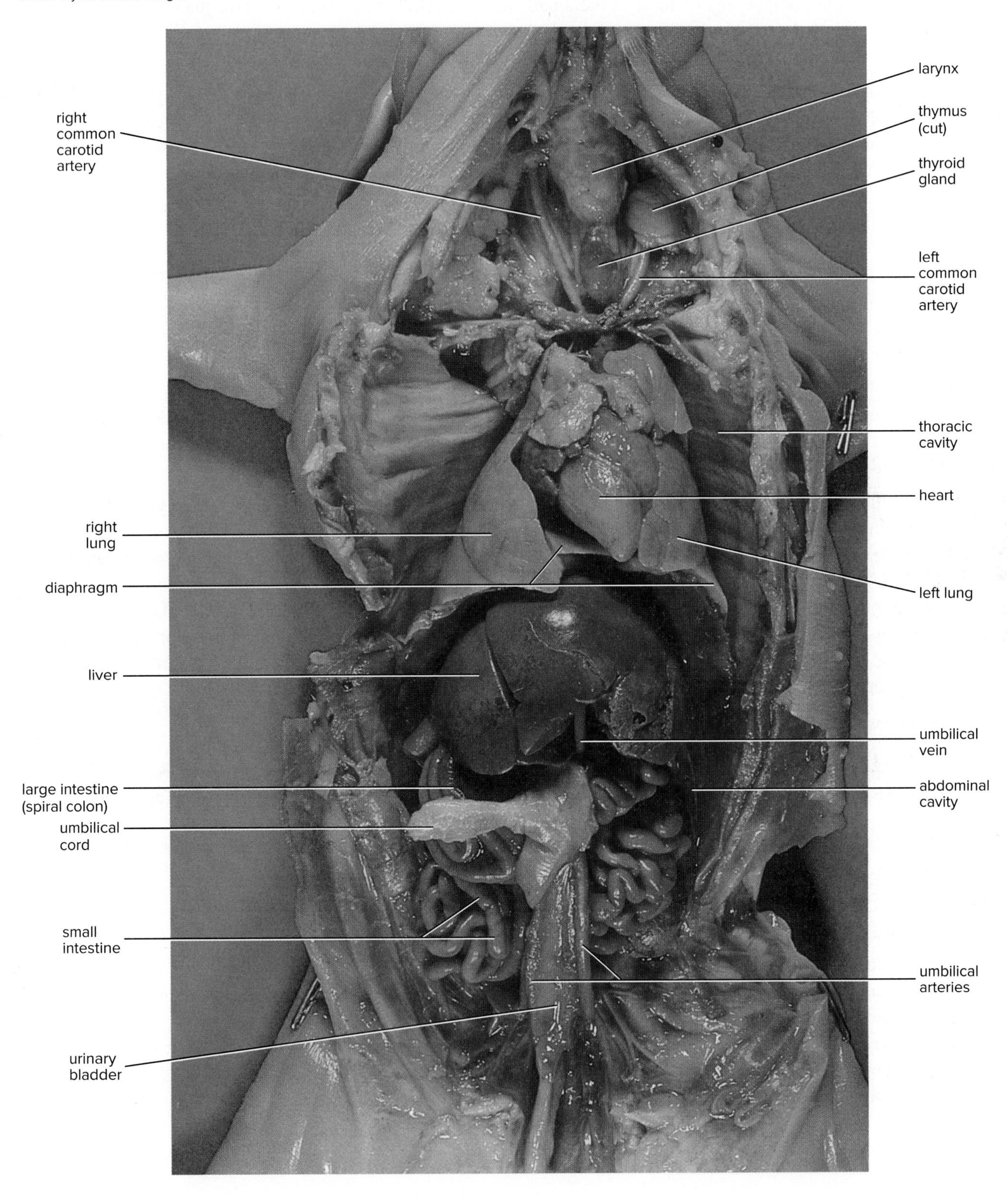

Gallbladder and Pancreas

The **gallbladder** stores and releases bile, which aids in the digestion of fat. The **pancreas** (see Fig. 7.5) is both an exocrine and an endocrine gland. As an exocrine gland, it produces and secretes pancreatic juice, which digests all the components of food in the small intestine. Both bile and pancreatic juice enter the duodenum by way of ducts. As an endocrine gland, the pancreas secretes the hormones insulin and glucagon into the bloodstream. Insulin and glucagon regulate blood glucose levels.

1. Locate the **bile duct,** which runs in the mesentery stretching between the liver and the duodenum. Find the gallbladder, embedded in the liver on the underside of the right lobe. It is a small, greenish sac.
2. Lift the stomach and locate the pancreas, the light-colored, diffuse gland lying in the mesentery between the stomach and the small intestine. The pancreas has a duct that empties into the duodenum of the small intestine.
3. What is the function of the gallbladder? ____

   ____
4. What is the function of the pancreas? ____

   ____

Large Intestine

The **large intestine** is the part of the digestive tract that absorbs water and prepares feces for defecation at the anus. The first part of the large intestine, called the **cecum,** has a projection called the vermiform (meaning wormlike) appendix.

1. Locate the distal (far) end of the small intestine, which joins the large intestine posteriorly, in the left side of the abdominal cavity (right side in humans). At this junction, note the cecum, a blind pouch.
2. Compare the large intestine of your pig to Figure 7.7. The organ does not have the same appearance in humans.
3. Follow the main portion of the large intestine, known as the **colon,** as it runs from the point of juncture with the small intestine into a tight coil (spiral colon), then out of the coil anteriorly, then posteriorly again along the midline of the dorsal wall of the abdominal cavity. In the pelvic region, the **rectum** is the last portion of the large intestine. The rectum leads to the **anus.**
4. The large intestine is a part of what system? ____
5. What is the function of the large intestine? ____
6. Sequence the organs of the digestive system to trace the path of food from the mouth to the anus. ________

   ____

Storage of Pigs

1. Before leaving the laboratory, place your pig in the plastic bag provided.
2. Expel excess air from the bag, and tie it shut.
3. Write your *name* and *section* on the tag provided, and attach it to the bag. Your instructor will indicate where the bags are to be stored until the next laboratory period.
4. Clean the dissecting tray and tools, and return them to their proper location.
5. Wipe off your goggles.
6. Wash your hands.

7.7 Human Anatomy

Humans and pigs are both mammals, and their organs are similar. A human torso model shows the exact location of the organs in humans (Fig. 7.7). You should learn to associate each human organ with its particular system. Six systems are color-coded in Figure 7.7.

Figure 7.7 Human internal organs.

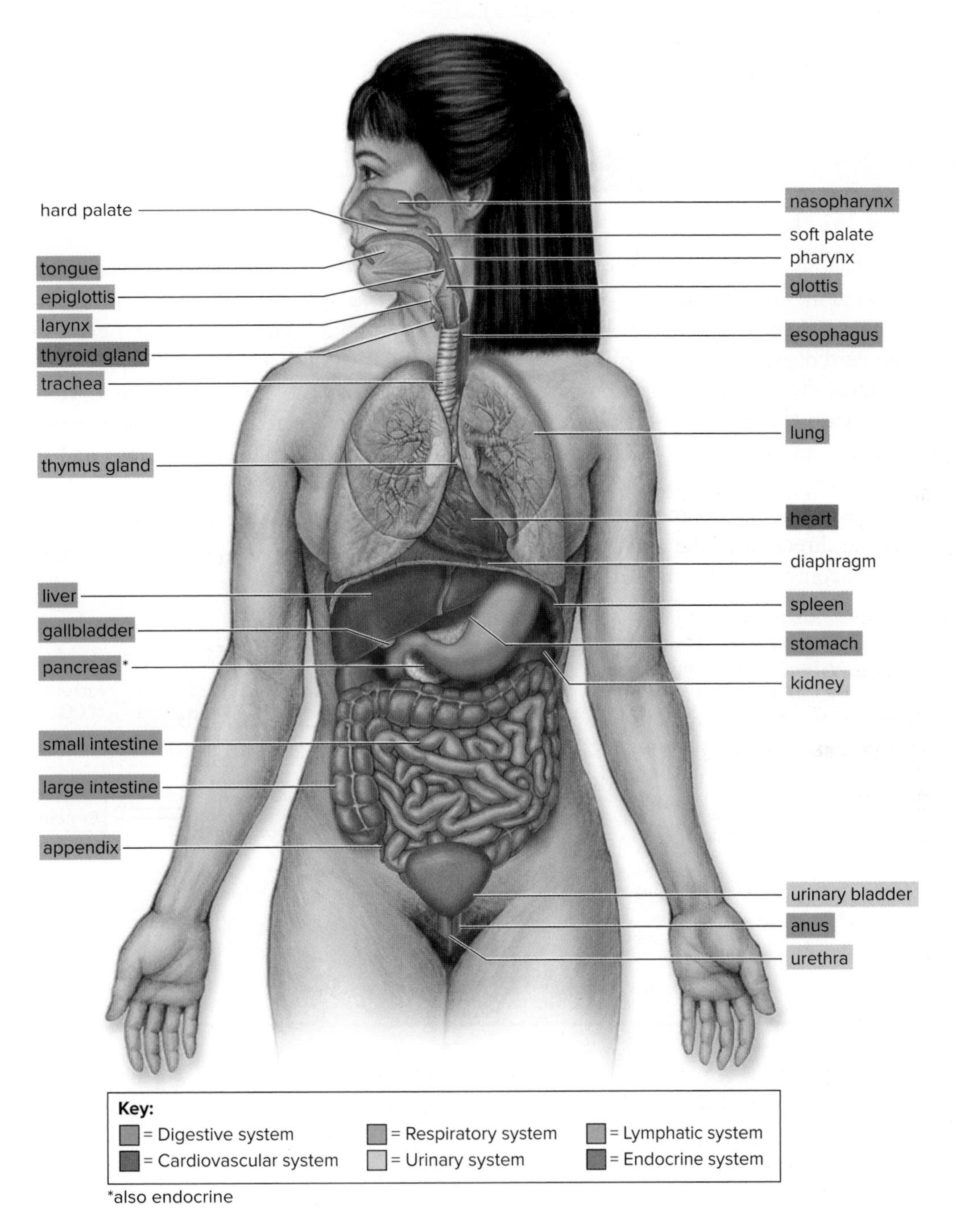

*also endocrine

Observation: Human Torso

1. Examine a human torso model, and using Figure 7.7 as a guide, locate the same organs just dissected in the fetal pig.
2. In your studies so far, have you seen any major differences between pig internal anatomy and human internal anatomy? ____

____

Application for Daily Living

The Systems Work Together

This laboratory has involved a number of different systems, including the respiratory system, the cardiovascular system, and the digestive system. Can one of these systems malfunction without negatively affecting the other systems? Likely not. For example, if you climb a high mountain and can't breathe adequately, the cardiovascular system will not be able to deliver oxygen to your cells (Fig. 7.8). Or, if you go on a hunger strike, the cardiovascular system will not be able to deliver glucose to your cells. Without oxygen and glucose, even your heart will not be able to continue beating. Blood vessels not only deliver oxygen and glucose to the heart; they also take these molecules to the lungs and the walls of the digestive tract.

In the body as in the environment, everything is connected to everything else. The systems work together under the direction of the nervous and endocrine systems as a harmonious whole so that your cells and you can continue to function as you should.

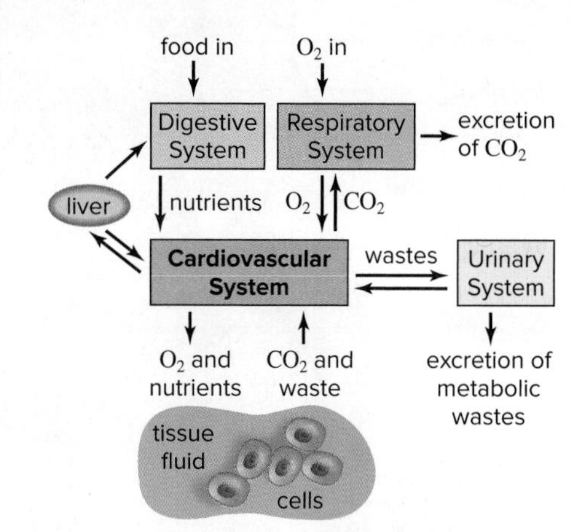

Figure 7.8 The systems work together. The cardiovascular system receives nutrients from the digestive tract and oxygen from the respiratory system, which it delivers to the cells. It also delivers CO_2 and other wastes from the cells to the respiratory system and urinary system respectively.

Laboratory Review 7

______________ 1. What directional term refers to the head end of the fetal pig?

______________ 2. If the urogenital opening is just anterior to the anus, is the fetal pig male or female?

______________ 3. What space is dorsal to the oral cavity and has three openings?

______________ 4. What is the name of the flap that closes during swallowing to prevent food and fluids from entering the trachea?

______________ 5. What is the structure that separates the thoracic cavity from the abdominal cavity?

______________ 6. What glandular organ is especially large in fetal pigs since their immune systems are still developing?

______________ 7. Is the trachea ventral or dorsal to the esophagus?

______________ 8. In what thoracic cavities are the lungs located?

______________ 9. What is the largest organ in the abdominal cavity?

______________ 10. What organ do secretions from the gallbladder and pancreas enter?

______________ 11. What is the name of the tissue that holds the small intestine together?

______________ 12. What is the name of the first part of the large intestine?

______________ 13. What organ produces digestive enzymes and hormones that regulate blood glucose?

______________ 14. To what organ system do the kidneys belong?

Thought Questions

15. Why would someone born without a thymus gland be likely to suffer from a greater number of infections?

16. Treatment for a person's laryngeal cancer might require surgical removal of the epiglottis. What would postsurgical therapy for this person involve?

17. Based on the location of the spleen, explain why it is damaged so often in car accidents, especially when the driver's side is struck.

LABORATORY

8

Cardiovascular System

Learning Outcomes

8.1 The Heart

- Name the four chambers of the heart and the blood vessels attached to these chambers.
- Trace the path of blood through the heart, naming the chambers and the valves in proper order.
- Describe the conduction system of the heart, including the nodes that control the heartbeat.

Prelab Question: Which side of the heart pumps O_2-poor blood and which side pumps O_2-rich blood?

8.2 Heartbeat and Blood Pressure

- Describe the heartbeat, including the sounds that occur when the heart beats.
- Relate blood pressure to the beat of the heart.
- Show blood pressure rises with exercise and offer an explanation.

Prelab Question: What is normal blood pressure?

8.3 The Blood and Blood Flow

- Distinguish between red blood cells and the various types of white blood cells.
- Explain the ABO system of blood typing and how blood type is determined.
- Describe the structure and function of arteries, veins, and capillaries.
- Name the major blood vessels and trace the path of blood in both the pulmonary and systemic circuits.

Prelab Question: Describe the differences between the pulmonary and systemic circuits.

***Application for Daily Living:* High Blood Pressure**

Introduction

The heart and blood vessels form the cardiovascular system. The heart is a double pump that (1) keeps the blood flowing in one direction—blood flows away from and then back to the heart; (2) keeps O_2-poor blood separate from O_2-rich blood; and (3) creates blood pressure, which moves the blood through the blood vessels. The blood vessels, on the other hand, (1) transport blood and its contents, (2) serve the needs of the body's cells by carrying out exchanges with them, and (3) direct blood flow to those systemic tissues that most require it at the moment.

We will make note of these functions again as we first study the structure and beat of the heart and how its beat brings about blood pressure. Following that, we concentrate on the composition of blood, how it is typed, and how it circulates in the body.

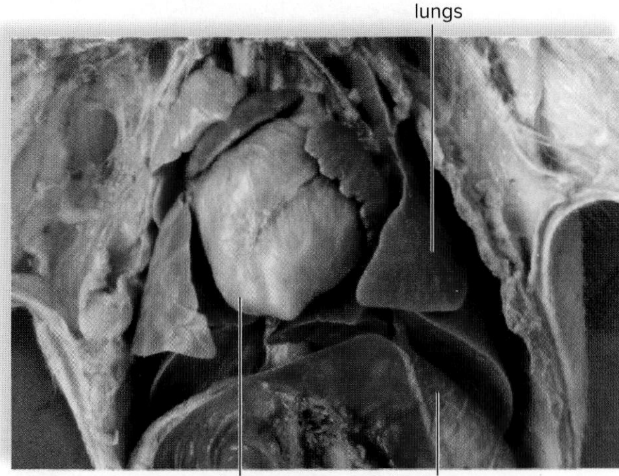

A thoracic cavity contains the heart and lungs.

8.1 The Heart

The heart has right and left sides divided by the **septum.** To tell the right from the left side, position Figure 8.1 so it corresponds to your body. There are four **chambers:** two upper, thin-walled atria and two lower, thick-walled ventricles. The heart has valves that keep the blood flowing in one direction; special muscles and tendons secure the valves to prevent backflow from the ventricles to the atria. The right side of the heart sends blood to the lungs, and the left side sends blood into the body. Therefore, the heart is called a double pump.

Observation: External Anatomy of the Heart

1. In a heart model and/or a sheep heart, identify the **right atrium** and its attached blood vessels, the superior and inferior **venae cavae** (Fig. 8.1). The superior vena cava and the inferior vena cava return blood from the head and body, respectively, to the right atrium. Is the blood that enters the right atrium O_2-poor or O_2-rich? ______________ Explain. ______________________________________

   ____

2. Identify the **right ventricle** and its attached blood vessel, the **pulmonary trunk.** The pulmonary trunk leaves the ventral side of the heart from the top of the right ventricle and then passes diagonally forward, before branching into the **right** and **left pulmonary arteries.**
3. Identify the **left atrium** and its attached blood vessels, the left and right **pulmonary veins.** The pulmonary veins return blood from the lungs to the left atrium. Is the blood that enters the left atrium O_2-poor or O_2-rich? ______________ Explain. ______________________________________

   ____

4. Identify the **left ventricle** and its attached blood vessel, the **aorta,** which arises from the anterior end of the left ventricle, just dorsal to the origin of the pulmonary trunk. The aorta soon bends to the animal's left as the aortic arch. The aorta carries blood to the body proper.

Figure 8.1 External heart anatomy.
a. Human heart. **b.** Sheep heart. © Eric Wise

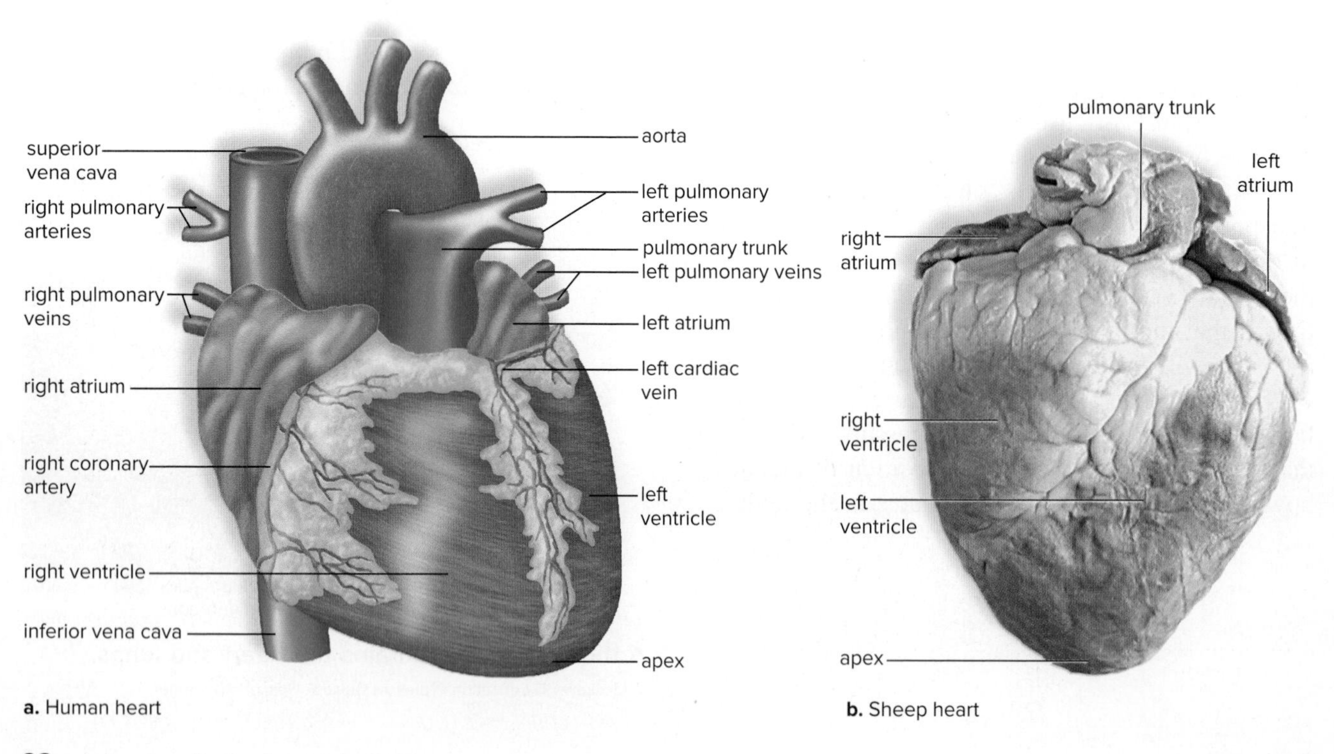

5. Identify the **coronary arteries** and the **cardiac veins,** which service the needs of the heart wall. The coronary arteries branch off the aorta as soon as it leaves the heart and appear on the surface of the heart. The cardiac veins, also on the surface of the heart, join and then enter the right atrium through the coronary sinus on the dorsal side of the heart.

Observation: Internal Anatomy of the Heart

Remove the ventral half of the human heart model and/or sheep heart. You are going to remove the top portion of the heart to achieve a view similar to Figure 8.2*b*. Position the heart as it appears in Figure 8.2. Using a sharp scalpel, cut through the left atrium and the left ventricle to the apex of the heart, being sure your cut is deep enough to reach the septum. Now position the heart so that the apex is uppermost. Using a sharp scalpel, cut from the apex of the heart through the right atrium, being sure your cut is deep enough to reach the septum of the heart.

> ⚠ **Latex gloves** Wear protective latex gloves when handling preserved animal organs. Use protective eyewear and exercise caution when using sharp instruments during this laboratory. Wash hands thoroughly upon completion of this laboratory.

1. Identify the four chambers of the heart in longitudinal section: right atrium, right ventricle, left atrium, and left ventricle. *Label Figure 8.2.*
2. Which ventricle is more muscular? ____________________

   Why is this appropriate? ____________________
3. Find the **right atrioventricular** (tricuspid) valve, located between the right atrium and the right ventricle.
4. Find the **left atrioventricular** (bicuspid or mitral) valve, located between the left atrium and the left ventricle.
5. Find the **pulmonary semilunar** valve, located in the base of the pulmonary trunk.
6. Find the **aortic semilunar** valve, located in the base of the aorta. What is the function of the

   heart valves? ____________________
7. Note the **chordae tendineae** ("heartstrings") that hold the atrioventricular valves in place while the heart contracts. These extend from the papillary muscles. The chordae tendineae prevent the atrioventricular valves from inverting into the atria when the ventricles contract.
8. If inspecting a sheep heart, insert a probe into the various vessels from the outside into the heart chambers. Identify the vessel and the chamber.

Figure 8.2 Internal heart anatomy.
a. Human heart. **b.** Sheep heart. © Eric Wise

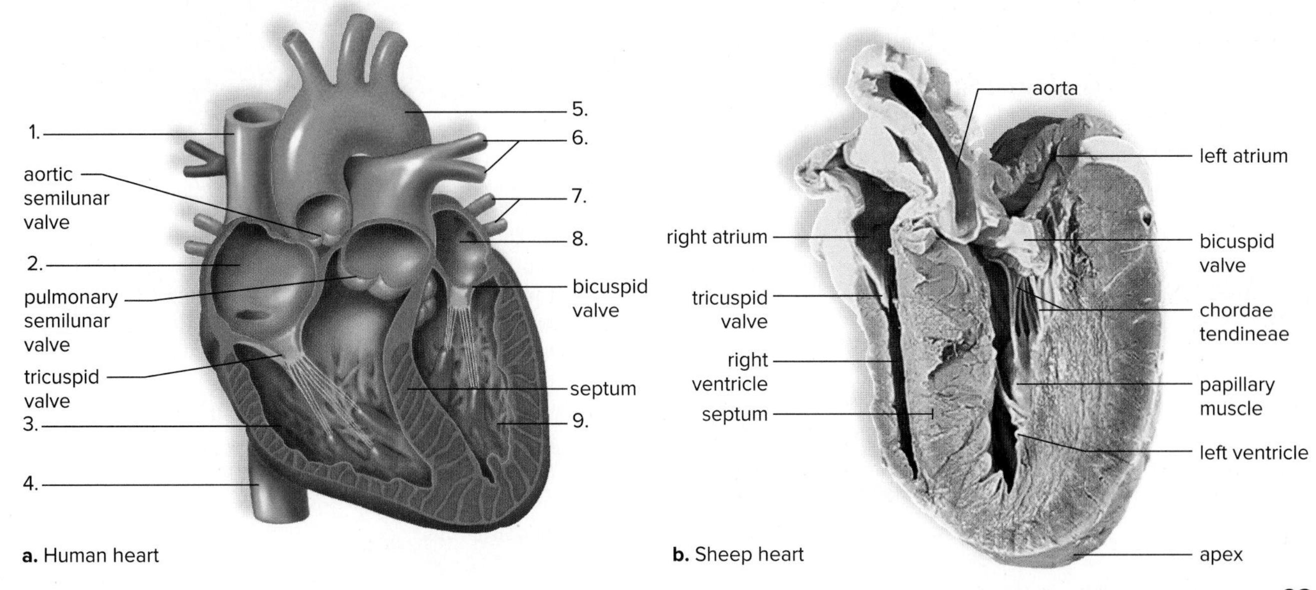

Summary

To demonstrate that O_2-poor blood is kept separate from O_2-rich blood, trace the path of blood from the right side of the heart to the aorta by filling in the following blanks.

Venae Cavae	***Lungs***
______________	______________
______________ valve	______________
______________	______________ valve
______________ valve	______________
______________	______________ valve
Lungs	Aorta

Which side of the heart (right or left) pumps O_2-poor blood? ______________

Which side of the heart (right or left) Which side pumps O_2-rich blood?

____

Physiology of the Heart

The heartbeat is controlled by nodal tissue. The SA node is the pacemaker of the heart because it initiates the heartbeat and sends an excitation impulse every 0.85 seconds, causing the atria to contract. After the impulse reaches the **AV node,** it passes into large fibers and, thereafter, spreads out by way of the smaller **Purkinje fibers** (Fig. 8.3*a*). These fibers signal the ventricles to contract.

With the contraction of any muscle, including the heart, electrolyte changes occur that can be detected by electrical recording devices. Therefore, it is possible to study the heartbeat by recording voltage changes that occur when the heart contracts. The record that results is called an electrocardiogram (ECG). The first wave in the electrocardiogram, called the P wave, occurs prior to the excitation and contraction of the atria. The second wave, the QRS wave, occurs prior to ventricular excitation and contraction. The third wave, the T wave, is caused by the recovery of the ventricles. (Atrial relaxation is not apparent in an ECG.) Examination of an ECG indicates whether the heartbeat has a normal or an irregular pattern (Fig. 8.3*b*).

1. Note the SA (sinoatrial) node in Figure 8.3*a*.
2. Explain Figure 8.3*b* by answering these questions:
   Why is an arrow drawn between the SA node and the P wave?

   ____

   Why is an arrow drawn between the AV node and the QRS wave?

   ____

   The voltage changes in an ECG are related to the

   ____.

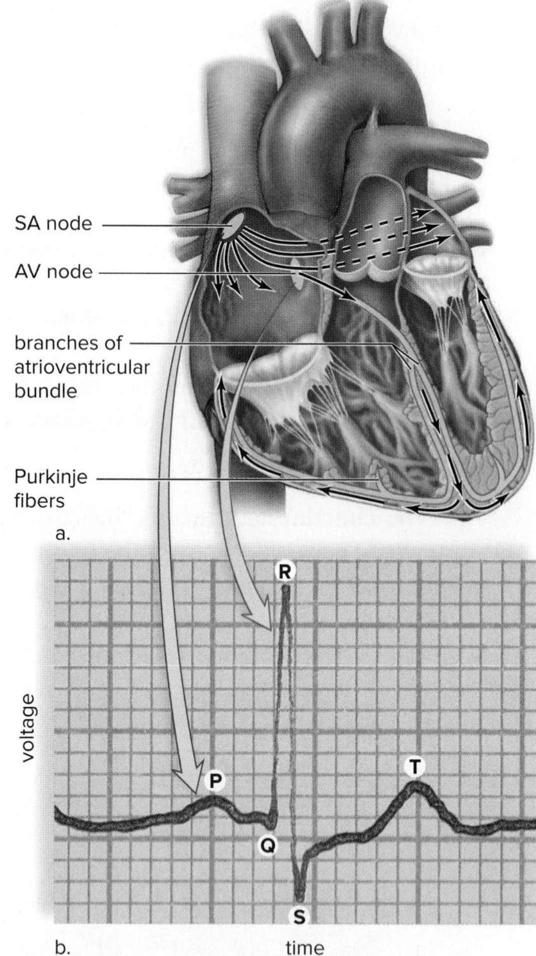

Figure 8.3 Control of the heartbeat.
a. The SA node sends out a stimulus that causes the atria to contract. When this stimulus reaches the AV node, it signals the ventricles to contract by way of the atrioventricular bundle and Purkinje fibers. **b.** A normal ECG indicates that the heart is functioning properly. The P wave indicates that the atria are about to contract; the QRS wave indicates that the ventricles are about to contract; and the T wave indicates that the ventricles are recovering from contraction.

Experimental Procedure: Electrocardiogram

1. Electrodes will be placed on the wrists and ankles, so the subject should remove any jewelry from these areas. The subject then lies down on a cot or table close to the electrocardiograph.
2. Clean the electrode placement areas with alcohol swabs and apply a small quantity of electrode cream to these areas.
3. Similarly spread electrode cream on the inner surfaces of four electrode plates before attaching one to each of the four sites on the subject. The electrode plate on the right ankle is the grounding system.
4. A selector switch allows various combinations of electrodes to be activated. Attach the leads from the electrodes to the corresponding cables of the lead selector switch.

   Lead I measures the potential difference between the right wrist and the left wrist.

   Lead II measures the potential difference between the right wrist and the left ankle.

   Lead III measures the potential difference between the left wrist and the left ankle.
5. Record the ECG for Lead I, Lead II, and Lead III for one minute each. Then remove the electrodes and clean the cream from the metal and skin of the subject.
6. Do the ECGs you recorded resemble that in Figure. 8.3? ____________________

   Explain. ____

8.2 Heartbeat and Blood Pressure

In this section, we will examine the heartbeat and blood pressure. The beat of the heart creates blood pressure.

Heartbeat

During a heartbeat, first the atria contract and then the ventricles contract. When a chamber contracts, it is called **systole;** when a chamber relaxes, it is called **diastole.** The atria and ventricles take turns being in systole:

Time	*Atria*	*Ventricles*
0.15 sec	Systole	Diastole
0.30 sec	Diastole	Systole
0.40 sec	Diastole	Diastole

Usually, there are two heart sounds with each heartbeat. The first sound (lub) is low and dull and lasts longer than the second sound. It is caused by the closure of the atrioventricular valves following atrial systole. The second sound (dub) follows the first sound after a brief pause. The sound has a snapping quality of higher pitch and shorter duration. The sound is caused by the closure of the semilunar valves following ventricle systole.

Experimental Procedure: Heartbeat at Rest

In the following procedure, you will work with a partner and use a stethoscope to listen to the heartbeat. It will not be necessary for you to count the number of beats per minute.

1. Obtain a stethoscope, and properly position the earpieces. They should point forward. Place the bell of the stethoscope on the left side of your partner's chest between the fourth and fifth ribs. This is where the apex (tip) of the heart is closest to the body wall.

2. Which of the two sounds (lub or dub) is louder? ___

3. Now switch, and your partner will determine your heartbeat.

Blood Pressure

Blood pressure is highest just after ventricular systole, and it is lowest during ventricular diastole.

Why? ___

We would expect a person to have lower blood pressure readings at rest than after exercise.

Why? ___

Experimental Procedure: Blood Pressure at Rest and After Exercise

A number of different types of digital blood pressure monitors are available, and your instructor will instruct you on how to use the type you will be using for this Experimental Procedure. The normal resting blood pressure readings for a young adult are 120/80 (systolic/diastolic), as displayed on the monitor shown in Figure 8.4.

You may work with a partner or by yourself. If working with a partner, each of you will assist the other in taking blood pressure readings. After you have noted the blood pressure readings, also note the pulse reading.

Blood Pressure at Rest

1. Reduce your activity as much as possible.
2. Use the blood pressure monitor to obtain several blood pressure readings, average them, and record your results in Table 8.1.

Blood Pressure After Exercise

1. Run in place for 1 minute.
2. Immediately use the blood pressure monitor to obtain a blood pressure reading, and record it in Table 8.1.

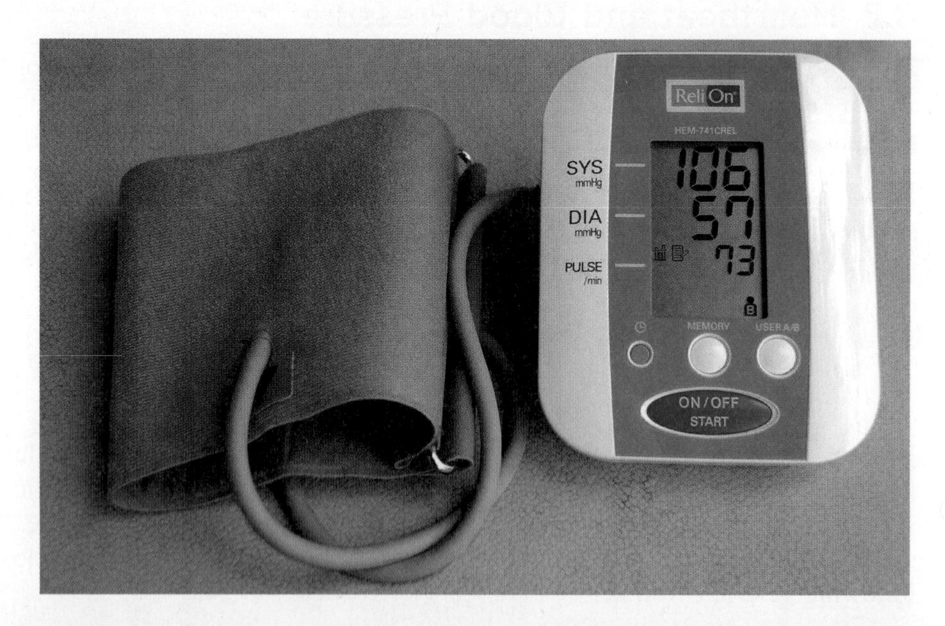

Figure 8.4 Measurement of blood pressure and pulse. There are many different types of digital blood pressure/pulse monitors now available. The one shown here uses a cuff to be placed on the arm. Others use a cuff for the wrist.

Table 8.1 Blood Pressure

	Blood Pressure at Rest	Blood Pressure After Exercise
Partner		
Yourself		

Conclusion: Blood Pressure

- Recall that in Laboratory 1, you determined that exercise increases the heart rate. Knowing this, offer an explanation for your results. ______________________________
- Under what conditions in everyday life would you expect the heart rate and the blood pressure to increase, even though you are not exercising? ______________________________
  When might this be an advantage? A disadvantage? ______________________________
  ____

8.3 The Blood and Blood Flow

Blood appears to be a somewhat viscous, homogeneous fluid. However, analysis shows it to be a fluid tissue, composed of plasma (the fluid portion) and formed elements (including the cells and platelets).

Red and White Blood Cells

There are two categories of blood cells: red blood cells (erythrocytes) and white blood cells (leukocytes). Red blood cells are smaller than white blood cells, and they lack a nucleus, which enables them to carry the maximum amount of oxygen. Red blood cells appear red because they contain the respiratory pigment hemoglobin. Each red blood cell lasts about 120 days in circulation.

White blood cells are larger than red blood cells, and they have a nucleus. The white blood cells are translucent if not stained; there are five different types (Fig. 8.5). White blood cells fight infection, and the white blood cell count is used by doctors to help diagnose diseases. The time white blood cells are in circulation is variable.

Figure 8.5 The white blood cells.
a. A neutrophil has a lobed nucleus with two to five lobes. **b.** An eosinophil has red-staining granules. **c.** A basophil has deep-blue-staining granules. **d.** A monocyte is the largest of the blood cells. **e.** A lymphocyte contains a large, round nucleus. **(a–c, e:** Magnification ×400; **d:** Magnification ×500) (a-b) © Ed Reschke; (c) © Biophoto Associates/Science Source; (d-e) © Ed Reschke

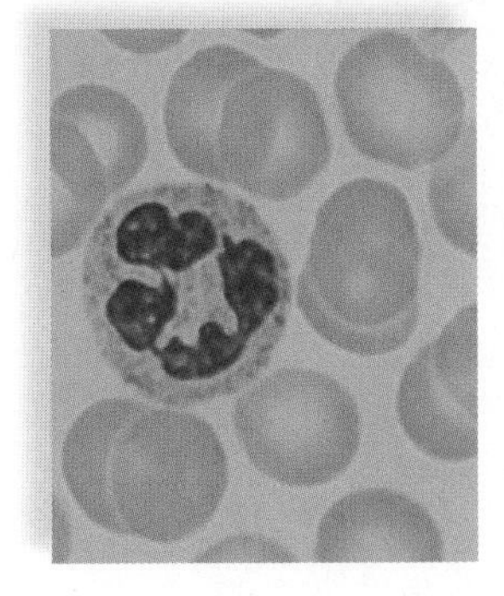

a. Neutrophil

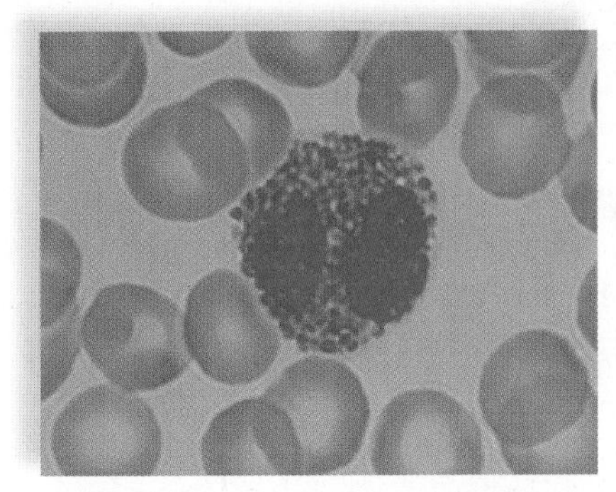

b. Eosinophil

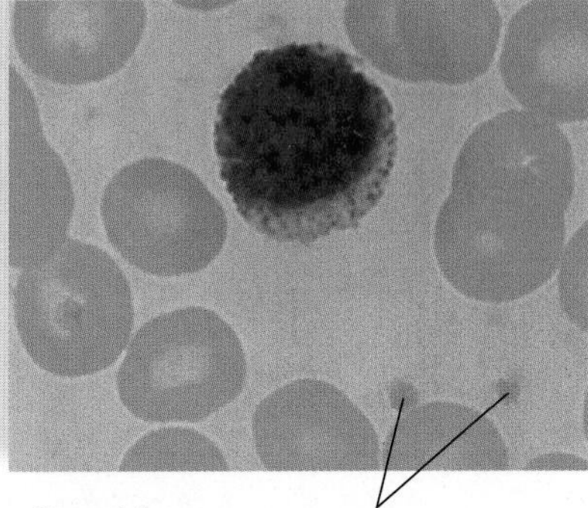

c. Basophil

d. Monocyte

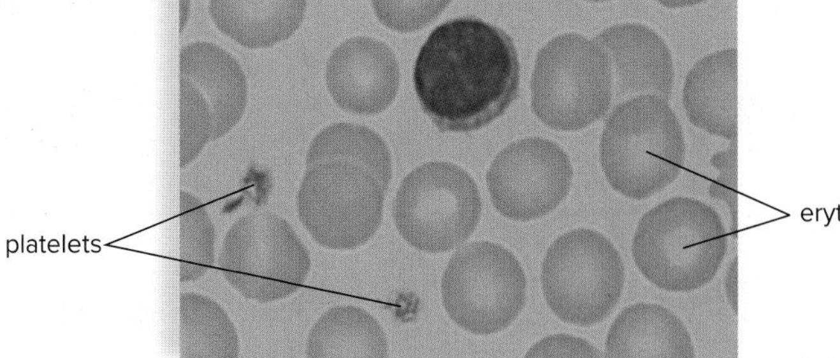

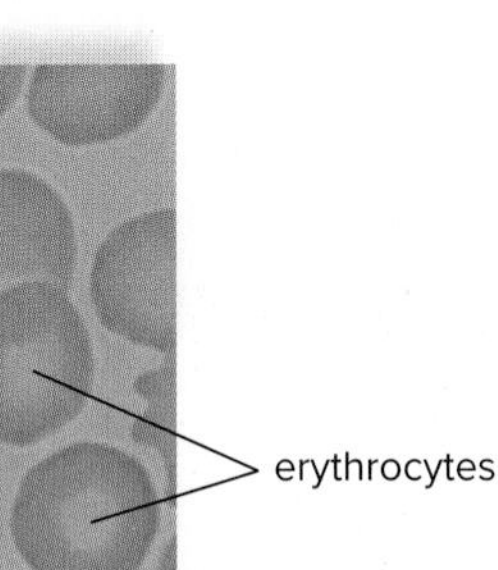

e. Lymphocyte

Observation: Blood Slide

1. Observe a prepared blood smear slide on high power, and note the biconcave (concave on both sides) red blood cells and the less numerous white blood cells. Also, differentiate the two types of cells according to their size and the presence or absence of a nucleus.
2. Complete Table 8.2.

Table 8.2 Slide of Blood

	Relative Number	Relative Size	Presence of Nucleus	Function
Red blood cells				
White blood cells				

Observation: White Blood Cell Slide

1. Observe the demonstration slides under oil immersion. With the help of Figure 8.5, identify the different types of white blood cells, which have been stained with Wright stain.
2. Identify the **granular leukocytes:**
   Neutrophils have cytoplasmic granules; the nucleus is multilobed with two to five connected parts. Neutrophils are, therefore, called polymorphonuclear leukocytes.
   Eosinophils have granules that stain deep red. The nucleus is bilobed.
   Basophils have granules that stain deep blue. The nucleus is bilobed.
3. Identify the **agranular leukocytes:**
   Monocytes are the largest of the white blood cells. The nucleus varies in shape.
   Lymphocytes are usually only slightly larger than red blood cells and typically have a relatively large, round nucleus surrounded by a thin rim of cytoplasm.

ABO Blood Typing

Red blood cells have surface molecules called antigens that indicate they belong to the person. In the ABO system, the presence or absence of an A antigen and/or a B antigen determines the blood type. In this system, there are four types of blood: A, B, AB, and O. Within the plasma, there are antibodies to the antigens that are not on the red blood cells as described in this chart.

Blood Type	Antigen on Red Blood Cells	Antibody in Plasma
A	A	Anti-B
B	B	Anti-A
AB	A, B	None
O	None	Anti-A and Anti-B

Blood type also indicates whether the person has an Rh factor, another type of antigen, on the red blood cells. This designation is attached to the ABO blood type as in A-positive or A-negative. It is customary to simply attach a positive or negative sign to A-negative and A-positive, which are compound words.

Experimental Procedure: ABO Blood Typing

Use the precautions given in the warning box, because the synthetic products used in this exercise can be harmful to the skin.

> ⚠ Wear protective laboratory clothing, latex gloves, and goggles. If the chemicals touch the skin, eyes or mouth, wash immediately. If inhaled, seek fresh air.

1. Obtain three testing plates, each of which contains three depressions; vials of persons 1, 2, and 3 blood; vials of anti-A serum, anti-B serum, and anti-Rh serum.
2. Using a wax pencil, number the plates so you know which plate is for which person #1, #2, or #3. Look carefully at the plate and notice the wells are designated as A, B, or Rh.
3. Being sure to close the cap to each vial in turn,
   Add a drop of person #1 blood to all three wells of plate #1—close the cap.
   Add a drop of anti-A (blue) to the well designated A—close the cap.
   Add a drop of anti-B (yellow) to the well designated B—close the cap.
   Add a drop of anti-Rh (clear) to the well designated Rh—close the cap.
4. Stir the contents of each well with a mixing stick of the correct color. After a few minutes, examine the wells for **agglutination,** i.e., granular appearances that indicate the blood type. (Rh-positive takes the longest to react.) If a person is AB-positive, which wells would show agglutination? ________________
5. Record the blood type results for each person here:

   Person #1 __________

   Person #2 __________

   Person #3 __________

Blood Flow

Blood must circulate to serve the body. The heart pumps the blood, which moves away from the heart in **arteries** and **arterioles** and returns to the heart in **venules** and **veins. Capillaries** connect arterioles to venules. Capillaries branch to form capillary beds that close when precapillary sphincters contract; in this way, capillaries direct the blood flow to regions that need it most (Fig. 8.6). When a capillary bed is closed, blood moves directly from the arteriole to the venule by way of the arteriovenous shunt.

Figure 8.6 Anatomy of blood vessels.
a. Photomicrograph of anatomy of veins. **b.** Blood flow. **c.** Capillary bed.
(a) © Ed Reschke; (c) © Biophoto Associates/Science Source

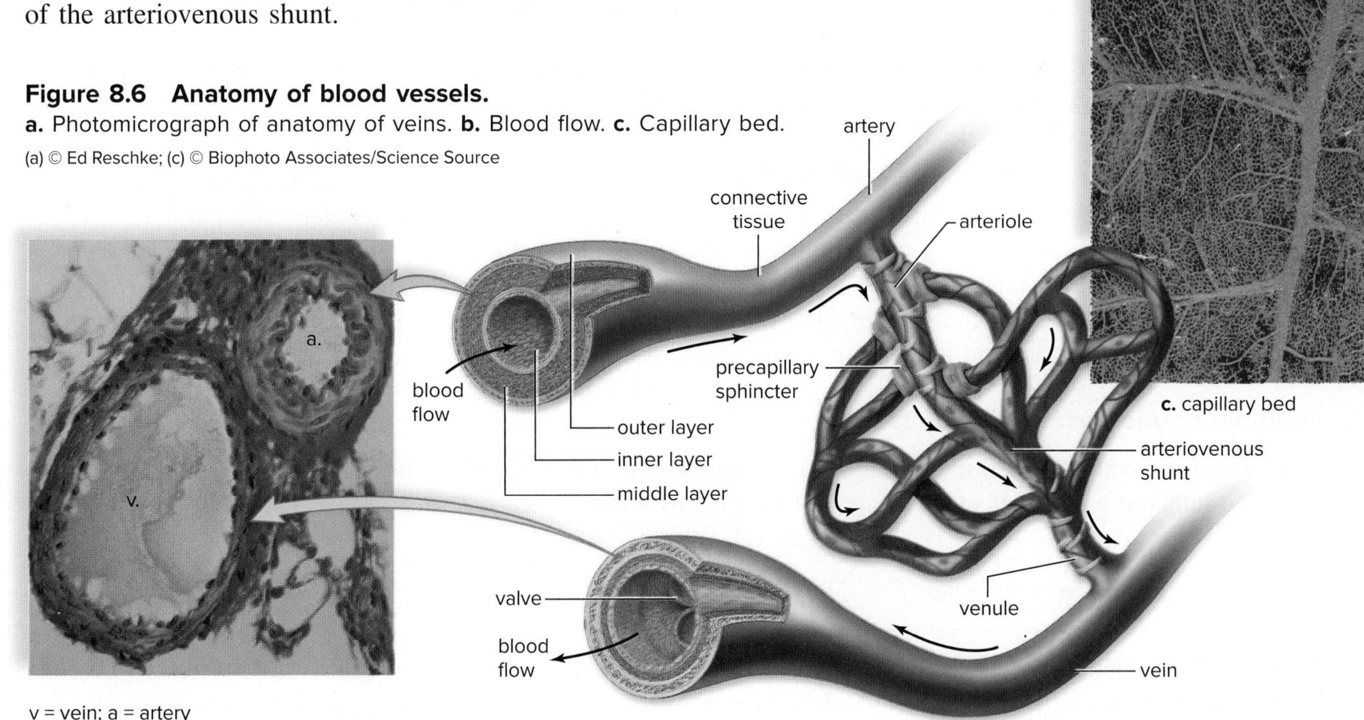

v = vein; a = artery
a. Photomicrograph

b. Blood flow

Observation: Blood Vessel Comparison

1. Obtain a microscope slide that shows an artery and a vein in cross section.
2. View the slide, under both low and high power; with the help of Figure 8.6*a*, determine which is the artery and which is the vein. Identify the **outer layer,** which contains many collagen and elastic fibers and often appears white in specimens. Identify the **middle layer,** the thickest layer, composed of

   smooth muscle and elastic tissue. Does this layer appear thicker in arteries than in veins? ____________

   Identify the **inner layer,** a wavy lining of simple squamous epithelial cells called the endothelium. In veins, the endothelium forms valves that keep the blood moving toward the heart. Veins are larger and more "flabby" than arteries. Why might that be based on the structure of their walls?

   ____

Experimental Procedure: Blood Flow

Observe blood flow through arterioles, capillaries, and venules, either in a video clip, in the tail of a goldfish, or in the webbed skin between the toes of a frog. Observe the pulse and the swiftly moving blood in the arterioles. Contrast this with the more slowly moving blood that circulates in the opposite direction in the venules. Many criss-crossing capillaries are visible. Look for blood cells floating in the bloodstream.

Do you see any? ____

Conclusions: Blood Flow

- Which type of blood vessel (arteries or veins) has thicker walls? _______________________ Why is this advantageous? ____
- Which type of blood vessel has thinner walls? _______________ Why is this advantageous?

  ____
- Which type of blood vessel has valves? _______________ Compare the direction of the flow of blood in the aorta to that in the inferior vena cava. Why is this advantageous for veins to have valves? ____

Path of Blood in Adult Humans

The right side of the heart pumps blood into the **pulmonary circuit**—to the lungs and back to the heart (Fig. 8.7). While the blood is in the lungs, it gives up carbon dioxide and gains oxygen. The left side of the heart, consisting of two chambers, pumps blood to the **systemic circuit**—throughout the whole body except the lungs. Blood in the systemic circuit gives up oxygen and gains carbon dioxide.

Figure 8.7 shows how to trace the path of blood in both the pulmonary and systemic circuits in adult humans. It will also assist you in learning the names of some of the major blood vessels.

Pulmonary and Systemic Circuits

1. Trace the path of blood in the pulmonary circuit from a chamber of the heart to the lungs, and then from the lungs to a chamber of the heart. Follow the arrows in Figure 8.7, and use the names of the

   blood vessels provided there. ____

   ____
2. Trace the path of blood in the systemic circuit from the heart to the kidneys, and then from the kidneys to the heart. ____

Figure 8.7 Diagram of the human cardiovascular system.
In the pulmonary circuit, the pulmonary arteries take O_2-poor blood to the lungs, and the pulmonary veins return O_2-rich blood to the heart. In the systemic circuit, the aorta branches into the various arteries that go to all other parts of the body. After blood passes through arterioles, capillaries, and venules, it enters various veins and then the superior and inferior venae cavae, which return it to the heart.

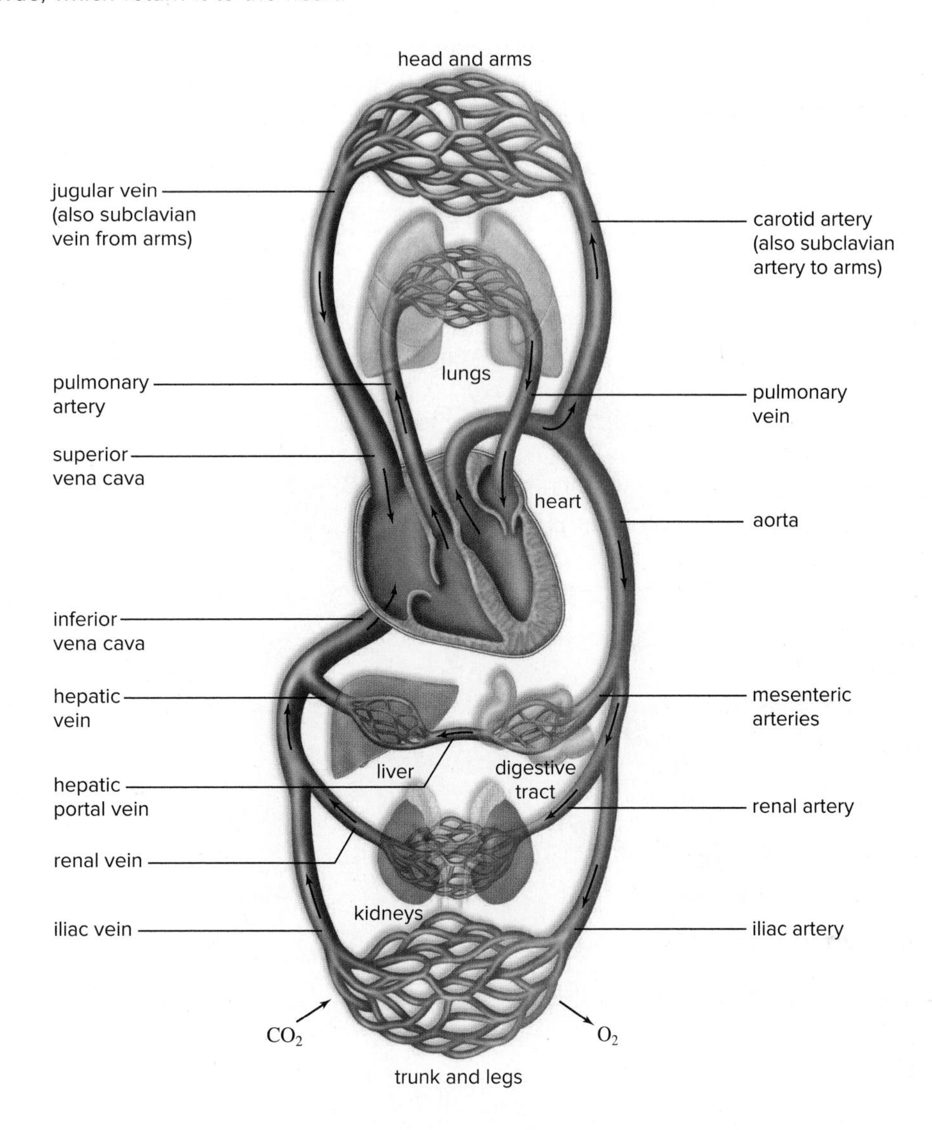

3. Is it true that all arteries carry O_2-rich blood and all veins carry O_2-poor blood? ____________________
Explain the color of the pulmonary blood vessels in Figure 8.7. ____

____

4. A portal system is a vein that begins in capillaries and ends in capillaries. For example, the hepatic portal vein begins in capillaries at the digestive tract and ends in capillaries at the liver. Trace the path of blood from the aorta to the inferior vena cava by way of the hepatic portal vein. ____________________

____

____

Names of Blood Vessels

Use Figures 8.7 and 8.8 to complete Table 8.3 by stating the name of the artery that takes blood to the body part and the name of the vein that takes blood away from the body part.

Figure 8.8 The major arteries and veins of the systemic circuit.
This illustration offers a more realistic representation of major blood vessels (arteries and veins) of the systemic circuit.

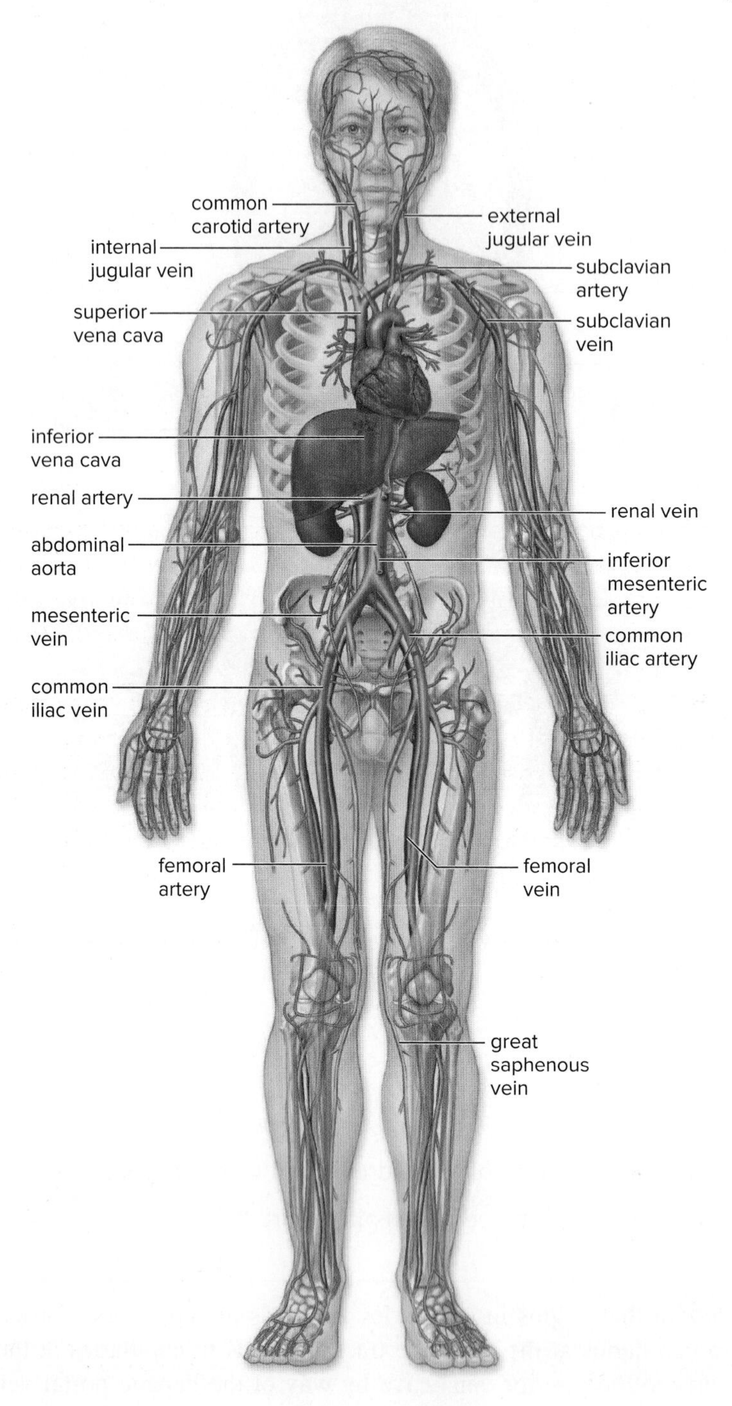

Table 8.3 Major Blood Vessels in the Systemic Circuit

Body Part	Artery	Vein
Heart		
Head		
Arms		
Kidney		
Legs		
Intestines		

Application for Daily Living

High Blood Pressure

Blood pressure is an indication of cardiovascular health because high blood pressure at rest is associated with clogged arteries that impede blood flow. Blood should flow freely to all parts of the body and especially to the heart because it has to pump the blood to keep it moving about the body. Unfortunately, the carotid arteries, which serve the heart, are the ones most apt to become clogged with fatty deposits called plaque.

Optimal blood pressure at rest is at or below 120/80. (These numbers are in mm Hg, a way to measure pressure, and systolic blood pressure is given first.) Prehypertension occurs when systolic pressures are between 120 and 139 and diastolic pressures are between 80 and 89 mm Hg. Hypertension, or high blood pressure, occurs when systolic pressure exceeds 140 and diastolic pressure exceeds 90, as in 150/95.

Age (over 55), male gender, and African-American ethnicity are all risk factors for developing plaque in the arteries. However, blood pressure can be improved through adequate exercise, good diet, loss of weight, not smoking, and reduced stress.

Laboratory Review 8

_______________ **1.** What type of blood cells are lymphocytes and monocytes?

_______________ **2.** Nutrients exit and wastes enter which type of blood vessel?

_______________ **3.** Which blood cells contain a respiratory pigment?

_______________ **4.** Which chamber of the heart receives venous blood from the systemic circuit?

_______________ **5.** Identify the vessel that conducts blood from the left ventricle.

_______________ **6.** The pulmonary trunk leaves which chamber?

_______________ **7.** Identify the artery that nourishes the heart tissue.

_______________ **8.** Which heart chamber pumps blood to the body proper?

_______________ **9.** Which is higher—systolic or diastolic pressure?

_______________ **10.** Identify the pacemaker region of the heart.

_______________ **11.** Does the pulmonary artery in adults carry O_2-rich or O_2-poor blood?

_______________ **12.** Which type of blood shows no reaction to anti-A and anti-B serum?

_______________ **13.** Identify the artery that serves the kidney.

_______________ **14.** Identify the arteries that take blood from the aorta to the legs.

_______________ **15.** What part of the human body is served by a subclavian vessel?

_______________ **16.** Identify the large abdominal vein that runs alongside the aorta and enters the right atrium.

_______________ **17.** Which type of blood vessel (artery or vein) has valves?

Thought Questions

18. What might be the significant result of nonfunctional or defective chordae tendineae?

19. During a heart attack, cardiac muscle cells are deprived of their blood supply, yet the atria are still receiving blood. Explain.

20. Evaluate the following statement: All arteries carry O_2-rich blood and all veins carry O_2-poor blood. Based on what you have learned in this laboratory, is this statement correct? Explain.

LABORATORY

9

Chemical Aspects of Digestion

Learning Outcomes

Introduction

- Sequence the organs of the digestive tract from the mouth to the anus.
- State the contribution of each organ, if any, to the process of chemical digestion.

9.1 Protein Digestion by Pepsin

- Associate the enzyme pepsin with the ability of the stomach to digest protein.
- Explain why stomach contents are acidic and how a warm body temperature aids digestion.

Prelab Question: How does an acidic pH aid pepsin?

9.2 Fat Digestion by Pancreatic Lipase

- Associate the enzyme lipase with the ability of the small intestine to digest fat.
- Explain why the emulsification process assists the action of lipase.
- Explain why a change in pH indicates that fat digestion has occurred.
- Explain the relationship between time and enzyme activity.

Prelab Question: What is the relationship between time and enzyme activity?

9.3 Starch Digestion by Pancreatic Amylase

- Associate the enzyme pancreatic amylase with the ability of the small intestine to digest starch.

Prelab Question: Why is a denatured enzyme unable to carry out digestion?

9.4 Requirements for Digestion

- Assuming a specific enzyme, list four factors that can affect the activity of all enzymes.
- Explain why the operative procedure that reduces the size of the stomach causes an individual to lose weight.

Which factors during digestion can affect the activity of enzymes?

Introduction

In this lab, we will examine the process of digestion by learning the organs associated with digestion (Fig. 9.1) and studying the action of digestive enzymes.

Enzymes are molecules (typically proteins) that catalyze chemical reactions. They are very specific, and each one usually participates in only one type of reaction. The active site of an enzyme has a shape that accommodates its substrate, and if an environmental factor such as a boiling temperature or a wrong pH alters this shape, the enzyme loses its ability to function well, if at all. We will have an opportunity to make these observations with controlled experiments. The box on page 106 reviews what is meant by a controlled experiment.

Planning Ahead Be advised that protein digestion requires 1½ hours and fat digestion requires 1 hour. Also a boiling water bath is required for starch digestion.

Figure 9.1 Organs of the digestive tract (right) and accessory organs (left).

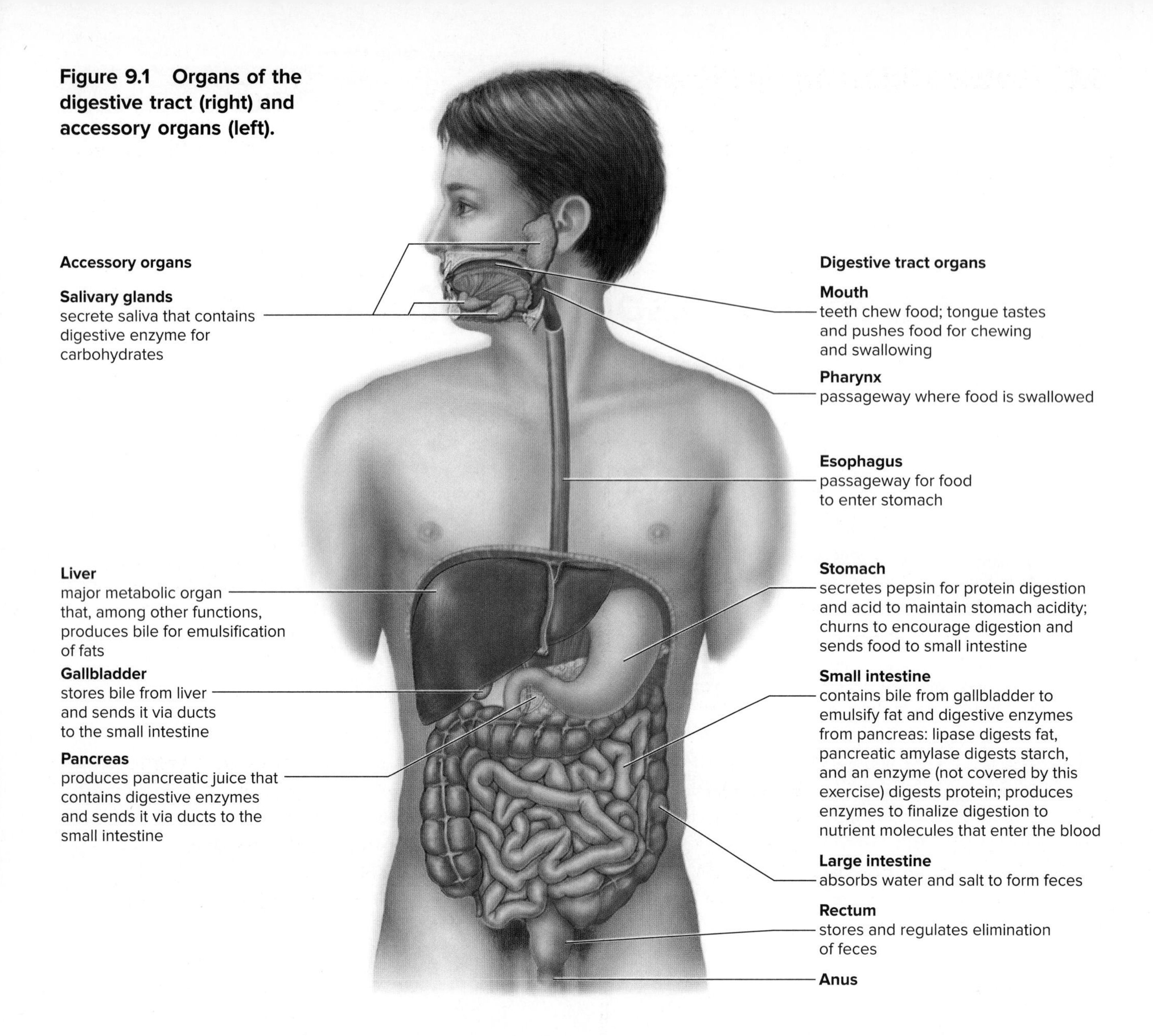

What Is a Control?

The experiments in today's laboratory have both a positive control and a negative control, *which should be saved for comparison purposes until the experiment is complete*. The **positive control** goes through all the steps of the experiment and does contain the substance being tested. Therefore, positive results are expected. The **negative control** goes through all the steps of the experiment, except it does not contain the substance being tested. Therefore, negative results are expected.

For example, if a test tube contains glucose (the substance being tested) and Benedict's reagent (blue) is added, a red color develops upon heating. This test tube is the positive control; it tests positive for glucose. If a test tube does not contain glucose and Benedict's reagent is added, Benedict's is expected to remain blue. This test tube is the negative control; it tests negative for glucose.

What benefit is a positive control? Positive controls give you a standard by which to tell if the substance being tested is present (or acting properly) in an unknown sample. Negative controls ensure that the experiment is giving reliable results; after all, if a negative control should happen to give a positive result, then the entire experiment may be faulty and unreliable.

9.1 Protein Digestion by Pepsin

Certain foods, such as meat and egg whites, are rich in protein. Egg whites contain albumin, which is the protein used in this Experimental Procedure. Protein is digested by **pepsin** in the stomach (Fig. 9.2), a process described by the following reaction:

$$\text{protein} + \text{water} \xrightarrow{\text{pepsin (enzyme)}} \text{peptides}$$

The stomach has a very low pH. Does this indicate that pepsin works effectively in an acidic or a basic environment? ________________________ This is the pH that allows the enzyme to maintain its normal shape so that it will combine with the substrate. A warm temperature causes molecules to move about more rapidly and increases the encounters between enzyme and substrate. Therefore you would hypothesize that the yield from this enzymatic reaction will be higher if the pH is ____________________ and the temperature is ____________________ (body temperature 37°C).

Test for Protein Digestion

Biuret reagent is used to test for protein digestion. If digestion has not occurred, biuret reagent turns purple, indicating that protein is present. If digestion has occurred, biuret reagent turns pinkish-purple, indicating that peptides are present.

> ⚠ **Biuret reagent** is highly corrosive. Exercise care in using this chemical. If any should spill on your skin, wash the area with mild soap and water. Follow your instructor's directions for its disposal.

Experimental Procedure: Protein Digestion

1. Label four clean test tubes (1 to 4). Using the designated graduated pipet, add 2 ml of the albumin solution to all tubes. Albumin is a protein.
2. Add 2 ml of the pepsin solution to tubes 1 to 3, as listed in Table 9.1.
3. Add 2 ml of 0.2% HCl to tubes 1 and 2. HCl simulates the acidic conditions of the stomach.
4. Add 2 ml of water to tube 3 and 4 ml of water to tube 4, as listed in Table 9.1.
5. Swirl to mix the tubes. Tube 2 remains at room temperature, but the other three are incubated for 1½ hours. Record the temperature for each tube in Table 9.1.
6. Remove the tubes from the incubator and place all four tubes in a tube rack. Add 2 ml of biuret reagent to all tubes and observe. Record your results in Table 9.1 as + or – to indicate digestion or no digestion.

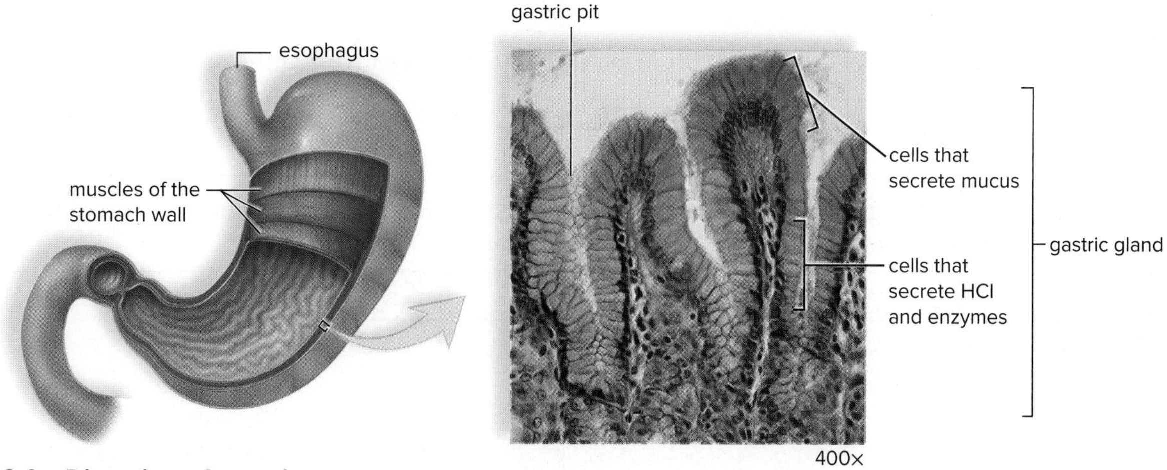

Figure 9.2 Digestion of protein.
Pepsin, produced by the gastric glands of the stomach, helps digest protein. © Ed Reschke

Table 9.1 Protein Digestion by Pepsin

Tube	Contents	Temperature	Digestion (+ or −)	Explanation
1	Albumin Pepsin HCl Biuret reagent			
2	Albumin Pepsin HCl Biuret reagent			
3	Albumin Pepsin Water Biuret reagent			
4	Albumin Water Biuret reagent			

Conclusions: Protein Digestion

- Explain your results in Table 9.1 by giving an explanation why digestion did or did not occur. To be complete, consider all the requirements for an enzymatic reaction as listed in Table 9.4. Now show here that tube 1 met all the requirements for digestion:

  Pepsin is the correct ____________.

  Albumin is the correct ____________.

  37°C is the optimum ____________.

  HCL provides the optimum ____________.

  1½ hours provides ____________ for the reaction to occur.

- Review "What Is a Control?" on page 106. Which tube was the negative control? ______________________

  Explain. ____

  ____

- If this control tube had given a positive result for protein digestion, what could you conclude about this experiment? ____

  ____

9.2 Fat Digestion by Pancreatic Lipase

Lipids include fats (e.g., butterfat) and oils (e.g., sunflower, corn, olive, and canola). Lipids are digested by **pancreatic lipase** in the small intestine (Fig. 9.3).

Figure 9.3 Emulsification and digestion of fat.
Bile from the liver (stored in the gallbladder) enters the small intestine, where lipase in pancreatic juice from the pancreas digests fat.

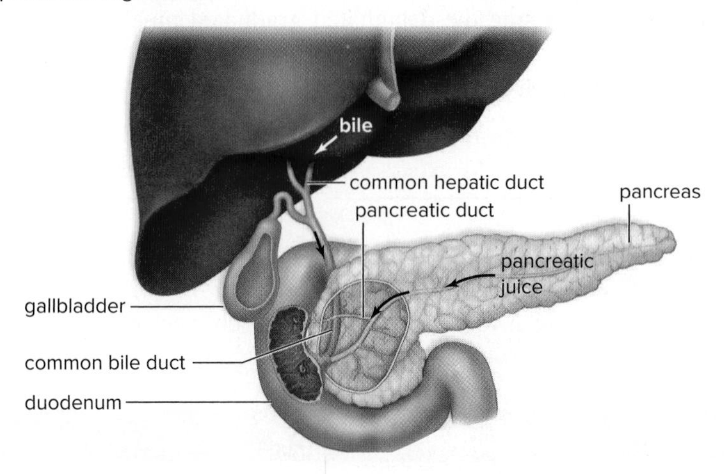

The following two reactions describe fat digestion:

1. $$\text{fat} \xrightarrow{\text{bile (emulsifier)}} \text{fat droplets}$$

2. $$\text{fat droplets} + \text{water} \xrightarrow{\text{lipase (enzyme)}} \text{glycerol} + \text{fatty acids}$$

With regard to the first step, consider that fat is not soluble in water; yet, lipase makes use of water when it digests fat. Therefore, bile is needed to emulsify fat—cause it to break up into fat droplets that disperse in water. The reason for dispersal is that bile contains molecules with two ends. One end (the nonpolar end) is soluble in fat, and the other end (the polar end) is soluble in water. Bile can emulsify fat because of this.

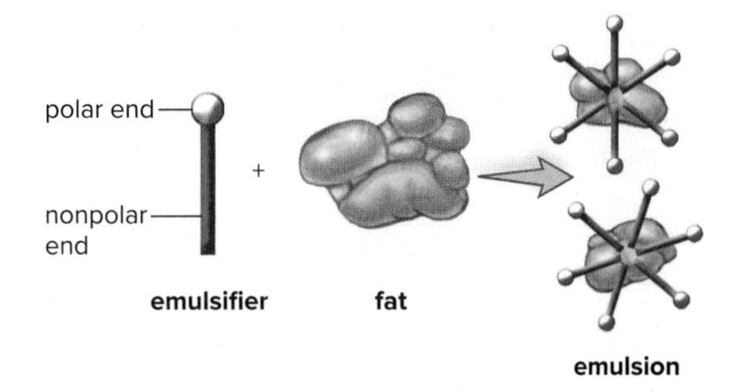

With regard to the second step, would the pH of the solution be lower before or after the enzymatic reaction? (*Hint:* Remember that an acid decreases pH and a base increases pH.) ____________________

____

Test for Fat Digestion

In the test for fat digestion, you will be using a pH indicator, which changes color as the solution in the test tube goes from basic conditions to acidic conditions. Phenol red is a pH indicator that is red in basic solutions and yellow in acidic solutions.

Experimental Procedure: Fat Digestion

1. Label three clean test tubes (1 to 3). Using the designated graduated pipet, add 1 ml of vegetable oil to all tubes.
2. Add 2 ml of phenol red solution to each tube. What role does phenol red play? ____________________
3. Add 2 ml of pancreatic lipase (pancreatin) to tubes 1 and 2 and 2 ml of water to tube 3, as listed in Table 9.2. What role does lipase play? ____________________
4. Add a pinch of bile salts to tube 1.
5. Record the initial color of all tubes in Table 9.2.
6. Incubate all three tubes at 37°C and check every 20 minutes.
7. Record any color change and how long it took to see this color change in Table 9.2.

Table 9.2 Fat Digestion by Pancreatic Lipase

Tube	Contents	Color		Time Taken	Explanation
		Initial	*Final*		
1	Vegetable oil Phenol red Pancreatin Bile salts				
2	Vegetable oil Phenol red Pancreatin				
3	Vegetable oil Phenol red Water				

Conclusions: Fat Digestion

- Explain your results in Table 9.2 by giving an explanation why digestion did or did not occur.
- What role did bile salts play in this experiment? ____________________

____
- What role did phenol red play in this experiment? ____________________
- Review "What Is a Control?" on page 106. Which test tube in this experiment could be considered a negative control? ____________________

____

9.3 Starch Digestion by Pancreatic Amylase

Starch is present in bakery products and in potatoes, rice, and corn. Starch is digested by **pancreatic amylase** in the small intestine, a process described by the following reaction:

$$\text{starch} + \text{water} \xrightarrow{\text{amylase (enzyme)}} \text{maltose}$$

1. If digestion *does not* occur, which will be present—starch or maltose? ______________________
2. If digestion *does* occur, which will be present—starch or maltose? ______________________

Tests for Starch Digestion

You will be using two tests for starch digestion:

1. If digestion has not taken place, the iodine test for starch will be positive (+) and a blue-black color will be observed. If digestion has occurred, the iodine test for starch will be negative (–) and the iodine will remain yellowish-brown.
2. If digestion has taken place, the Benedict's test for sugar (maltose) will be positive (+) and a color change ranging from green to red will be observed. If digestion has not taken place, the Benedict's test for sugar will be negative (–) and the solution will remain blue.

   To test for sugar, add five drops of Benedict's reagent to each test tube. Place the tube in a boiling water bath for a few minutes, and note any color changes. Boiling the test tube is necessary for the Benedict's reagent to react.

> ⚠ **Benedict's reagent** is highly corrosive. Use protective eyewear when performing this experiment. Exercise care in using this chemical. If any should spill on your skin, wash the area with mild soap and water. Follow your instructor's directions for disposal of this chemical.

Experimental Procedure: Starch Digestion

1. Label six clean test tubes (1 to 6).
2. Using the designated graduated transfer pipet, add 1 ml of pancreatic amylase solution to tubes 1 to 4 and 1 ml of water to tubes 5 and 6.
3. Test tubes 1 and 2 immediately

   Tube 1 Shake the starch solution and add 1 ml of starch solution. Immediately add five drops of iodine to test for starch. Put this tube in a test tube rack and record your results in Table 9.3.

   Tube 2 Shake the starch solution and add 1 ml of starch solution. Immediately test for sugar with the Benedict's test following the preceding directions. Put this tube in a test tube rack and record your results in Table 9.3.
4. Shake the starch solution and add 1 ml of starch solution to tubes 3 to 6. Allow the tubes to stand for 30 minutes.

   Tubes 3 and 5 After the 30 minutes have passed, test for starch using the iodine test. Place these tubes in the test tube rack and record your results in Table 9.3.

   Tubes 4 and 6 After the 30 minutes have passed, test for sugar with the Benedict's test following the preceding directions. Place these tubes in the test tube rack and record your results in Table 9.3.
5. Examine all your tubes in the test tube rack and decide whether digestion occurred (+) or did not occur (–). Complete Table 9.3.

Table 9.3 Starch Digestion by Amylase

Tube	Contents	Time*	Type of Test	Results	Explanation
1	Pancreatic amylase Starch	0	Iodine	+	
2	Pancreatic amylase Starch				
3	Pancreatic amylase Starch				
4	Pancreatic amylase Starch				
5	Water Starch				
6	Water Starch				

* Enter either 0 for immediately or T for after 30 minutes.

Conclusions: Starch Digestion

- Considering tubes 1 and 2, this experimental procedure showed that __________ must pass for digestion to occur.
- Considering tubes 5 and 6, this experimental procedure showed that an active __________ must be present for digestion to occur.
- Why would you not recommend doing the test for starch and the test for sugar on the same tube? __________

  ____

  ____

- Which test tubes served as a negative control in this experiment? ____________________

  Explain your answer. ____

  ____

Absorption of Sugars and Other Nutrients

Figure 9.4 shows that the folded lining of the small intestine has many fingerlike projections called villi. The small intestine not only digests food; it also absorbs the products of digestion, such as sugars from carbohydrate digestion, amino acids from protein digestion, and glycerol and fatty acids from fat digestion at the villi.

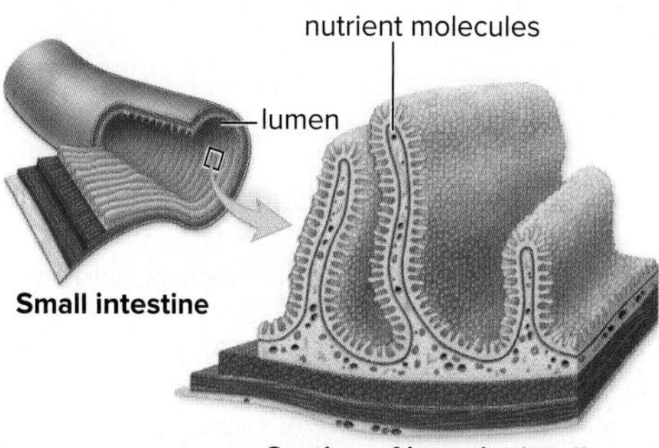

Figure 9.4 Anatomy of the small intestine. Nutrients enter the bloodstream across the much-convoluted walls of the small intestine.

9.4 Requirements for Digestion

Explain in Table 9.4 how each of the requirements listed influences effective digestion.

Table 9.4 Requirements for Digestion

Requirement	Explanation
Specific enzyme	
Specific substrate	
Warm temperature	
Specific pH	
Time	
Fat emulsifier	

To lose weight, some obese individuals undergo an operation in which (1) the stomach is reduced to the size of a golf ball, and (2) food bypasses the duodenum (first 2 feet) of the intestine. Answer these questions to explain how this operation would affect the requirements for digestion.

1. How is the amount of substrate reduced? ______________________________
2. How is the amount of digestive enzymes reduced? ______________________________
3. How is time reduced? ______________________________
4. What makes the pH of the small intestine higher than before? ______________________________
5. How is fat emulsification reduced? ______________________________
6. How does surgery to reduce obesity sometimes result in malnutrition? ______________________________

Laboratory Review 9

_______________ **1.** What part of an enzyme is specific for and accommodates its substrate?

_______________ **2.** What two environmental conditions affect the ability of an enzyme to function properly?

_______________ **3.** Where are carbohydrates first digested by an enzyme?

_______________ **4.** What organ produces the lipase and amylase that perform digestion in the small intestine?

_______________ **5.** What is the name given to a sample that contains the factor being tested and that goes through all the steps of the experiment?

_______________ **6.** Does pepsin function in an acidic or a basic pH?

_______________ **7.** What results from the digestion of proteins by pepsin?

_______________ **8.** What reagent was used to test for the digestion of proteins?

_______________ **9.** What molecule emulsifies fats in the small intestine?

_______________ **10.** What organ stores bile?

_______________ **11.** What component of foods is digested by pancreatic amylase?

_______________ **12.** What molecule is present after pancreatic amylase digests starch?

_______________ **13.** What two tests were done to test for starch digestion?

_______________ **14.** What structures absorb the products of digestion in the small intestine?

Thought Questions

15. Salivary amylase digests starch to maltose in the mouth while chewing takes place. Why is it beneficial to have the enzyme that digests starch, amylase, in two different locations?

16. How would the digestion of fat be impacted by the removal of the gallbladder? Explain.

17. Explain why consuming excess antacids might affect protein digestion. How could you investigate the impact of consuming antacids on protein digestion?

LABORATORY

10

Energy Requirements and Ideal Weight

Learning Outcomes

10.1 Average Daily Energy Intake

- Keep a food diary and use it to calculate your average daily energy intake.

Prelab Question: What unit of measurement is used for daily energy intake?

10.2 Average Daily Energy Requirement

- Keep a physical activity diary and use it to calculate average daily energy required for physical activity.
- Calculate the daily energy required for your basal metabolism.
- Calculate the daily energy required for your specific dynamic action.
- Calculate your total average daily energy requirement.

Prelab Question: Daily energy requirement has what three components?

10.3 Comparison of Average Daily Energy Intake and Average Daily Energy Requirement

- Predict whether weight gain or weight loss will occur after energy intake is compared to energy required.

Prelab Question: What two energy measurements determine whether you gain weight or lose weight?

10.4 Ideal Weight

- Calculate ideal weight based on body mass index (BMI).
- Calculate ideal weight based on body composition.
- Make a recommendation regarding daily energy intake and daily energy required for achieving or maintaining ideal weight.

Prelab Question: What are two methods for calculating ideal weight?

***Application for Daily Living:* Losing Weight**

Introduction

This laboratory helps you determine your average daily energy (kcal) requirements, your average daily energy (kcal) intake, and your ideal weight. In addition to the energy content of food, other nutritional aspects are also important. Minerals and vitamins, along with specific types of biological molecules, contribute to a healthy diet. So, while the focus of this lab is on energy in our food, there is more to diet than just calories.

Before you come to the laboratory, it is necessary for you to keep two daily diaries for 3 days. The physical activity diary is a record of the energy you expend for physical activity. Physical activity is only a portion of your daily energy requirement. You also need energy for **basal metabolism** (energy needed when the body is resting) and **specific dynamic action** (energy needed to process food), also known as thermal effect of food.

The food diary is a record of the energy you take in as food. If the energy you expend and the energy you take in are in balance, you will neither gain nor lose weight. If your ideal weight is different from your present weight, you will be able to determine whether to increase or decrease your physical activity and food intake to attain an ideal weight.

10.1 Average Daily Energy Intake

To complete this laboratory, you must calculate an average daily energy intake. The first method given is preferred because it makes use of your actual intake. However, when appropriate, a hypothetical method is also provided.

Personal Diet

Follow these directions.

1. Keep a written food diary of your own design for 3 days. Your diary should list all food and beverages consumed throughout an entire 3-day period, including snacks and alcoholic beverages.
2. Make sure your food diary contains a column for recording the energy in kcal (kilocalories)* of the food you eat. Table 10.1 may be helpful to you in this regard. Also, food packages often contain kcal information, as do various calorie books and computer programs. For example, after eating two slices of bread, your diary might look like this:

Day of Week ____________			
Type of Food	**Portion Size**	**kcal**	**Total kcal**
Bread	Two slices	70/one slice	140

Continue in this manner until you have listed all the foods you eat in 1 day. Do the same for 2 more days.

3. Add up your total kcal intake for each day, and record the information here:

   Day 1 energy intake ____________ kcal

   Day 2 energy intake ____________ kcal

   Day 3 energy intake ____________ kcal

   Total ____________ kcal

4. Divide the total by 3 to calculate the average daily energy intake.

Av. energy intake/day = ____________ kcal

*The amount of heat required to raise 1 kg (kilogram) of water 1°C.

Table 10.1 Nutrient and Energy Content of Foods

Breads
1 slice of bread or any of the following:
¾ cup ready-to-eat cereal
⅓ cup corn
1 small potato
(1 bread = 15 g carbohydrate, 2 g protein, and 70 kcal)

Milk
1 cup skim milk or: (2% milk—add 5 g fat)
1 cup skim-milk yogurt, plain (Whole milk—add 8 g fat)
1 cup buttermilk
½ cup evaporated skim milk or milk dessert
(1 milk = 12 g carbohydrate, 8 g protein, and 80 kcal)

Vegetables
½ cup greens
½ cup carrots
½ cup beets
(1 vegetable = 5 g carbohydrate, 1 g protein, and 25 kcal)

Fruits
½ small banana or
1 small apple
½ cup orange juice or ½ grapefruit
(1 fruit = 10 g carbohydrate and 40 kcal)

Meats (lean)
1 oz. lean meat or:
1 oz. chicken meat without the skin
1 oz. any fish
¼ cup canned tuna or 1 oz. low-fat cheese
(1 oz. low-fat meat = 7 g protein, 3 g fat, and 55 kcal)

Meats (medium fat)
1 oz. pork loin
1 egg
¼ cup creamed cottage cheese
(1 medium-fat meat = 7 g protein,
5½ g fat, and about 80 kcal)

Meats (high fat)
1 oz. high-fat meat is like:
1 oz. country-style ham
1 oz. cheddar cheese
A small hot dog (frankfurter)
(1 high-fat meat = 7 g protein, 8 g fat, and 100 kcal)

Peanut butter
Peanut butter is like a meat in terms of its protein content but is very high in fat. It is estimated as (2 tbsp peanut butter = 7 g protein, 15½ g fat, and about 170 kcal)

Fats
1 tsp butter or margarine
1 tsp any oil
1 tbsp salad dressing
1 strip crisp bacon
5 small olives
10 whole Virginia peanuts
(1 fat = 5 g fat and 45 kcal)

Legumes (beans and peas)
Legumes are like meats because they are rich in protein and iron but are lower in fat than meat. They contain much starch. They can be treated as (½ cup legumes = 15 g carbohydrate, 9 g protein, 3 g fat, and 125 kcal)

Miscellaneous Foods

	g (grams)			*kcal*		*g (grams)*			*kcal*
	Protein	**Fat**	**Carbohydrate**			**Protein**	**Fat**	**Carbohydrate**	
Ice cream (1 cup)	5	14	32	274	Beer (1 can)	1	0	14	60
Cake (1 piece)	3	1	32	149	Soft drink	0	0	37	148
Doughnuts (1)	3	11	22	199	Soup (1 cup)	2	3	15	95
Pie	3	15	51	351	Coffee and tea	0	0	0	0
Caramel candy (1 oz.)	1	3	22	119					

Hypothetical Diet

Suppose you have decided to eat your meals at the fast-food restaurants listed in the table.

1. Circle the restaurant and the items you have chosen for breakfast, lunch, and dinner. Choose a different restaurant for each meal.
2. Record the number of kcal for the items chosen.
3. Add the number of kcal.
4. This will be your average energy intake/day.

Fast-Food Menus and kcal

Restaurant/Menu	kcal	Restaurant/Menu	kcal
Taco Bell (www.tacobell.com)		**McDonald's (www.mcdonalds.com)**	
Steak Baja chalupa	390	Egg McMuffin	300
Soft beef taco supreme	240	Sausage McMuffin w/egg	450
Chicken Baja gordita	320	Ham and egg cheese bagel	550
Spicy chicken soft taco	170	Hash browns	130
Nachos Bellgrande	760	Hotcakes (w/margarine & syrup)	350
Soft drink (small)	150	Breakfast burrito	290
Subway (www.subway.com)		6 pc. chicken nugget	280
Meatball marinara	560	Barbeque sauce	50
6" roasted chicken breast	310	Hamburger	250
6" roast beef	290	Quarter pounder w/cheese	510
Soft drink (small)	150	Big Mac	540
Burger King (www.burgerking.com)		Filet-O-Fish	380
Croissan'wich w/sausage, egg & cheese	520	Chicken McGrill w/mayo	420
Croissan'wich w/egg & cheese	320	Med. french fries	380
Egg'wich™ w/Canadian bacon & egg	380	Large french fries	500
French toast sticks	390	Hot fudge sundae	340
Lg. hash brown rounds	390	Vanilla milk shake	570
Double Whopper w/Cheese and mayo	1150	Soft drink (small)	150
Whopper w/cheese and mayo	760		
BK big fish	710	**Papa John's (www.papajohns.com)**	
King-sized fries w/salt	600	Cheese pizza	283
Med.-sized fries w/salt	360	All-meat pizza	390
King-sized onion rings	550	Garden special pizza	280
Med.-sized onion rings	320	The works pizza	342
Dutch apple pie	340	Cheese sticks	180
Hershey's sundae pie	310	Breadsticks	140
Soft drink (small)	150	Garlic sauce	235
Vanilla milk shake (medium)	720	Pizza sauce	25
		Nacho cheese sauce	60
		Soft drink (small)	150

Av. energy intake/day = ________________ kcal

10.2 Average Daily Energy Requirement

As stated at the beginning of this laboratory, the body needs energy for three purposes: (1) energy for physical activity, (2) energy to support basal metabolism, and (3) energy for the specific dynamic action of processing food. We will calculate the daily energy requirement for each of these.

Calculating Average Daily Energy Required for Physical Activity

You will need to know your average daily energy requirement for physical activity before coming to this laboratory.

1. Keep a physical activity diary for 3 days. Make three copies of Table 10.2, one for each day.
2. Use Scale 3 (left and right sides) in Figure 10.1 to convert your body weight to kg (kilograms). Place this value on the appropriate line in your physical activity diary (Table 10.2).
3. Consult Table 10.3 and fill in column 4 of your physical activity diary.
4. Fill in column 5 of your physical activity diary by using a calculator and this formula:

$$\underset{\text{(min)}}{\text{time spent}} \times \text{energy cost} \times \underset{\text{(kg)}}{\text{body weight}} = \underset{\text{kcal}}{\text{total energy expended}}$$

Therefore, if 100 minutes were spent doing word processing, and the body weight is 77.3 kg, the kcal expended for typing would be

$$100 \text{ min} \times 0.01 \text{ kcal/kg min} \times 77.3 \text{ kg} = 77.3 \text{ kcal}$$

5. Total the number of kcal expended for each of the 3 days, and divide this total by 3 to get the average (av.) daily energy required for physical activity:

Energy for physical activity (day 1)	________________	kcal (see page 120)
Energy for physical activity (day 2)	________________	kcal
Energy for physical activity (day 3)	________________	kcal
Total energy for physical activity/3 days	________________	kcal
Av. energy required for physical activity/day	________________	kcal

A completed food diary and physical activity diary are necessary before you begin the laboratory.

Table 10.2 Physical Activity Diary for One Day

Time of Day	Activity	Time Spent (min)	Factor (kcal/kg min)	Weight (kg)	Total Energy Expended (kcal)

Total energy expended ________________

Table 10.3 Energy Cost for Activities (Exclusive of Basal Metabolism and Specific Dynamic Action)

Type of Activity	Energy Cost (kcal/kg min)	Type of Activity	Energy Cost (kcal/kg min)
Sitting or standing still	0.010	Heavy exercise	0.070
Studying		Fast dancing	
Writing		Walking uphill	
Word processing		Jogging	
TV watching		Fast swimming	
Eating		Severe exercise	0.110
Very light activity	0.020	Tennis	
Driving car		Racquetball	
Walking slowly		Running	
Light exercise	0.025	Aerobic dancing	
Light housework		Soccer	
Walking at moderate speed		Very severe exercise	0.140
Carrying books or packages		Wrestling	
Moderate exercise	0.040	Boxing	
Fast walking		Racing	
Slow dancing		Rowing	
Slow bicycling		Full-court basketball	
Golf			

Calculating BMR/Day

The **basal metabolic rate (BMR)** is the rate at which kcal are spent for basal metabolism. Basal metabolism is the minimum amount of energy the body needs at rest in the fasting state. The beating of the heart, breathing, maintaining body temperature, and sending nerve impulses are some of the activities that maintain life. BMR varies according to a person's body surface area, age, and gender. You will multiply your body surface area by a basal metabolic rate constant to arrive at your BMR/hour. This hourly rate multiplied by 24 hr/day will give you the BMR/day.

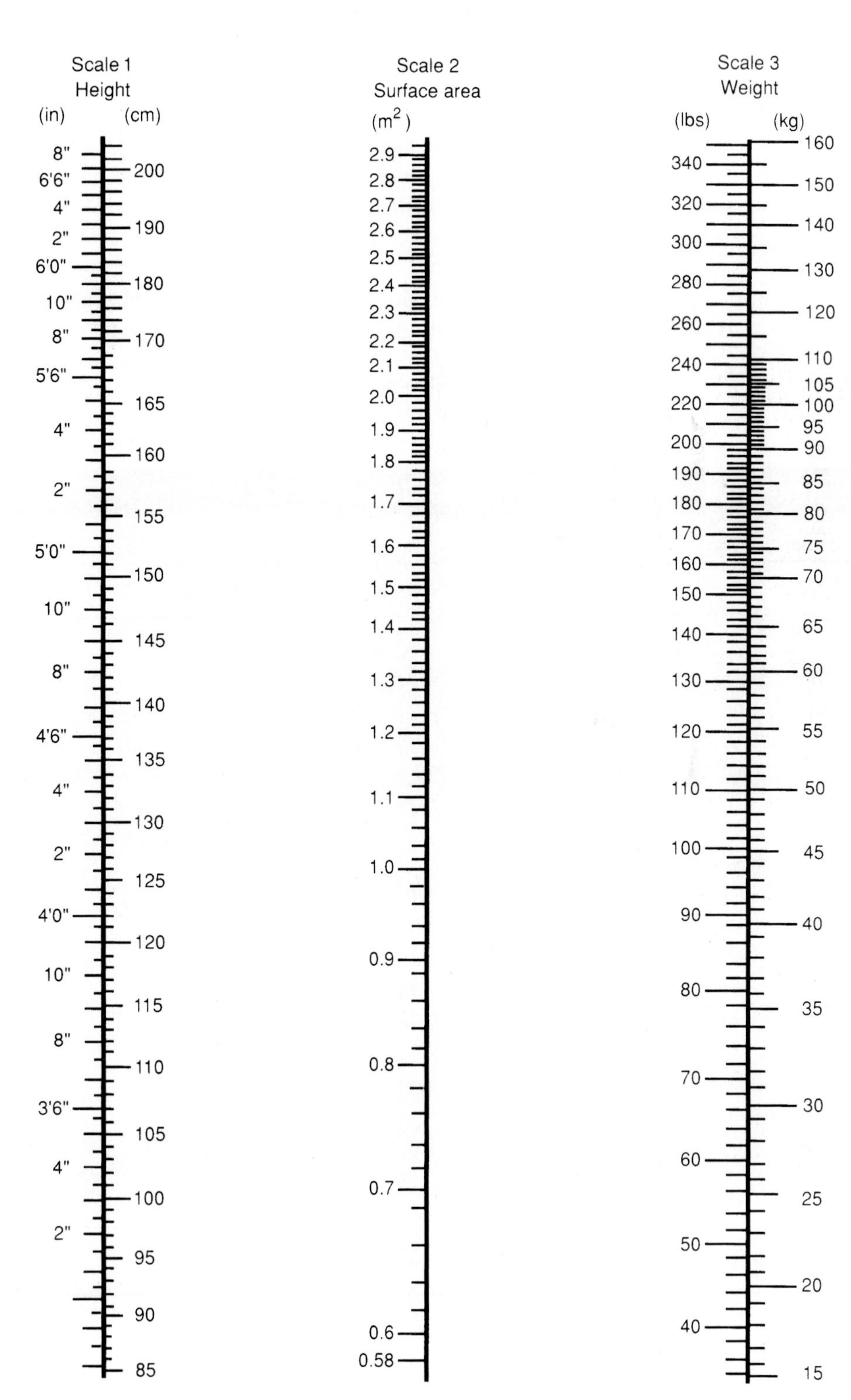

Figure 10.1 Method to estimate body surface area from height and weight. A straight line is drawn from the subject's height (Scale 1) to the subject's weight (Scale 3). The point at which the line intersects Scale 2 is the subject's body surface area in m^2 (meters squared).

Procedure for Calculating BMR/Day

Body Surface Area

1. Use a scale to determine your weight and a measuring device (e.g., measuring tape) to determine your height. It is assumed that you are fully clothed and wearing shoes with a 1-inch heel.
   Be honest about your weight. There is no need to tell anyone else.
2. Consult Figure 10.1 and with a ruler, draw a straight line from your height to your weight. The point where that line crosses the middle column shows your surface area in m^2 (squared meters). For example, a person who is 6 ft tall and weighs 170 lb has a body surface area of 1.99 m^2.

   What is your body surface area? ______________________________

BMR/Day

1. Consult Table 10.4 to find the BMR constant for your age and sex. Multiply your surface area by this factor to calculate your BMR/hr. For example, a 17-year-old male has a BMR constant of 41.5 kcal/m^2 hr. If his surface area is 1.99 m^2, his BMR is 1.99 $m^2 \times$ 41.5 kcal/m^2 = 82.6 kcal/hr.

   What is your BMR/hr? ______________________________

2. Multiply your BMR/hr by 24 to obtain the total number of kcal you need for BMR/day. For example, if the BMR is 82.6 kcal/hr, then the daily BMR rate is 24 hr × 82.6 kcal/hr = 1,982 kcal/day.

   What is your BMR/day? ______________________________

Table 10.4 Basal Metabolic Rate Constants

	BMR (kcal/m^2 hr)			BMR (kcal/m^2 hr)	
Age	**Males**	**Females**	**Age**	**Males**	**Females**
10	47.7	44.9	29	37.7	35.0
11	46.5	43.5	30	37.6	35.0
12	45.3	42.0	31	37.4	35.0
13	44.5	40.5	32	37.2	34.9
14	43.8	39.2	33	37.1	34.9
15	42.9	38.3	34	37.0	34.9
16	42.0	37.2	35	36.9	34.8
17	41.5	36.4	36	36.8	34.7
18	40.8	35.8	37	36.7	34.6
19	40.5	35.4	38	36.7	34.5
20	39.9	35.3	39	36.6	34.4
21	39.5	35.2	40–44	36.4	34.1
22	39.2	35.2	45–49	36.2	33.8
23	39.0	35.2	50–54	35.8	33.1
24	38.7	35.1	55–59	35.1	32.8
25	38.4	35.1	60–64	34.5	32.0
26	38.2	35.0	65–69	33.5	31.6
27	38.0	35.0	70–74	32.7	31.1
28	37.8	35.0	75+	31.8	

Calculating Energy Required for SDA

The **specific dynamic action** (**SDA**) refers to the amount of energy needed to process food. For example, muscles that move food along the digestive tract and glands that make digestive juices use up energy. To calculate the amount of average energy you require for daily SDA, (1) add the average energy required for daily physical activity and the energy for daily BMR. (2) Multiply the total by 10% to obtain an estimated average daily SDA. For example, if 1,044 kcal/day is required for physical activity and 1,882 kcal/day is required for BMR, then an average SDA = 293 kcal/day.

What is your **average SDA/day**? ______________ kcal

Calculating Average Daily Energy Requirement

Total the amounts you have calculated for average daily physical activity, daily BMR, and average daily SDA. This is your average (av.) energy requirement/day.

Av. physical activity/day ______________ kcal

BMR/day ______________ kcal

Av. SDA/day ______________ kcal

Av. energy requirement/day = ______________ kcal

10.3 Comparison of Average Daily Energy Intake and Average Daily Energy Requirement

Figure 10.2 illustrates that if the average daily energy intake is the same as the average daily energy requirement, weight is likely to remain the same.

1. Compare your average daily energy intake to your average daily energy requirement: ______________________________
2. Our methods of calculation are only approximate. If your two figures are within 20% of each other, you will most likely neither lose nor gain weight. If the two figures are not within 20% of each other, are you apt to lose weight or gain weight? ______________________________

   Explain. ______________________________

   ____

Figure 10.2 Comparison of average daily energy intake and average daily energy requirement. If average daily energy intake is the same as the average daily energy requirement, the person's weight stays the same.

10.4 Ideal Weight

Three methods of determining your ideal weight are explained next. Your instructor will choose which one you are to use.

Ideal Weight Based on Height and Weight Tables

Tables 10.5 and 10.6 are constructed by life insurance companies from data on thousands of people, at an age when bone and muscle growth are complete. Such tables are not highly accurate. Determine your ideal weight

Table 10.5 Weight and Height Table for Males

Height		Small Frame	Medium Frame	Large Frame
(ft)	(in)	(lb)	(lb)	(lb)
5	2	128–134	131–141	138–150
5	3	130–136	133–143	140–153
5	4	132–138	135–145	142–156
5	5	134–140	137–148	144–160
5	6	136–142	139–151	146–164
5	7	138–145	142–154	149–168
5	8	140–148	145–157	152–172
5	9	142–151	148–160	155–176
5	10	144–154	151–163	158–180
5	11	146–157	154–166	161–184
6	0	149–160	157–170	164–188
6	1	152–164	160–174	168–192
6	2	155–168	164–178	172–197
6	3	158–172	167–182	176–202
6	4	162–176	171–187	181–207

Table 10.6 Weight and Height Table for Females

Height		Small Frame	Medium Frame	Large Frame
(ft)	(in)	(lb)	(lb)	(lb)
4	10	102–111	109–121	118–131
4	11	103–113	111–123	120–134
5	0	104–115	113–126	122–137
5	1	106–118	115–129	125–140
5	2	108–121	118–132	128–143
5	3	111–124	121–135	131–147
5	4	114–127	124–138	134–151
5	5	117–130	127–141	137–155
5	6	120–133	130–144	140–159
5	7	123–136	133–147	143–163
5	8	126–139	136–150	146–167
5	9	129–142	139–153	149–170
5	10	132–145	142–156	152–173
5	11	135–148	145–159	155–176
6	0	138–151	148–162	158–179

Weight at ages 25–59 based on lowest mortality. Weight according to frame (indoor clothing weighing 3 lb is allowed for women; 5 lb for men), shoes with 1-inch heels.

range. A person more than 10% over the weight in the tables is overweight; a person 20% over is obese. Similarly, a person 10% below the weight in the table is underweight. What is your ideal weight based on these height and weight tables?

Ideal weight ______________ lb

Ideal Weight Based on Body Mass Index (BMI)

Your **body mass index** (**BMI**) is a measure of your weight relative to your height (Table 10.7). A BMI of 19–25 is considered ideal. Below 19 a person is considered underweight, and above 25 a person is considered overweight. An obese person has a BMI of 30 or greater.

Table 10.7 Body Mass Index

Height (Inches)	19	20	21	22	23	24	25	26	27	28	29	30	31	32	33	34	35
	Body Weight (Pounds)																
	Ideal							Not Ideal									
58	91	96	100	105	110	115	119	124	129	134	138	143	148	153	158	162	167
59	94	99	104	109	114	119	124	128	133	138	143	148	153	158	163	168	173
60	97	102	107	112	118	123	128	133	138	143	148	153	158	163	168	174	179
61	100	106	111	116	122	127	132	137	143	148	153	158	164	169	174	180	185
62	104	109	115	120	126	131	136	142	147	153	158	164	169	175	180	186	191
63	107	113	118	124	130	135	141	146	152	158	163	169	175	180	186	191	197
64	110	116	122	128	134	140	145	151	157	163	169	174	180	186	192	197	204
65	114	120	126	132	138	144	150	156	162	168	174	180	186	192	198	204	210
66	118	124	130	136	142	148	155	161	167	173	179	186	192	198	204	210	216
67	121	127	134	140	146	153	159	166	172	178	185	191	198	204	211	217	223
68	125	131	138	144	151	158	164	171	177	184	190	197	203	210	216	223	230
69	128	135	142	149	155	162	169	176	182	189	196	203	209	216	223	230	236
70	132	139	146	153	160	167	174	181	188	195	202	209	216	222	229	236	243
71	136	143	150	157	165	172	179	186	193	200	208	215	222	229	236	243	250
72	140	147	154	162	169	177	184	191	199	206	213	221	228	235	242	250	258
73	144	151	159	166	174	182	189	197	204	212	219	227	235	242	250	257	265
74	148	155	163	171	179	186	194	202	210	218	225	233	241	249	256	264	272
75	152	160	168	176	184	192	200	208	216	224	232	240	248	256	264	272	279
76	156	164	172	180	189	197	205	213	221	230	238	246	254	263	271	279	287

Procedure to Determine Ideal Weight Based on Body Mass Index (BMI)

1. Consult Table 10.7 and find your height in inches.
2. Go across that row to your weight.
3. Read the number at the top of that column. What is your BMI? ______________
4. Is your BMI within the normal range (19–25)? ______________ What is your ideal weight range based on BMI? ______________

Comparison

How does this weight range compare to your desirable weight range from Table 10.5 or Table 10.6?

____

Ideal Weight Based on Body Composition

Body composition refers to the lean body weight plus the body fat. Ideal weight based on body composition is preferred because then you know how much of your weight is due to fat. In adult males, lean body weight should equal about 84% of weight, and body fat should be only 16% of weight. In adult females, lean body weight should equal about 77% of weight, and body fat should be only 23% of weight. Well-conditioned male athletes, such as marathon runners and swimmers, usually have 10–12% body fat, whereas football players can have as high as 19–20% body fat. At no time should the percentage of body fat drop below 5% in males and 10% in females.

To calculate **lean body weight (LBW)**, it is necessary to subtract the amount of your body fat from your present weight. The amount of body fat is to be determined by taking skin-fold measurements, because the fat attached to the skin is roughly proportional to total body fat. The more areas of the body measured, the more accurate the estimate of body fat will be. In this laboratory, we will take only those measurements that permit you to remain fully clothed.

Procedure to Determine Lean Body Weight

1. Work in pairs. Obtain calipers designed to measure skin folds. All measurements should be done on the right side in mm (millimeters). Firmly grasp the fold of skin between the left thumb and the other four fingers and lift. Pinch and lift the fold several times to make sure no musculature is grasped. Hold the skin fold firmly, and place the contact side of the calipers 1/2 inch below the thumb and fingers; do not let go of the fold. Close the calipers on the skin fold, and do the measurements noted in items 2–6 following. Take the reading to the nearest 1/2 mm. Release the grip on the caliper and release the fold. To make sure the reading is accurate, repeat the measurement a few times. If the second measurement is within 1–2 mm of the first, it is reliable. Record two readings.
2. **Triceps (females only).** The triceps skin fold is measured on the back of the upper arm, halfway between the elbow and the tip of the shoulder, while the arm is hanging loosely at the subject's side. Grasp the skin fold parallel to the long axis of the arm and measure as described in step 1. Record in #7 below. Record two readings.
3. **Ilium—hip (males and females).** Fold the skin diagonally just above the top of the hip bone on an imaginary line that would divide the body into front and back halves. Measure and record as before. Record two readings.
4. **Abdomen (males and females).** Fold the skin vertically 1 inch to the right of the navel. Measure and record as before. Record two readings.
5. **Chest (males only).** Fold the skin diagonally, midway between the nipple and the armpit. Measure and record as before. Record two readings.
6. **Axilla—side (males only).** Fold the skin vertically at the level of nipple on an imaginary line that would divide the body into front and back halves. Measure and record as before. Record two readings.
7. Total any four of your skin-fold measurements.

	Females	Males
Triceps	________ ________ mm	
Ilium	________ ________ mm	________ ________ mm
Abdomen	________ ________ mm	________ ________ mm
Chest		________ ________ mm
Axilla		________ ________ mm
Total of any four skin-fold measurements	________________ mm	________ ________ mm

8. Consult Table 10.8 (males) and Table 10.9 (females) to determine percentage (%) body fat from the sum of four skin-fold measurements and your age.

   What is your % body fat? ________________ %

Table 10.8 Percentage Fat Estimates for Males

Sum of Four Skin Folds (mm)	Age								
	18–22	23–27	28–32	33–37	38–42	43–47	48–52	53–57	58–older
8–12	1.9%	2.5%	3.2%	3.8%	4.4%	5.0%	5.7%	6.3%	6.9%
13–17	3.3%	3.9%	4.5%	5.1%	5.7%	6.4%	7.0%	7.6%	8.2%
18–22	4.5%	5.2%	5.8%	6.4%	7.0%	7.7%	8.3%	8.9%	9.5%
23–27	5.8%	6.4%	7.1%	7.7%	8.3%	8.9%	9.5%	10.2%	10.8%
28–32	7.1%	7.7%	8.3%	8.9%	9.5%	10.2%	10.8%	11.4%	12.0%
33–37	8.3%	8.9%	9.5%	10.1%	10.8%	11.4%	12.0%	12.6%	13.2%
38–42	9.5%	10.1%	10.7%	11.3%	11.9%	12.6%	13.2%	13.8%	14.4%
43–47	10.6%	11.6%	11.9%	12.5%	13.1%	13.7%	14.4%	15.0%	15.6%
48–52	11.8%	12.4%	13.0%	13.6%	14.2%	14.9%	15.5%	16.1%	16.7%
53–57	12.9%	13.5%	14.1%	14.7%	15.4%	16.0%	16.6%	17.2%	17.9%
58–62	14.0%	14.6%	15.2%	15.8%	16.4%	17.1%	17.7%	18.3%	18.9%
63–67	15.0%	15.6%	16.3%	16.9%	17.5%	18.1%	18.8%	19.4%	20.0%
68–72	16.1%	16.7%	17.3%	17.9%	18.5%	19.2%	19.8%	20.4%	21.0%
73–77	17.1%	17.7%	18.3%	18.9%	19.5%	20.2%	20.8%	21.4%	22.0%
78–82	18.0%	18.7%	19.3%	19.9%	20.5%	21.0%	21.8%	22.4%	23.0%
83–87	19.0%	19.6%	20.2%	20.8%	21.5%	22.1%	22.7%	23.3%	24.0%
88–92	19.9%	20.5%	21.2%	21.8%	22.4%	23.0%	23.6%	24.3%	24.9%
93–97	20.8%	21.4%	22.1%	22.7%	23.3%	23.9%	24.8%	25.2%	25.8%
98–102	21.7%	22.6%	22.9%	23.5%	24.2%	24.8%	25.4%	26.0%	26.7%
103–107	22.5%	23.2%	23.8%	24.4%	25.0%	25.6%	26.3%	26.9%	27.5%
108–112	23.4%	24.0%	24.6%	25.2%	25.8%	26.5%	27.1%	27.7%	28.3%
113–117	24.1%	24.8%	25.4%	26.0%	26.6%	27.3%	27.9%	28.5%	29.1%
118–122	24.9%	25.5%	26.2%	26.8%	27.4%	28.0%	28.6%	29.3%	29.9%
123–127	25.7%	26.3%	26.9%	27.5%	28.1%	28.8%	29.4%	30.0%	30.6%
128–132	26.4%	27.0%	27.6%	28.2%	28.8%	29.5%	30.1%	30.7%	31.3%
133–137	27.1%	27.7%	28.3%	28.9%	29.5%	30.2%	30.8%	31.4%	32.0%
138–142	27.7%	28.3%	29.0%	29.6%	30.2%	30.8%	31.4%	32.1%	32.7%
143–147	28.3%	29.0%	29.6%	30.2%	30.8%	31.5%	32.1%	32.7%	33.3%
148–152	29.0%	29.6%	30.2%	30.8%	31.4%	32.7%	32.7%	33.3%	33.9%
153–157	29.5%	30.2%	30.8%	31.4%	32.0%	32.7%	33.3%	33.9%	34.5%
158–162	30.1%	30.7%	31.3%	31.9%	32.6%	33.2%	33.8%	34.4%	35.1%
163–167	30.6%	31.2%	31.9%	32.5%	33.1%	33.7%	34.3%	35.0%	35.6%
168–172	31.1%	31.7%	32.4%	33.0%	33.6%	34.2%	34.8%	35.5%	36.1%
173–177	31.6%	32.2%	32.8%	33.5%	34.1%	34.7%	35.3%	35.9%	36.6%
178–182	32.0%	32.7%	33.3%	33.9%	34.5%	35.2%	35.8%	36.4%	37.0%
183–187	32.5%	33.1%	33.7%	34.3%	34.9%	35.6%	36.2%	36.8%	37.4%
188–192	32.9%	33.5%	34.1%	34.7%	35.3%	36.0%	36.6%	37.2%	37.8%
193–197	33.2%	33.8%	34.5%	35.1%	35.7%	36.3%	37.0%	37.8%	38.2%
198–202	33.6%	34.2%	34.8%	35.4%	36.1%	36.7%	37.3%	37.9%	38.5%
203–207	33.9%	34.5%	35.1%	35.7%	36.4%	37.0%	37.6%	38.2%	38.9%

Table 10.9 Percentage Fat Estimates for Females

Sum of Four Skin Folds (mm)	Age								
	18–22	23–27	28–32	33–37	38–42	43–47	48–52	53–57	58–older
8–12	8.8%	9.0%	9.2%	9.4%	9.5%	9.7%	9.9%	10.1%	10.3%
13–17	10.8%	10.9%	11.1%	11.3%	11.5%	11.7%	11.8%	12.0%	12.2%
18–22	12.6%	12.8%	13.0%	13.2%	13.4%	13.5%	13.7%	13.9%	14.1%
23–27	14.5%	14.6%	14.8%	15.0%	15.2%	15.4%	15.6%	15.7%	15.9%
28–32	16.2%	16.4%	16.6%	16.8%	17.0%	17.1%	17.3%	17.5%	17.7%
33–37	17.9%	18.1%	18.3%	18.5%	18.7%	18.9%	19.0%	19.2%	19.4%
38–42	19.6%	19.8%	20.0%	20.2%	20.3%	20.5%	20.7%	20.9%	21.1%
43–47	21.2%	21.4%	21.6%	21.8%	21.9%	22.1%	22.3%	22.5%	22.7%
48–52	22.8%	22.9%	23.1%	23.3%	23.5%	23.7%	23.8%	24.0%	24.2%
53–57	24.2%	24.4%	24.6%	24.8%	25.0%	25.2%	25.3%	25.5%	25.7%
58–62	25.7%	25.9%	26.0%	26.2%	26.4%	26.6%	26.8%	27.0%	27.1%
63–67	27.1%	27.2%	27.4%	27.6%	27.8%	28.0%	28.2%	28.3%	28.5%
68–72	28.4%	28.6%	28.7%	28.9%	29.1%	29.3%	29.5%	29.7%	29.8%
73–77	29.6%	29.8%	30.0%	30.2%	30.4%	30.6%	30.7%	30.9%	31.1%
78–82	30.9%	31.0%	31.2%	31.4%	31.6%	31.8%	31.9%	32.1%	32.3%
83–87	32.0%	32.2%	32.4%	32.6%	32.7%	32.9%	33.1%	33.3%	33.5%
88–92	33.1%	33.3%	33.5%	33.7%	33.8%	34.0%	34.2%	34.4%	34.6%
93–97	34.1%	34.3%	34.5%	34.7%	34.9%	35.1%	35.2%	35.4%	35.6%
98–102	35.1%	35.3%	35.5%	35.7%	35.9%	36.0%	36.2%	36.4%	36.6%
103–107	36.1%	36.2%	36.4%	36.6%	36.8%	37.0%	37.2%	37.3%	37.5%
108–112	36.9%	37.1%	37.3%	37.5%	37.7%	37.9%	38.0%	38.2%	38.4%
113–117	37.8%	37.9%	38.1%	38.3%	39.2%	39.4%	39.6%	39.8%	40.0%
118–122	38.5%	38.7%	38.9%	39.1%	39.4%	39.6%	39.8%	40.0%	40.5%
123–127	39.2%	39.4%	39.6%	39.8%	40.0%	40.1%	40.3%	40.5%	40.7%
128–132	39.9%	40.1%	40.2%	40.4%	40.6%	40.8%	41.0%	41.2%	41.3%
133–137	40.5%	40.7%	40.8%	41.0%	41.2%	41.4%	41.6%	41.7%	41.9%
138–142	41.0%	41.2%	41.4%	41.6%	41.7%	41.9%	42.1%	42.3%	42.5%
143–147	41.5%	41.7%	41.9%	42.0%	42.2%	42.4%	42.6%	42.8%	43.0%
148–152	41.9%	42.1%	42.3%	42.8%	42.6%	42.8%	43.0%	43.2%	43.4%
153–157	42.3%	42.5%	42.6%	42.8%	43.0%	43.2%	43.4%	43.6%	43.7%
158–162	42.6%	42.8%	43.0%	43.1%	43.3%	43.5%	43.7%	43.9%	44.1%
163–167	42.9%	43.0%	43.2%	43.4%	43.6%	43.8%	44.0%	44.1%	44.3%
168–172	43.1%	43.2%	43.4%	43.6%	43.8%	44.0%	44.2%	44.3%	44.5%
173–177	43.2%	43.4%	43.6%	43.8%	43.9%	44.1%	44.3%	44.5%	44.7%
178–182	43.3%	43.5%	43.7%	43.8%	44.0%	44.2%	44.4%	44.6%	44.8%

9. Multiply your present body weight in pounds (lb) by your present % body fat to determine how much of your body weight is fat. For example, if a male weighs 180 lb and his percentage body fat is 20%, then 180 × .20 = 36 lb. His body fat is 36 lb.

How much of your body weight is fat? ________________ lb

10. Subtract the weight of fat from your present body weight. This is lean body weight (LBW). For example, if a male weighs 180 lb and his body fat is 36 lb, then 180 – 36 = 144. His LBW is 144 lb.

What is your **LBW?** ________________________ lb

11. Calculate ideal weight based on body composition.

For males, 84% of ideal weight should be LBW. Therefore, ideal weight = LBW/0.84. What is your ideal weight based on body composition? For example, if a male has a LBW of 144 lb, his ideal weight is 144/.84 = 171 lb.

Male ideal weight = ________________ lb

(At this weight, your body will contain 16% fat, the amount generally recommended for males.)

For females, 77% of ideal weight should be LBW. Therefore, ideal weight = LBW/0.77. What is your ideal weight based on body composition?

Female ideal weight = ________________ lb

(At this weight, your body will contain 23% fat, the amount generally recommended for females.)

Comparison

How does your ideal weight based on body composition compare to your ideal weight based on BMI?

____

____

Conclusions: Ideal Body Weight

- What is your average daily energy requirement? ______________________ kcal
- What is your average daily energy intake? ______________________ kcal
- What is your ideal weight? (Take an average of any you have calculated.) ____________ lb

For your information, 1 lb of fat represents 3,500 kcal. Therefore, if you want to lose 1 lb of fat, you must either reduce your intake by 3,500 kcal or increase your activity by 3,500 kcal. Assuming the same amount of physical activity, you could lose 1 lb per week by reducing your kcal intake by 500 kcal per day or gain 1 lb per week by increasing your kcal intake by 500 kcal per day.

- What is your recommendation for maintaining or achieving your ideal weight? ____________________

____

Application for Daily Living

Losing Weight

Before turning to more drastic measures, an overweight or obese person should first attempt to lose weight by lowering his or her caloric intake and increasing caloric output through exercise. Fad diets and pills lack lasting value. The effort to control weight is worth it because two serious illnesses are associated with obesity: type 2 diabetes and cardiovascular disease.

When a person has type 2 diabetes, the pancreas produces insulin but the cells do not respond to it. Therefore, glucose builds up in the blood and spills over into the urine. How might diet contribute to the occurrence of type 2 diabetes? Simple sugars in foods, such as candy and ice cream, immediately enter the bloodstream, as do sugars from the digestion of starch within white bread and potatoes. This apparently leads to a high blood fatty acid level, insulin resistance, and type 2 diabetes.

Cardiovascular disease due to arteries blocked by plaque is another condition seen in obese individuals. Plaque contains saturated fats and cholesterol. Therefore, limiting cholesterol (present in cheese, egg yolks, shrimp, and lobster) in the diet may be helpful. You should also be aware that beef, dairy foods, and coconut oil are rich sources of saturated fat. Further, processed foods made with or fried in partially hydrogenated oils (e.g., vegetable shortening and stick margarine) are sources of trans fats that contribute to plaque formation.

Laboratory Review 10

________________ 1. What is a listing of all foods eaten for a day?

________________ 2. What does *kcal* mean?

________________ 3. Daily energy requirement includes energy for physical activity, SDA, and what else?

________________ 4. Give an example of a basal metabolism activity.

________________ 5. SDA refers to energy needed for what activity?

________________ 6. With age, the BMR decreases. What is the implication for daily energy intake?

________________ 7. Why does a tall, thin person have a higher BMR than a short, stout person?

________________ 8. Daily energy requirement must be in balance with what to not gain weight?

________________ 9. If the average energy intake/day equals the average energy requirement/day, the person will not ________________.

________________ 10. Ideal weight includes lean body weight and what else?

________________ 11. Generally speaking, a male should have no more than what percentage body fat?

________________ 12. Calculation of ideal weight is considered to be the most accurate if it is based on your ________________.

________________ 13. How many kcal are equal to a pound of fat?

________________ 14. In which sex is it natural to have a higher percentage of body fat?

Thought Questions

15. Why do males typically have a higher basal metabolic rate than females?

16. Why is energy needed when the body is at rest?

17. Why is energy needed to process food?

18. Why should someone monitor his or her average daily energy intake when attempting to maintain average body weight?

LABORATORY

11

Homeostasis

Learning Outcomes

Introduction

- Define homeostasis and the internal environment.

11.1 Kidneys

- Understand kidney and nephron structure and blood supply.
- State the three steps in urine formation and how they relate to the parts of a nephron.
- Perform a urinalysis and explain how the results are related to kidney functions.

Prelab Question: What substance will be present in urine if a person has diabetes mellitus?

11.2 Lungs

- Describe the mechanics of breathing and the role of the alveoli in gas exchange.
- Explain how the excretion of CO_2 helps stabilize the pH of the blood.
- Measure respiratory volumes (e.g., tidal volume) and explain their relationship to homeostasis.

Prelab Question: When CO_2 exits blood at the alveoli, how does the pH of blood change?

11.3 Liver

- Describe the anatomy of the liver, including the path of blood from the intestines, through the liver and to the heart.
- Compare the glucose level in the mesenteric artery, the hepatic portal vein, and the hepatic vein before and after eating.
- State two ways the liver contributes to homeostasis.

Prelab Question: What hormone causes the liver to store glucose as glycogen?

***Application for Daily Living:* The Liver Is a Vital Organ**

Introduction

Homeostasis refers to the dynamic equilibrium of the body's internal environment. The internal environment of humans consists of blood and tissue fluid. To meet their needs, cells take nutrients (e.g., glucose) and O_2 from the blood and return waste products, including CO_2, to the blood by way of tissue fluid. Tissue fluid in turn exchanges molecules with the blood. Homeostasis also involves adjusting blood pH, ionic concentrations, and blood volume. All internal organs contribute to homeostasis, but this laboratory specifically examines the contributions of the kidneys, lungs, and the liver.

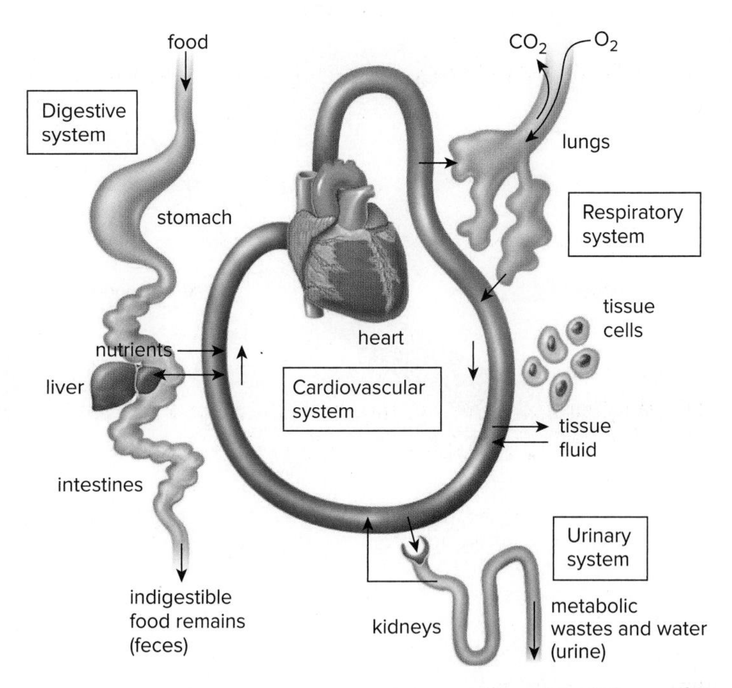

Negative Feedback

> **Planning Ahead** Starting the water bath needed for Section 11.3 now may save you valuable time.

Two systems in the body, the endocrine (hormone) system and the nervous system, work together to regulate homeostasis, the relative constancy of the internal environment (blood and tissue fluid). Both systems operate by utilizing negative feedback. For example, when you eat a candy bar and your blood glucose rises, an endocrine gland called the pancreas senses the rise in blood sugar and releases the hormone insulin (Fig. 11.1). Insulin causes the liver to store glucose as glycogen, maintaining blood sugar level at a normal level. The drop in blood sugar level is **negative feedback** that causes the pancreas to stop secreting insulin. Similarly, when blood pressure lowers, the brain sends out impulses to the smooth muscles of arterioles to constrict, raising blood pressure. This rise in blood pressure is negative feedback that causes the brain to stop sending nerve impulses to the arterioles; they relax and the blood pressure lowers.

As you study the operation of the kidneys, lungs, and the liver in this laboratory, realize that negative feedback ultimately is what enables these organs to maintain homeostasis.

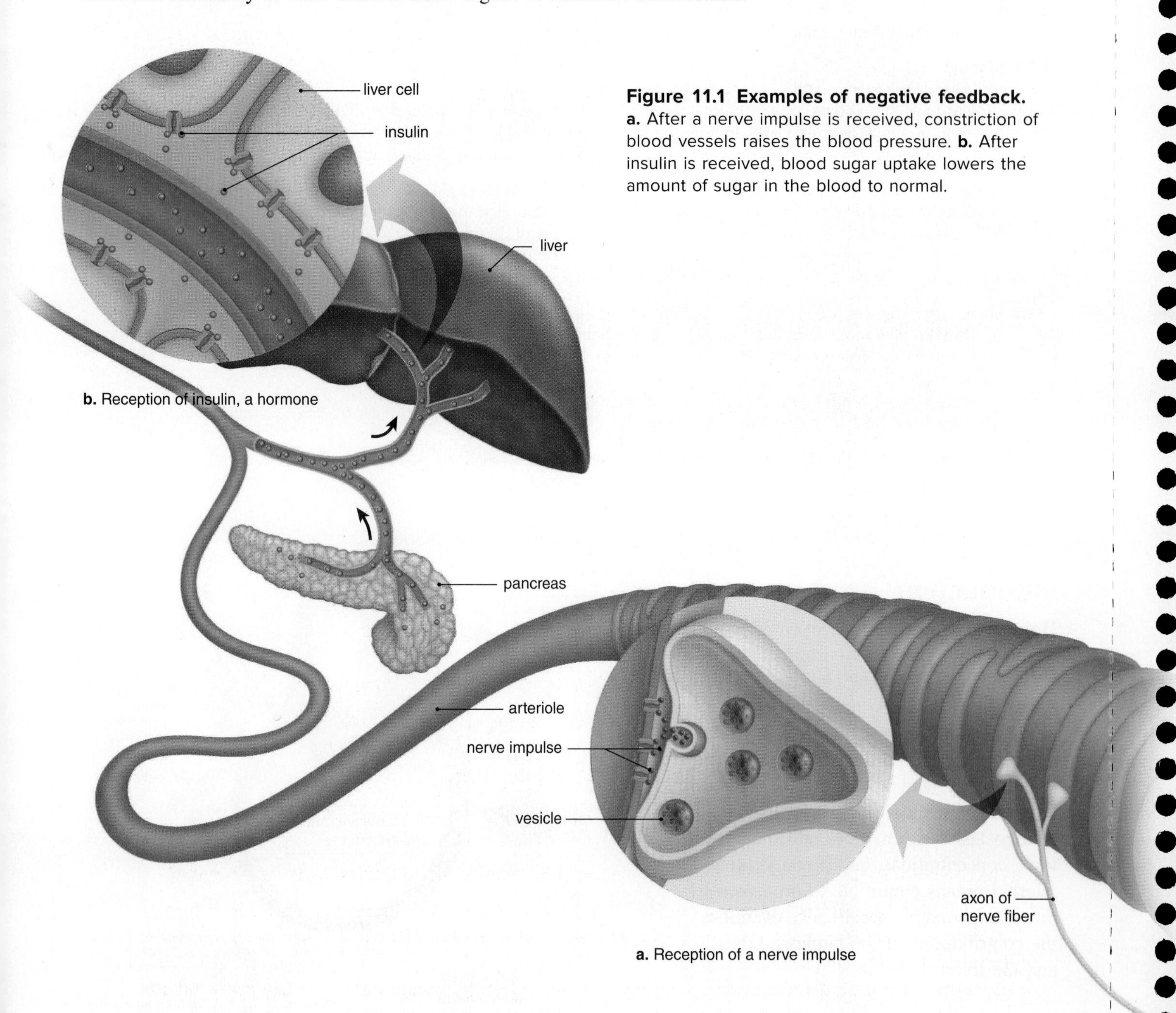

Figure 11.1 Examples of negative feedback. **a.** After a nerve impulse is received, constriction of blood vessels raises the blood pressure. **b.** After insulin is received, blood sugar uptake lowers the amount of sugar in the blood to normal.

11.1 Kidneys

Kidneys are bean-shaped organs that lie along the dorsal wall of the abdominal cavity. Figure 11.2 shows the macroscopic and microscopic structure of a kidney. The macroscopic structure of a kidney is due to the placement of over 1 million **nephrons.** Nephrons are tubules that do the work of producing urine.

Figure 11.2 Longitudinal section of a kidney.
a. The kidneys are served by the renal artery and renal vein. **b.** Macroscopically, a kidney has three parts: renal cortex, renal medulla, and renal pelvis. **c.** Microscopically, each kidney contains over a million nephrons.

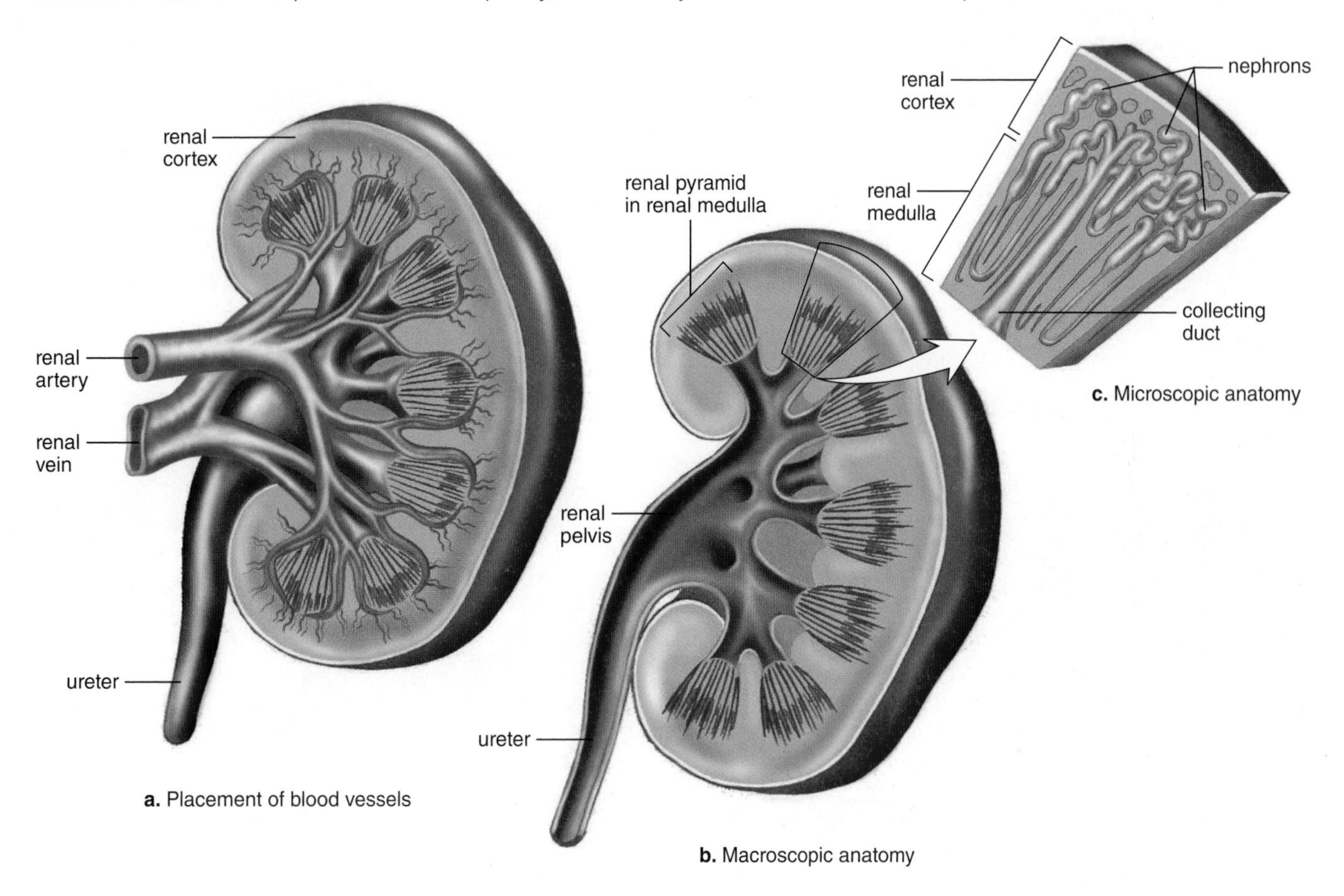

Observation: Kidney Structure

Study a model of a kidney, and with the help of Figure 11.2, locate the following:

1. **Renal cortex:** a granular region that contains most regions of the nephrons.
2. **Renal medulla:** contains the renal pyramids consisting of the loops of nephrons and collecting ducts.
3. **Renal pelvis:** where urine is received from the collecting ducts.

Observation: Nephron Structure

Study the nephron model and with the help of Figure 11.3, identify the following parts of a nephron:

1. **Glomerular capsule** (Bowman's capsule): closed end of the nephron pushed in on itself to form a cuplike structure; the inner layer has pores that allow **glomerular filtration** to occur; substances move from the blood to inside the nephron.
2. **Proximal convoluted tubule:** The inner layer of this region has many microvilli that allow tubular reabsorption to occur; substances move from inside the nephron to the blood.

3. **Loop of the nephron:** Nephron narrows to form a U-shaped portion. Functions in water reabsorption.
4. **Distal convoluted tubule:** second convoluted section that lacks microvilli and functions in **tubular secretion;** substances move from blood to inside nephron.

Several nephrons enter one collecting duct. The **collecting ducts** also function in water reabsorption, and they conduct urine to the pelvis of a kidney.

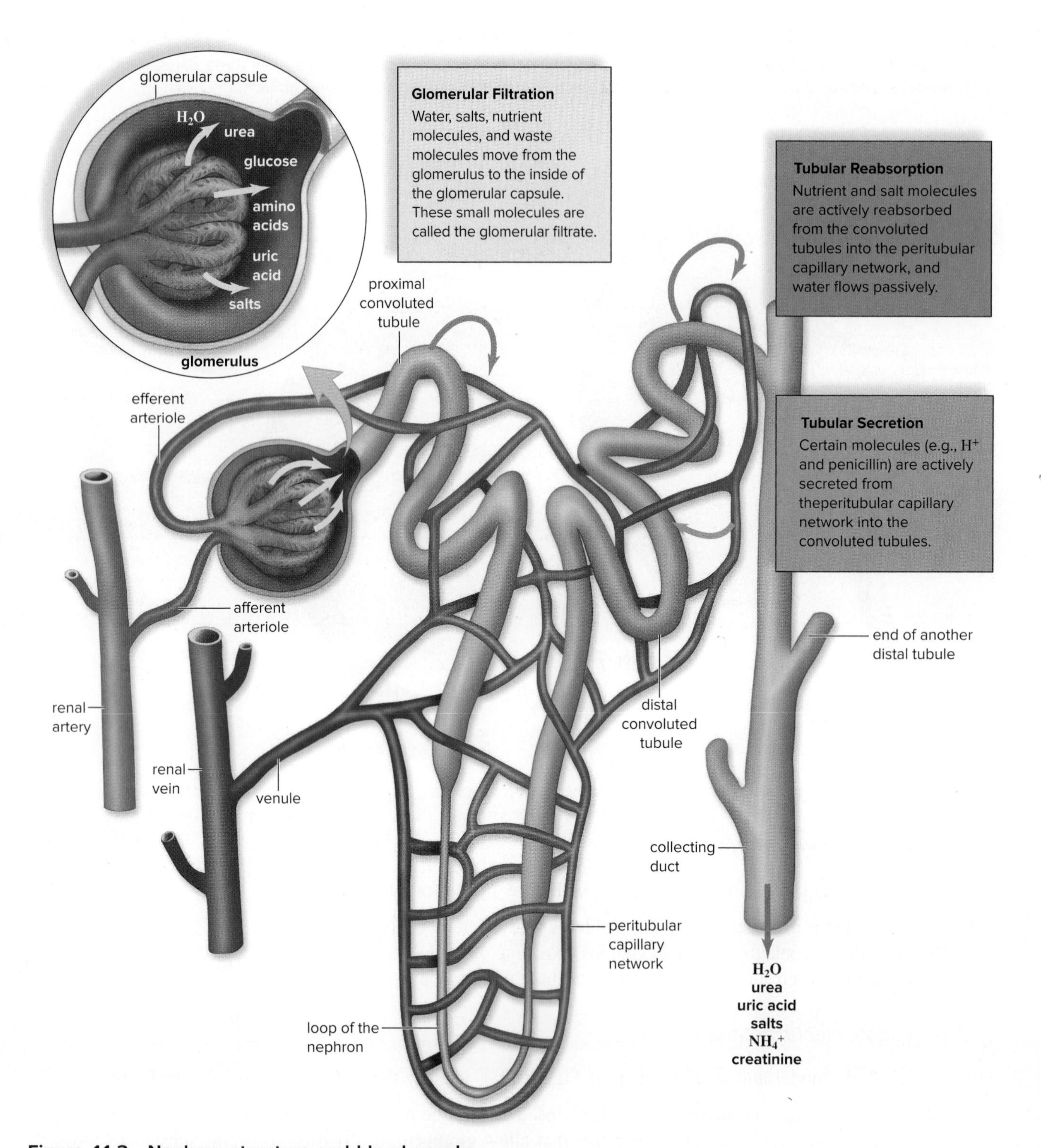

Figure 11.3 Nephron structure and blood supply.
The three main processes in urine formation are described in boxes and color coded to arrows that show the movement of molecules out of or into the nephron at specific locations. In the end, urine is composed of the substances within the collecting duct (see brown arrow).

Observation: Circulation About a Nephron

Study a nephron model and with the help of Figure 11.3 and Table 11.1, trace the path of blood about a nephron:

1. **Afferent arteriole:** small vessel that conducts blood from the renal artery to a nephron.
2. **Glomerulus:** capillary network that exists inside the glomerular capsule; small molecules move from inside the capillary to the inside of the glomerulus during glomerular filtration.
3. **Efferent arteriole:** small vessel that conducts blood from the glomerulus to the peritubular capillary network.
4. **Peritubular capillary network:** surrounds the proximal convoluted tubule, the loop of the nephron, and the distal convoluted tubule.
5. **Venule:** takes blood from the peritubular capillary network to the renal vein.

Table 11.1 Circulation of Blood Around a Nephron

Name of Structure	Significance
Afferent arteriole	Brings arteriolar blood toward glomerular capsule
Glomerulus	Capillary tuft enveloped by glomerular capsule
Efferent arteriole	Takes arteriolar blood away from glomerular capsule
Peritubular capillary network	Capillary bed that envelops the rest of the nephron
Venule	Takes venous blood away from the nephron

Kidney Function

The kidneys produce urine and in doing so help maintain homeostasis in several ways. Urine formation requires three steps: glomerular filtration, tubular reabsorption, and tubular secretion (see Fig. 11.3).

Glomerular Filtration

1. Blood entering the glomerulus contains blood cells, proteins, glucose, amino acids, salts, urea, and water. Blood cells and proteins are too large to pass through the glomerular wall and enter the filtrate.
2. Blood pressure causes small molecules of glucose, amino acids, salts, urea, and water to exit the blood and enter the glomerular capsule. The fluid in the glomerular capsule is called the **filtrate**.
3. In the list that follows, *draw an arrow* from left to right for the small substances that leave the glomerulus and become a part of the filtrate.

Glomerulus	Glomerular Filtrate
Cells	
Proteins	
Glucose	
Amino acids	
Urea	
Water and salts	

4. Complete the second and third columns in Table 11.2, page 137. Use an X to indicate that the substance is at the locations noted.

Tubular Reabsorption

1. When the filtrate enters the proximal convoluted tubule, it contains glucose, amino acids, urea, water, and salts. Some water and salts remain in the nephron but enough are passively reabsorbed into the peritubular capillary to maintain blood volume and blood pressure. Use this information to state a way kidneys help maintain homeostasis. ______________________________

____

2. The cells that line the proximal convoluted tubule are also engaged in active transport and usually completely reabsorb nutrients (glucose and amino acids) into the peritubular capillary. What would happen to cells if the body lost all its nutrients by way of the kidneys? ______________________________

____

3. Which of the filtrate substances is reabsorbed the least and will become a part of urine? __________

   Urea is a nitrogenous waste. State here another way kidneys contribute to homeostasis. __________

____

4. In the list that follows, *draw an arrow* from left to right for all those molecules passively reabsorbed into the blood of the peritubular capillary. Use darker arrows for those that are reabsorbed completely by active transport.

Proximal Convoluted Tubule	**Peritubular Capillary**
Water and salts	
Glucose	
Amino acids	
Urea	

Tubular Secretion

During tubular secretion, certain substances—for example, penicillin and histamine—are actively secreted from the peritubular capillary into the fluid of the tubule. Also, hydrogen ions (H^+) and ammonia (NH_3) are secreted as NH_4^+ as necessary.

The excretion of H^+ in this way raises the pH of the blood. State here a third way the kidneys contribute to homeostasis. ______________________________

We learned earlier that the lungs help raise the pH of the blood by excreting CO_2 but only the kidneys can excrete H^+.

Summary: Kidney Function and Homeostasis

For each substance listed at the left in Table 11.2, place an X in the last column if you expect the substance to be present in urine.

Answer the following questions.

1. The presence of urea in the urine illustrates which of the kidney's functions? ______________

   Do the kidneys make urea? ______________ What organ makes urea? ______________

   What do the kidneys produce? ______________________________

2. The presence of NH_4^+ in the urine illustrates which of the kidney's functions? ______________

____

3. Regulation of the blood's water and salt content by the kidneys helps maintain ______________

______________________________ within normal limits.

Table 11.2 Urine Constituents			
Substance	**In Blood of Glomerulus**	**In Filtrate**	**In Urine**
Protein (albumin)			
Glucose and amino acids			
Urea			
Water and salts			
NH_4^+			

Urinalysis: A Diagnostic Tool

Urinalysis can indicate whether the kidneys are functioning properly or whether an illness such as diabetes mellitus is present. The procedure is easily performed with a Chemstrip test strip, which has indicator spots that produce specific color reactions when certain substances are present in the urine.

Experimental Procedure: Urinalysis

A urinalysis has been ordered, and you are to test the urine. (In this case, you will be testing simulated urine.)

Assemble Supplies

1. Obtain three Chemstrip urine test strips, each of which tests for leukocytes, pH, protein, glucose, ketones, and blood, as noted in Figure 11.4.
2. The color key on the diagnostic color chart or on the Chemstrip vial label will explain what any color changes mean in terms of the pH level and amount of each substance present in the urine sample. You will use these color blocks to read the results of your test.
3. Obtain three "specimen containers of urine" marked 1 through 3. Among them is a normal specimen and two that indicate the patient has an illness.

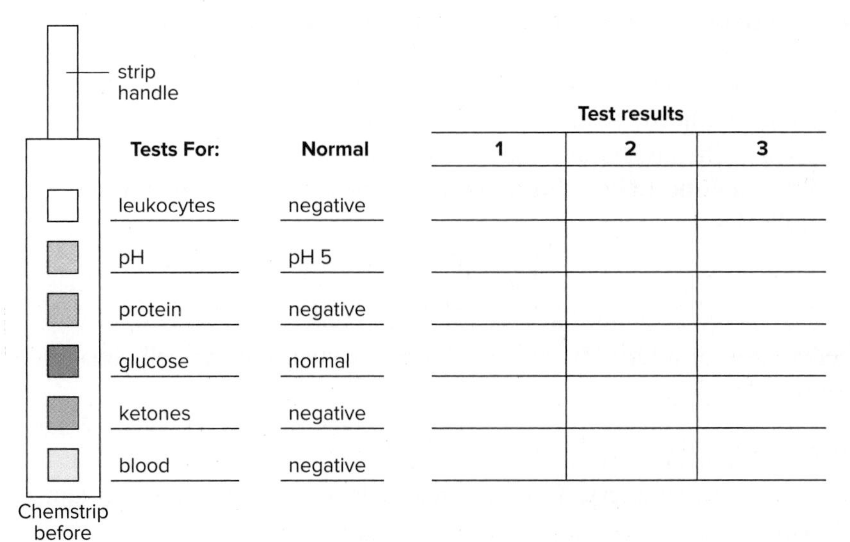

Tests For:	Normal	Test results 1	2	3
leukocytes	negative			
pH	pH 5			
protein	negative			
glucose	normal			
ketones	negative			
blood	negative			

Figure 11.4 Urinalysis test. A Chemstrip test strip can help determine illness in a patient by detecting substances in the urine. The normal test results are given for comparative purposes.

Test the Specimen

1. Be sure the chemically treated patches on the test strip are totally immersed. Briefly (no longer than 1 second) dip a test strip into the first specimen of urine.
2. Draw the edge of the strip along the rim of the specimen container to remove excess urine.
3. Turn the test strip on its side, and tap once on a piece of absorbent paper to remove any remaining urine and to prevent the possible mixing of chemicals.
4. After 60 seconds, read the results as follows: Hold the strip close to the color blocks on the diagnostic color chart (Fig. 11.4) or vial label, and match carefully, ensuring that the strip is properly oriented to the color chart. Enter the test results in Figure 11.4. Use a negative symbol (−) for items that are not present in the urine, a plus symbol (+) for those that are present, and a number for the pH.
5. Test the other two specimens. Add these results to Figure 11.4.

Conclusion: Urinalysis

- State below if the urinalysis is normal or indicates a urinary tract infection (leukocytes, blood, and possibly protein in the urine) or the patient has diabetes mellitus (glucose in the urine).

  Test results 1 ______________________________

  Test results 2 ______________________________

  Test results 3 ______________________________

- The hormone insulin promotes the uptake of glucose by cells. When glucose is in the urine, either the pancreas is not producing insulin (diabetes mellitus type 1) or cells are resistant to insulin (diabetes mellitus type 2). Ketones (acids) are also in the urine because the cells are metabolizing fat instead of glucose. Explain why cells are metabolizing fat. ______________________________

  ____

  Why is the pH of urine lower than normal? ______________________________

  ____

- If urinalysis shows that proteins are excreted instead of retained in the blood, would capillary exchange in the tissues be normal? __________ Why or why not? ______________________________

  ____

11.2 Lungs

Air moves from the nasal passages to the trachea, bronchi, bronchioles, and finally, lungs. The right and left lungs lie in the thoracic cavity on either side of the heart. A **lung** is a spongy organ consisting of irregularly shaped air spaces called **alveoli** (sing., *alveolus*). The alveoli are lined with a single layer of squamous epithelium and are supported by a mesh of fine, elastic fibers. The alveoli are surrounded by a rich network of tiny blood vessels called **pulmonary capillaries.**

Observation: Lung Structure

1. Observe a prepared slide of a stained section of a lung. In stained slides, the nuclei of the cells forming the thin alveolar walls appear purple or dark blue (Fig. 11.5*a*).
2. Look for areas with groups of red- or orange-colored, disk-shaped **erythrocytes** (red blood cells). When these appear in strings, you are looking at capillary vessels in side view.
3. In some part of the slide, you may even observe an artery. Thicker, circular or oval structures with a lumen (cavity) are cross sections of **bronchioles,** tubular pathways through which air reaches the air spaces.

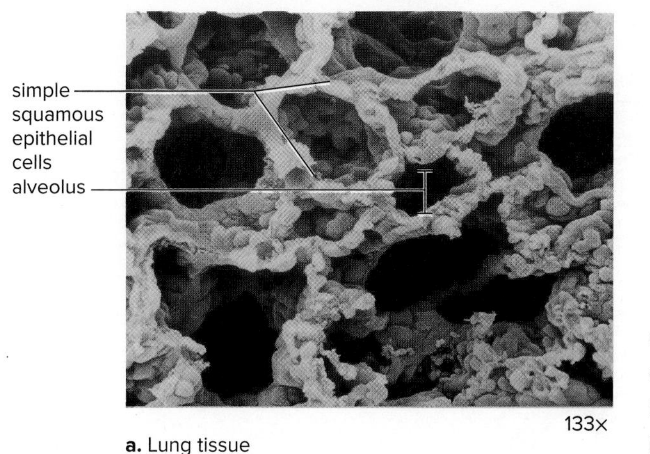

a. Lung tissue

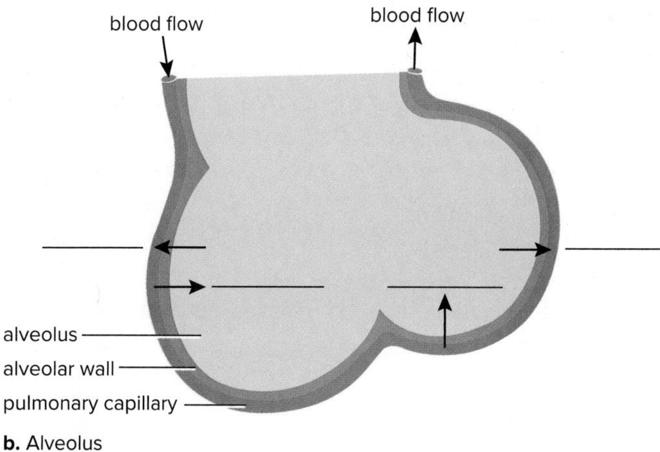

b. Alveolus

Figure 11.5 Gas exchange in the lungs.
a. A photomicrograph shows that the lungs contain many air sacs called alveoli. The alveoli are surrounded by pulmonary capillaries. **b.** During gas exchange, carbon dioxide leaves the blood and enters the alveoli; oxygen leaves the alveoli and enters the blood. (a) © Science Photo Library/Alamy RF

Lung Function

1. Lungs carry out gas exchange.

During gas exchange in the lungs, carbon dioxide (CO_2) leaves the blood within the pulmonary capillaries, and oxygen (O_2) leaves the alveoli and enters the blood. *Show gas exchange in Figure 11.5*b *by writing O_2 or CO_2 on the lines provided.* After the lungs take up O_2, it is transported by hemoglobin (Hb) inside red blood cells as HbO_2. In the tissues, Hb releases O_2 (Fig. 11.6). Why do cells require oxygen?__________________

____

2. The lungs help maintain the pH of the blood.

Carbon dioxide is carried in the blood as bicarbonate ions:

$$CO_2 + H_2O \longrightarrow \underset{\text{carbonic acid}}{H_2CO_3} \longrightarrow \underset{\text{bicarbonate ion}}{HCO_3^-} + \underset{\text{hydrogen ions}}{H^+}$$

Hydrogen ions increase the acidity of blood. Is blood more acidic when it is carrying carbon dioxide? ____________________ Explain your answer. __________________________________

____

As CO_2 leaves the blood, this reaction reverses:

$$HCO_3^- + H^+ \longrightarrow H_2CO_3 \longrightarrow CO_2 + H_2O$$

Is blood less acidic when the carbon dioxide exits? ________________ Explain your answer. ____________

____

During breathing cessation, it is the increase in acidity rather than the depletion of oxygen that first stimulates the urge to breathe. Chemoreceptors located in the aorta and carotid arteries detect changes in the blood's acidity and stimulate the respiratory center, which triggers contractions in the diaphragm and thoracic muscles, causing alterations in the rate and depth of breathing.

Figure 11.6 Exchange of gases at the pulmonary capillaries and at the systemic capillaries.

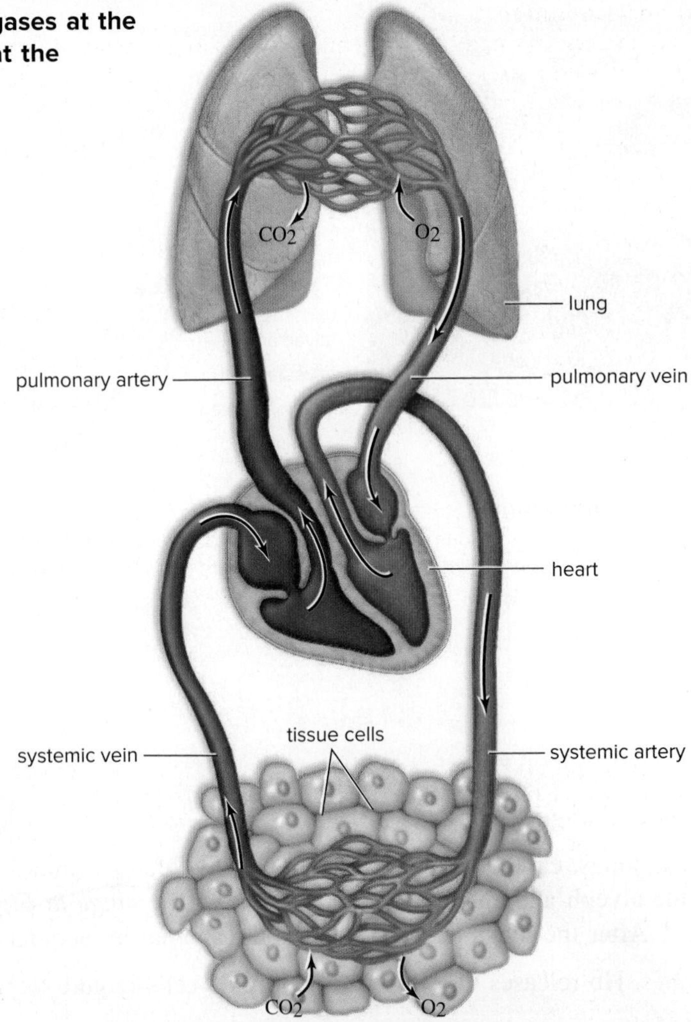

Summary: Lung Function and Homeostasis

1. Gas exchange assists homeostasis because it supplies cells with ________________ needed for ______________________________ and rids the body of ________________, a metabolic waste.
2. When the lungs excrete CO_2, they help maintain the ________________ of the blood. This is the second way that the lungs help maintain homeostasis.

Human Respiratory Volumes

Breathing in, called **inspiration** or inhalation, is the active part of breathing because that's when contraction of rib cage muscles causes the rib cage to move up and out, and contraction of the diaphragm causes the diaphragm to lower. Due to an enlarged thoracic cavity, air is drawn into the lungs. Breathing out, called **expiration** or exhalation, occurs when relaxation of these same muscles causes the thoracic cavity to resume its original capacity. Now air is pushed out of the lungs (Fig. 11.7).

Figure 11.7 Inspiration and expiration.
a. Inspiration occurs after the rib cage moves up and out and the diaphragm moves down. Air rushes in because of the expanded thoracic cavity. **b.** Expiration occurs as the rib cage moves down and in and the diaphragm moves up. As the thoracic cavity gets smaller, air is pushed out.

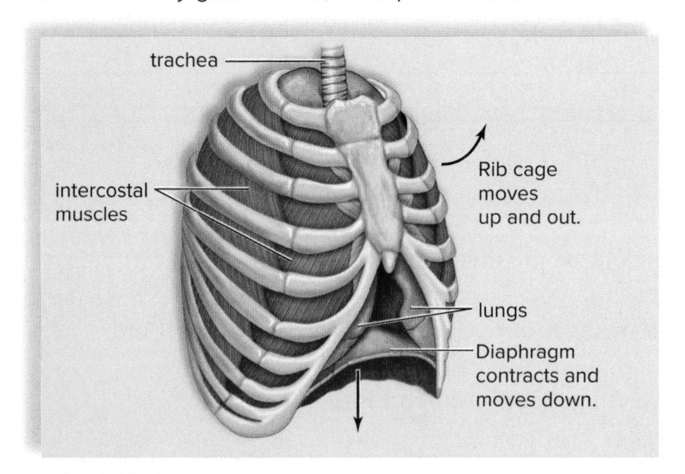

a. Inspiration

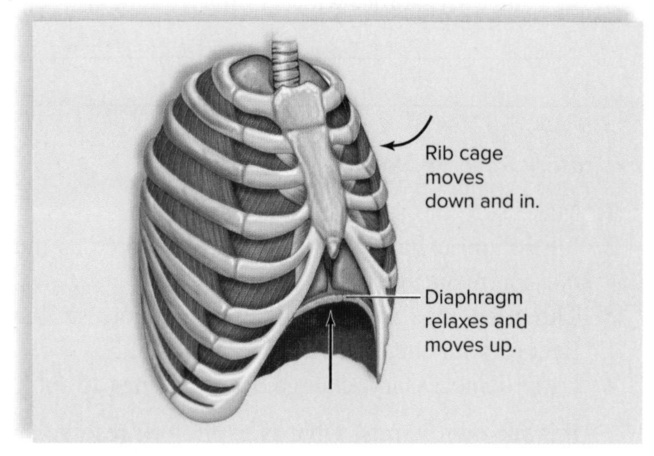

b. Expiration

Experimental Procedure: Human Respiratory Volumes

During this Experimental Procedure you will be working with a spirometer, an instrument that measures the amount of exhaled air (Fig. 11.8). Normally, about 500–600 ml of air move into and out of the lungs with each breath. This is called the **tidal volume** (**TV**). You can inhale deeply after a normal breath and more air will enter the lungs; this is the **inspiratory reserve volume** (**IRV**). You can also force more air out of your lungs after a normal breath; this is the **expiratory reserve volume** (**ERV**). **Vital capacity** is the volume of air that can be forcibly exhaled after forcibly inhaling.

Figure 11.8 Nine-liter student wet spirometer.
© Phipps & Bird, Inc., Richmond, VA

Tidal Volume (TV)

1. When it's your turn to use the spirometer, install a new disposable mouthpiece and set the spirometer to zero.
2. Inhale normally, then exhale normally (with no extra effort) through the mouthpiece of the spirometer. Record your measurement in Table 11.3.

Table 11.3 Measurements of Lung Volumes

Tidal Volume (TV)	Expiratory Reserve Volume (ERV)	Vital Capacity (VC)	Inspiratory Reserve Volume (IRV)
1st	1st	1st	———
2nd	2nd	2nd	———
3rd	3rd	3rd	———
Average ml	Average ml	Average ml	Calculated value = ml

3. Three readings are needed, so twice more set the spirometer to zero and repeat the same procedure. Record your measurements in Table 11.3.
4. Later, if necessary, change your readings to milliliters (ml), and calculate your average TV in ml.

   In your own words, what is tidal volume?

   ____

   ____

Expiratory Reserve Volume (ERV)

1. Make sure the spirometer is set to zero.
2. Inhale and exhale normally and then force as much air out as possible into the spirometer. Record your measurement in Table 11.3.
3. Three readings are needed, so twice more set the spirometer to zero and repeat the same procedure. Record your measurements in Table 11.3.
4. Later, if necessary, change your readings to ml, and calculate your average ERV.

   In your own words, what is expiratory reserve volume? ____

   ____

Vital Capacity (VC)

1. Make sure the spirometer is set to zero.
2. Inhale as much as possible and then exhale as much as possible into the spirometer.
3. Three readings are needed, so twice more set the spirometer to zero and repeat the same procedure. Record your measurements in Table 11.3.
4. Later, if necessary, change your readings to ml, and calculate your average VC.

   In your own words, what is vital capacity? ____

   ____

Inspiratory Reserve Volume (IRV)

It will be necessary for us to calculate IRV because a spirometer only measures exhaled air, not inhaled air.

Explain. ____

From having measured vital capacity (VC) you can see that VC = TV + IRV + ERV. To calculate IRV, simply subtract the average TV + the average ERV from the value you recorded for the average VC:

IRV = VC − (TV + ERV) = __________ ml. Record your IRV in Table 11.3.

Conclusions: Human Respiratory Volumes

- Vital capacity varies with age, sex, and height; however, typically for men vital capacity is about 5,200 ml and for women it is about 4,000 ml. How does your vital capacity compare to the typical values for your gender? ______________________ If smaller than normal, are you a smoker or is there any health reason why it would be smaller? If larger than normal, are you a sports enthusiast or do you play a musical instrument that involves inhaling and exhaling deeply? ______________________
- Diffusion alone accounts for pulmonary gas exchange. Therefore, how does good lung ventilation assist gas exchange? ____

  ____

11.3 Liver

The **liver,** which is the largest organ in the body, lies mainly to the right under the diaphragm. The liver has two main **lobes** and the lobes are further divided into **lobules,** which contain the cells of the liver, called **hepatic cells** (Fig. 11.9*a*).

Observation: Liver Structure

Study a model of the liver, and identify the following:

1. **Right and left lobes:** The liver has two lobes. The right lobe is larger than the left lobe. Each lobe has many **lobules.** Each lobule has many cells.
2. **Hepatic artery:** the blood vessel that transports O_2-rich blood to the liver.
3. **Hepatic portal vein:** the blood vessel that transports blood containing nutrients from the intestine to the liver (Fig. 11.9*b*).
4. **Hepatic veins:** the blood vessels that transport O_2-poor blood out of the liver to the inferior vena cava.

Liver Function

The liver has many functions in homeostasis, and this lab will study two of these.

1. The liver produces urea.

The liver removes amino groups (—NH_2) from amino acids and converts them to urea. **Urea** is the primary nitrogenous end product of metabolism that is excreted by the kidneys. What remains of the amino acid (a hydrocarbon) can be used by the body in a number of ways.

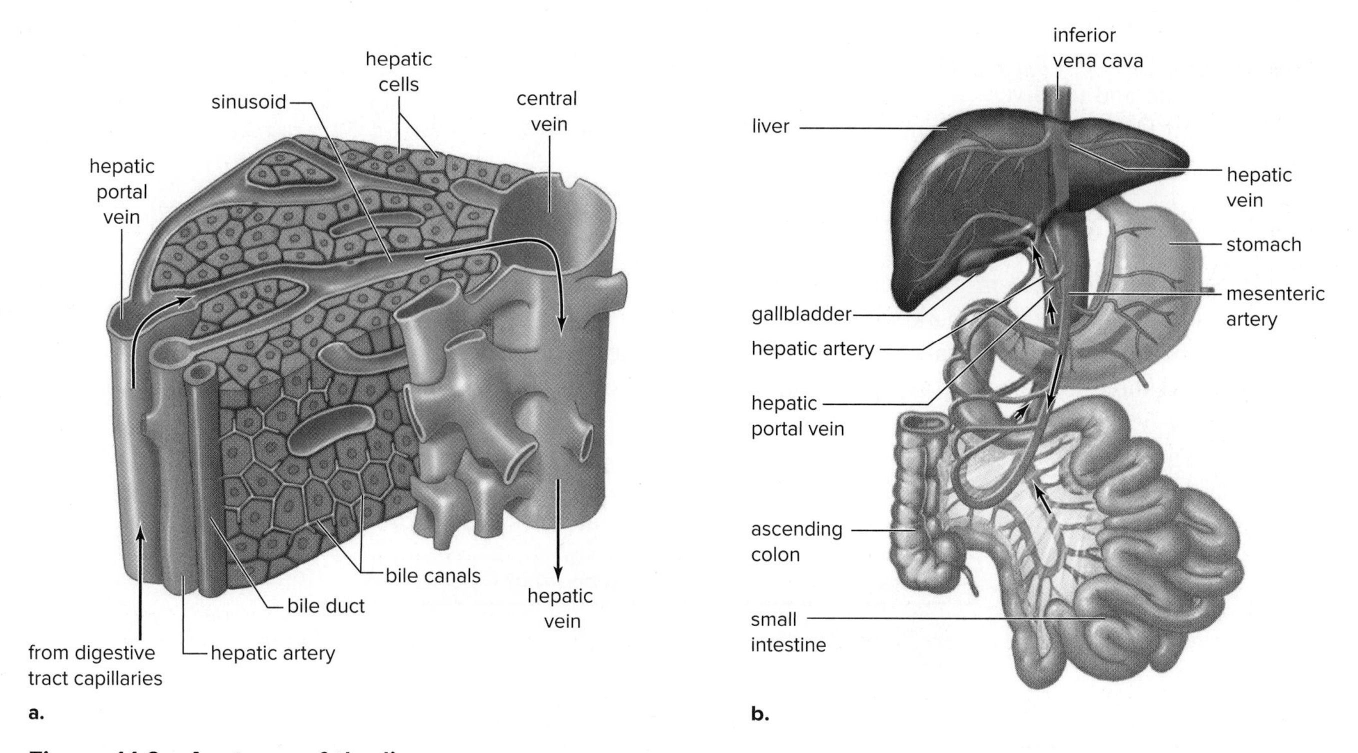

Figure 11.9 Anatomy of the liver.
a. Liver lobule. **b.** The hepatic portal vein lies between the digestive tract and the liver. The hepatic veins enter the inferior vena cava, which returns blood to the heart.

In the chemical formula for urea that follows, circle the portions that would have come from amino groups:

$$NH_2{-}\overset{\displaystyle O}{\overset{\|}{C}}{-}NH_2$$

2. The liver regulates the blood glucose level.

After you eat, blood glucose level rises. This increase is detected by the pancreas, which in response releases the hormone insulin. Insulin causes the liver to store excess glucose as glycogen. Before your next meal, the drop in blood sugar causes the pancreas to release glucagon, a hormone that promotes the liver to break down glycogen and release glucose.

Complete the following equation by writing *glucose* and *glycogen* on the appropriate sides of the arrows.

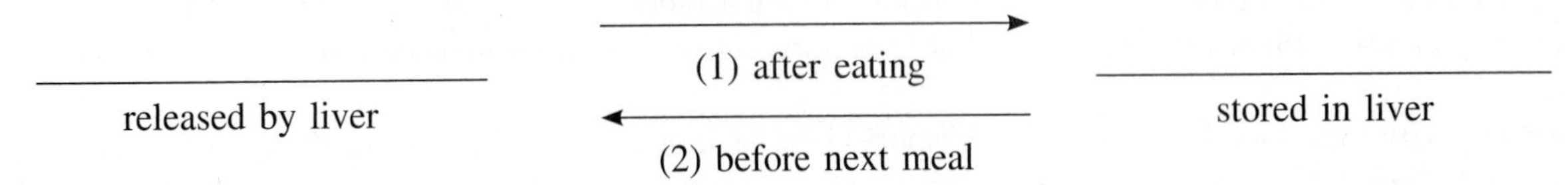

If glucose is excreted in the urine, instead of being stored, the individual has the medical condition called **diabetes mellitus,** commonly known as diabetes. In type 1 diabetes, the pancreas is no longer making insulin; in type 2 diabetes, the plasma membrane receptors are unable to bind properly to insulin. In type 1 diabetes, but not type 2, ketones (strong organic acids), a breakdown product of fat metabolism, also appear in the urine.

Experimental Procedure: Blood Glucose Level After Eating

Study the diagram of the human cardiovascular system in Figure 8.7 (page 101), and trace the path of blood from the mesenteric artery to the vena cava via the intestine and liver. Simulated serum samples have been prepared to correspond to these blood vessels in a person who ate a short time ago:

A_1: Serum from a mesenteric artery. The mesenteric arteries take blood from the aorta to the intestine.

B_1: Serum from the hepatic portal vein, which lies between the intestine and the liver.

C_1: Serum from the hepatic vein, which takes blood from the liver to the vena cava.

> ⚠ **Benedict's reagent** Use protective eyeware when performing this experiment. Benedict's reagent is highly corrosive. Exercise care in using this chemical. If any should spill on your skin, wash the area with mild soap and water. Follow your instructor's directions for disposal of this chemical.

1. With a wax pencil, label three test tubes A_1, B_1, and C_1, and mark them at 1 cm and 2 cm.
2. Fill tube A_1 to the 1 cm mark with serum A_1 and to the 2 cm mark with Benedict's reagent.
3. Fill tube B_1 to the 1 cm mark with serum B_1 and to the 2 cm mark with Benedict's reagent.
4. Fill tube C_1 to the 1 cm mark with serum C_1 and to the 2 cm mark with Benedict's reagent.
5. Place all three test tubes in the water bath at the same time. Heat the tubes in the same boiling water bath for 5 minutes.
6. Note any color change in the test tubes, and record the color and your conclusions in Table 11.4. The tube that shows color first has the most glucose, and so forth. Use the following chart to assist you in making your conclusions:

Color Change	Amount of Glucose
Color is still blue.	None
Green	Very low
Yellow-orange	Moderate
Orange	High
Orange-red	Very high

Table 11.4 Blood Glucose Level After Eating		
Test Tubes	**Color (After Heating)**	**Conclusion**
A_1 (mesenteric artery)		
B_1 (hepatic portal vein)		
C_1 (hepatic vein)		

Conclusions: Blood Glucose Level After Eating

- Which blood vessel—a mesenteric artery, the hepatic portal vein, or the hepatic vein—contains the most glucose after eating? ______
- Why do you suppose that the hepatic vein does not contain as much glucose as the hepatic portal vein after eating? ______

______

Experimental Procedure: Blood Glucose Level Before Next Meal

Simulated serum samples have been prepared to correspond to these blood vessels in a person who has not eaten for some time:

A_2: Serum from a mesenteric artery
B_2: Serum from the hepatic portal vein
C_2: Serum from the hepatic vein

1. With a wax pencil, label three test tubes A_2, B_2, and C_2, and mark them at 1 cm and 2 cm.
2. Fill tube A_2 to the 1 cm mark with serum A_2 and to the 2 cm mark with Benedict's reagent.
3. Fill tube B_2 to the 1 cm mark with serum B_2 and to the 2 cm mark with Benedict's reagent.
4. Fill tube C_2 to the 1 cm mark with serum C_2 and to the 2 cm mark with Benedict's reagent.
5. Heat the tubes in the same boiling water bath for 5 minutes.
6. Note any color change in the test tubes, and record the color and your conclusions in Table 11.5. Use the following list to assist you in making your conclusions:

Color Change	Amount of Glucose
Color is still blue.	None
Green	Very low
Yellow-orange	Moderate
Orange	High
Orange-red	Very high

Table 11.5 Blood Glucose Level Before Next Meal		
Test Tubes	**Color (After Heating)**	**Conclusion**
A_2 (mesenteric artery)		
B_2 (hepatic portal vein)		
C_2 (hepatic vein)		

Conclusions: Blood Glucose Level Before Next Meal

- Which blood vessel—a mesenteric artery, the hepatic portal vein, or the hepatic vein—contains the most glucose before your next meal? ______________________
- Why do you suppose that the hepatic vein now contains more glucose than the hepatic portal vein?

____

Summary: Liver Function and Homeostasis

We studied two ways the liver contributes to homeostasis.

1. The liver produces ______________, which is excreted by the ______________. Urea is the way we excrete nitrogen from the body; it is a metabolic ______________.
2. The liver stores glucose as ______________ after eating and releases glucose in between eating so that the concentration of glucose in the blood stays relatively constant at 0.1%. What hormone promotes storage of glucose by the liver? ______________ What hormone promotes breakdown of glycogen to glucose by the liver?

Conclusion: Homeostasis

1. As noted at the beginning of this laboratory, homeostasis is the dynamic equilibrium of the body's internal environment, the blood and tissue fluid surrounding tissue cells. The lungs and kidneys have boundaries that interact with the external environment to refresh blood. The liver also regulates blood content. Fill in the following chart to show the activities of the kidneys, lungs, and liver.

Processes	Kidneys	Lungs	Liver
Gas exchange	______	O_2 enters and CO_2 exits the blood	______
pH maintenance	a.	e.	______
Glucose level	b.	______	g.
Waste removal	c.	f.	h.
Blood volume	d.	______	______

2. Which of these organs contributes most to homeostasis? ______________________
3. As described throughout this lab, homeostasis is maintained when a change in the internal environment values triggers a response that restores the normal condition. Complete Table 11.6 to show how the kidneys, lungs, and liver specifically respond to changes in the internal environment. Under "Response," also include any hormones involved.

Table 11.6 Changes in Internal Environment		
Change	**Organ (kidneys, lungs, or liver)**	**Response**
Decrease in blood glucose level		
Decrease in blood volume and pressure		
Increase in blood CO_2		

Application for Daily Living

The Liver Is a Vital Organ

The liver is a vital organ because we can't live without one. The liver is the gatekeeper to the blood. After the molecules enter the blood at the digestive tract, they go first to the liver, which performs these functions.

- As we studied in this lab, the liver maintains glucose in the blood so that cells always have a source of energy. After we eat, the liver stores glucose as glycogen and then it breaks down glycogen gradually so that the blood concentration of glucose stays within a normal range.
- The liver removes and breaks down poisons, i.e., chemicals that can harm the body. The liver destroys old enzymes and hormones that no longer function properly, drugs such as an antibiotics and pain killers that have done their job, and drugs of abuse that damage the brain.
- The liver produces several kinds of blood proteins whose functions are varied. Among these are proteins needed to make the blood clot and if these proteins are missing, a simple cut can cause severe bleeding.
- The liver produces bile, which is sent to the digestive tract, where it emulsifies fat. Without bile, eating ice cream, the icing on a cake, or a juicy steak can make you ill.

It is important to realize that your behavior can affect your liver. The liver ordinarily does a marvelous job but it can be overwhelmed. Alcohol is a poison the liver valiantly breaks down, but if you drink heavily for a long time, the liver can't keep up and develops cirrhosis, characterized by scar tissue that has taken place of working liver cells. Another cause of liver failure is hepatitis, an infection that can be sexually transmitted or acquired by using dirty needles to shoot drugs into the body. When a liver is severely damaged, only a liver transplant can prevent death. The need for a healthy liver provides a good reason for a healthy lifestyle.

Laboratory Review 11

_________________ 1. Equilibrium in the body's internal environment is called ______.

_________________ 2. What type of feedback helps organs maintain homeostasis?

_________________ 3. When molecules leave the glomerulus, they enter what portion of the nephron?

_________________ 4. Name a substance that is in the glomerular filtrate but not in the urine.

_________________ 5. Name the process by which molecules move from the proximal convoluted tubule into the blood.

_________________ 6. H^+ is excreted in combination with what molecule?

_________________ 7. What are the air spaces in the lungs called?

_________________ 8. What molecule is removed by the lungs?

_________________ 9. When we exhale, the diaphragm relaxes and moves in what direction?

_________________ 10. When measuring tidal volume, should a student exhale normally or maximally?

_________________ 11. Vital capacity is expected to have a higher or lower volume than tidal volume?

_________________ 12. What blood vessel lies between the intestines and the liver?

_________________ 13. In what form is glucose stored in the liver?

_________________ 14. The liver removes the amino group from amino acids to form what molecules?

Thought Questions

15. Smoking cigarettes causes emphysema, in which the alveoli burst. Why would you expect the patient to have an abnormally low vital capacity and be tired?

16. Which systemic blood vessel would you expect to have a high glucose content immediately after eating? ____________ Explain your answer.

17. In what ways do the kidneys aid homeostasis?

LABORATORY

12

Musculoskeletal System

Learning Outcomes

12.1 Anatomy of a Long Bone

- Locate and identify the portions of a long bone, and associate particular tissues with each portion.
- Identify significant features of compact bone, spongy bone, and hyaline cartilage.

Prelab Question: Which portion of long bone and which type of bone contain osteons?

12.2 The Skeleton

- Locate and identify the bones of the appendicular and axial human skeletons.

Prelab Question: To which division of the skeleton should you associate the limbs?

12.3 The Skeletal Muscles

- Locate and identify selected human skeletal muscles.
- Illustrate types of joint movements.
- Give examples of antagonistic pairs of muscles and the actions involved.
- Distinguish between isometric and isotonic contractions.

Prelab Question: Why do muscles work in antagonistic pairs?

12.4 Mechanism of Skeletal Muscle Fiber Contraction

- Describe the structure of skeletal muscle.
- Describe an experiment that demonstrates the role of ATP and ions in the contraction of sarcomeres.

Prelab Question: In general, what is the role of ATP in muscle contraction?

Application for Daily Living: **Bone Marrow Transplants**

Introduction

The human skeletal system consists of the bones (206 in adults) and joints, along with the cartilage and ligaments that occur at the joints. The muscular system contains three types of muscles: smooth, cardiac, and skeletal. The term **musculoskeletal system** recognizes that contraction of skeletal muscles causes the bones to move.

In humans, the skeletal muscles are most often attached across a joint (Fig. 12.1). The biceps brachii muscle has two origins, and the triceps brachii has three origins on the humerus and scapula. Find the tendon of insertion of the biceps brachii muscle by feeling on the anterior surface of your elbow while contracting your biceps muscle. Feel for the bone at your posterior elbow. This is the ulna, the site of insertion of the tendon for the triceps brachii muscle.

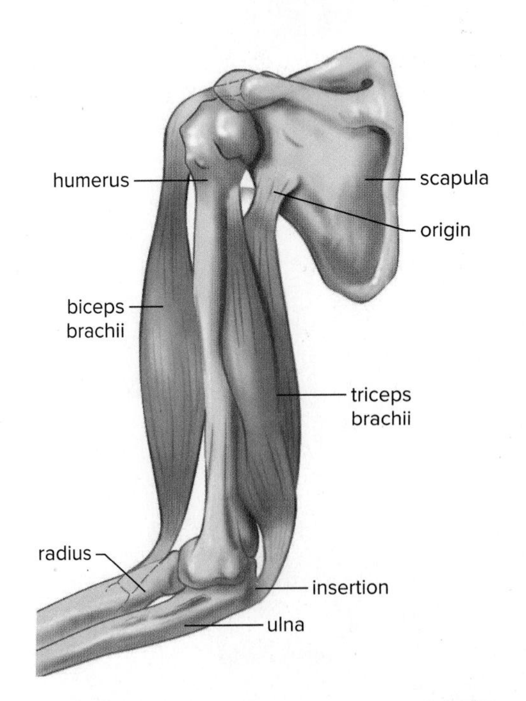

Figure 12.1 Muscular action.
Muscles, such as these muscles of the arm (which have their origin on the scapula and their insertion on the bones of the forearm), cause bones to move.

Muscles work in antagonistic pairs. For example, when the biceps brachii contracts, the bones of the forearm are pulled upward, while the triceps brachii relaxes; when the triceps brachii contracts, the bones of the forearm are pulled downward, while the biceps brachii relaxes.

12.1 Anatomy of a Long Bone

Although the bones of the skeletal system vary considerably in shape as well as in size, a long bone, such as the human femur, illustrates the general principles of bone anatomy (Fig. 12.2).

Figure 12.2 Anatomy of a bone from the macroscopic to the microscopic level.
A long bone is encased by the periosteum except at the ends, where it is covered by hyaline (articular) cartilage (see micrograph, *top left*). Spongy bone located at the ends may contain red bone marrow. The medullary cavity contains yellow bone marrow and is bordered by compact bone, shown in the enlargement and micrograph (*right*).
(hyaline, bone): © Ed Reschke; (osteocyte): © Biophoto Associates/Science Source

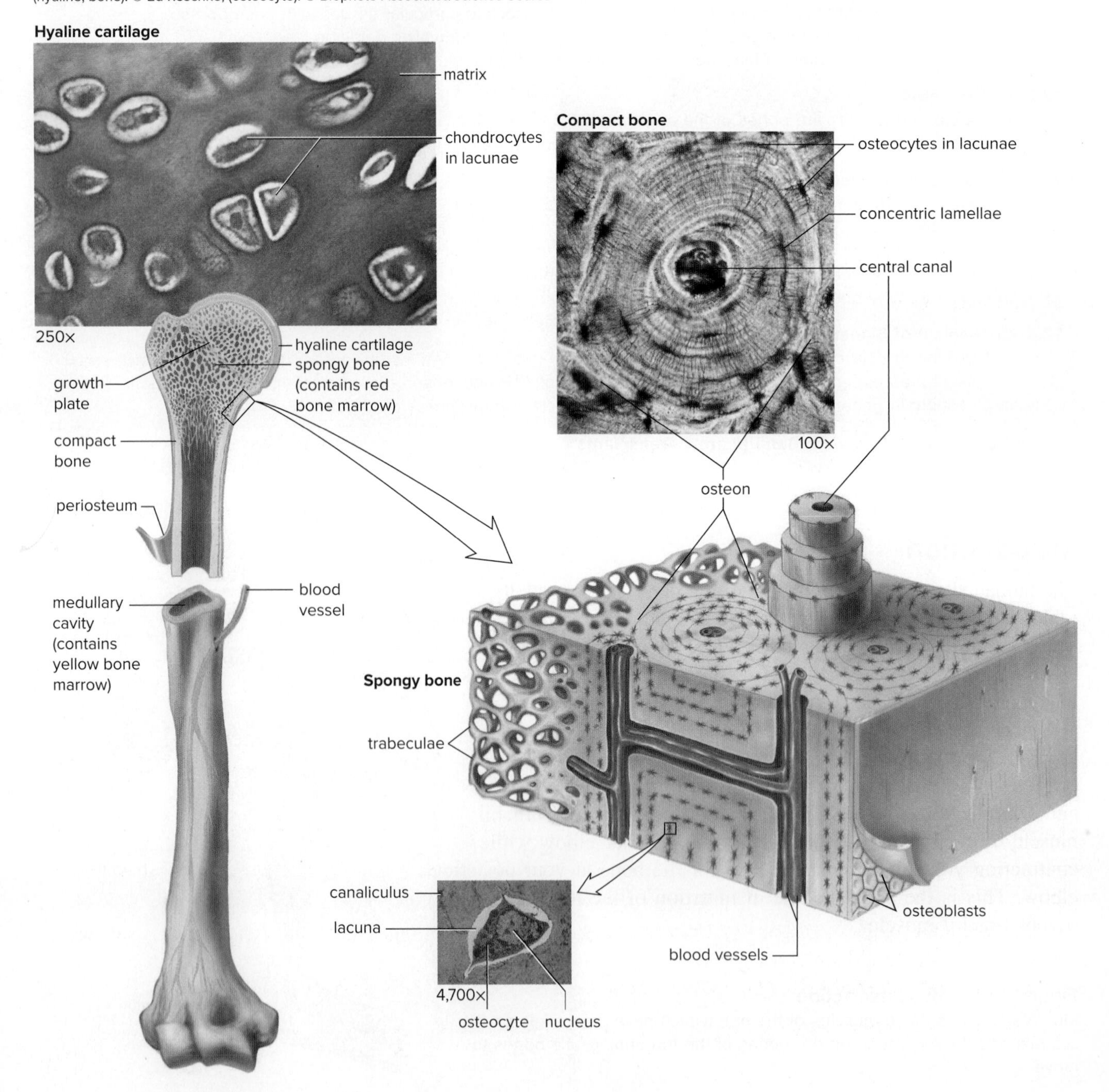

Observation: Anatomy of a Long Bone

Examine the exterior and a longitudinal section of a long bone or a model of a long bone, and with the help of Figure 12.2, identify the following:

1. **Periosteum:** tough, fibrous connective tissue covering continuous with the ligament and tendons that anchor bones; the periosteum allows blood vessels to enter the bone and service its cells
2. Expanded portions at each end of the bone (epiphysis) that contain spongy bone
3. Extended portion, or shaft (diaphysis) of a long bone that lies between the epiphyses; walls of diaphysis are compact bone
4. **Hyaline (articular) cartilage:** layer of cartilage where the bone articulates with (meets) another bone; decreases friction between bones during movement
5. **Medullary cavity:** cavity located in the diaphysis that stores yellow marrow, which contains a large amount of fat.

   Label the diaphysis and the epiphysis (twice) in Figure 12.2. Which of these contains the growth line where a long bone can grow in length? Diaphysis

Observation: Tissues of a Long Bone

The medullary cavity is bounded at the sides by **compact bone** and at the ends by **spongy bone.** Beyond a thin shell of compact bone is the layer of articular cartilage. **Red marrow,** a specialized tissue that produces all types of blood cells, occurs in the spongy bone of the skull, ribs, sternum, and vertebrae and in the ends of the long bones.

1. Examine a prepared slide of compact bone, and with the help of Figure 12.2, identify
   a. **Osteons:** cylindrical structural units.
   b. **Lamellae:** concentric rings of matrix.
   c. **Matrix:** nonliving material maintained by osteocytes; contains mineral salts (notably calcium salts) and protein.
   d. **Lacunae:** cavities between the lamellae that contain osteocytes (bone cells)
   e. **Central canal:** canal in the center of each osteon; Figure 12.2 shows that there are blood vessels in a central canal.
   f. **Canaliculi:** tiny channels that contain the processes of cells; these processes allow nutrients to pass between the osteocytes (sing., canaliculus).

   Describe how an osteocyte located near a central canal can pass nutrients to osteocytes located far from the central canal. The central canal contains blood vessels, Canaliculus which process cells that allow nutrients to pass between osteocyte

2. Examine a prepared slide of spongy bone, and with the help of Figure 12.2, identify
   a. **Trabeculae:** bony bars and plates made of mineral salts and protein
   b. **Lacunae:** cavities scattered throughout the trabeculae that contain osteocytes
   c. **Red bone marrow:** within large spaces separated by the trabeculae

   What activity occurs in red bone marrow? produces all types of blood cells.

3. Examine a prepared slide of hyaline cartilage, and with the help of Figure 12.2, identify
   a. **Lacunae:** cavities in twos and threes scattered throughout the matrix, which contain chondrocytes (cells that maintain cartilage)
   b. **Matrix:** material more flexible than bone because it consists primarily of protein

   Seniors tend to have joints that creak. What might be the matter? chondrocytes

12.2 The Skeleton

The human skeleton is divided into axial and appendicular components (Fig. 12.3). The **axial skeleton** is the main longitudinal portion and includes the skull, the vertebral column, the sternum, and the ribs. The **appendicular skeleton** includes the bones of the appendages and their supportive pectoral and pelvic (shoulder and hip) girdles.

Figure 12.3 The axial and appendicular skeletons.
a. Axial skeleton bones are colored blue. **b.** Bones of the appendicular skeleton are tan.

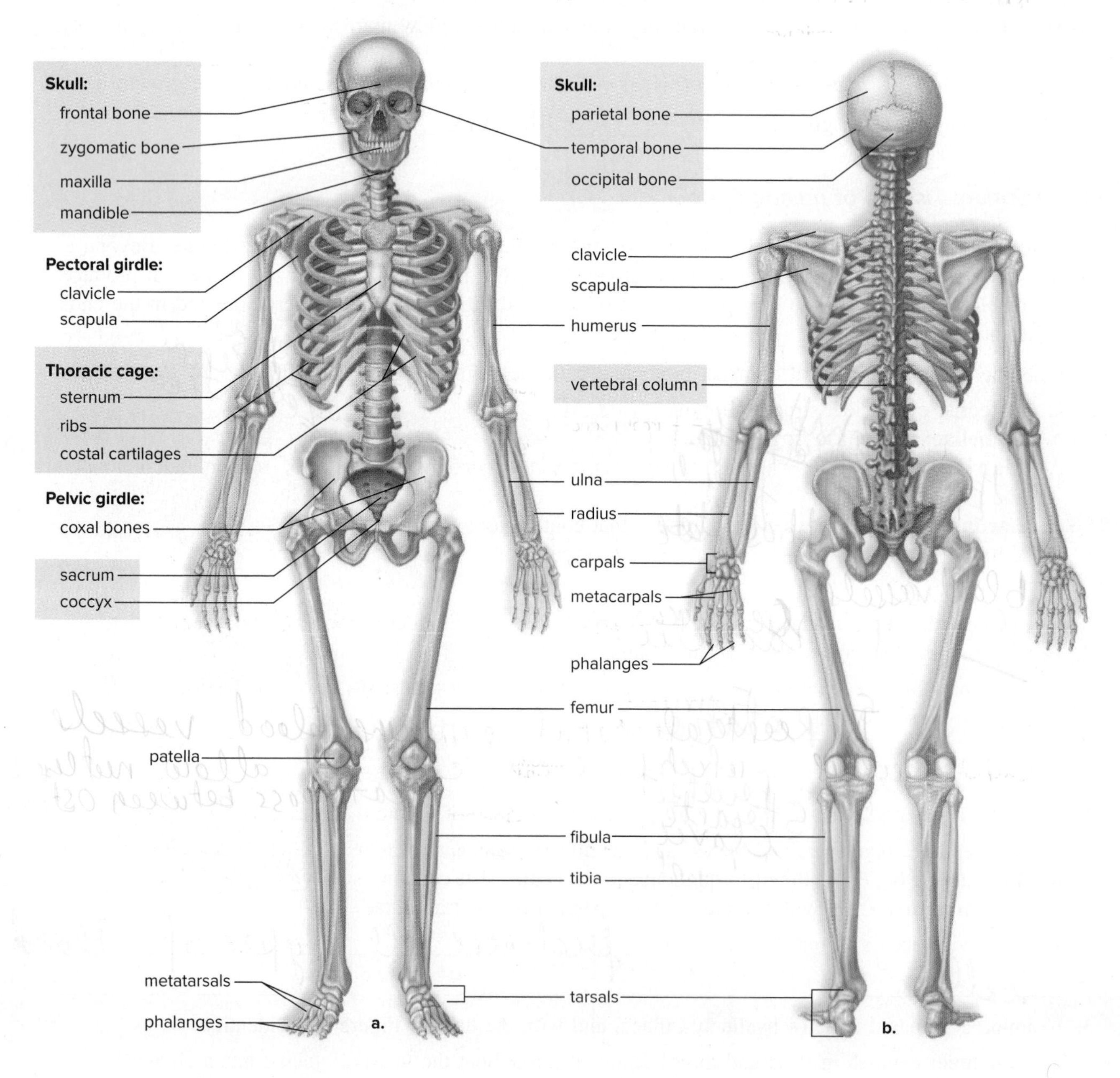

Observation: Axial Skeleton

Examine a human skeleton, and with the help of Figure 12.4, identify the **foramen magnum,** a large opening through which the spinal cord passes, and the following bones:

1. The **skull** is composed of many bones fused together at fibrous joints called **sutures.** Note the following in the cranium:
   a. **Frontal bone:** forms forehead
   b. **Parietal bones:** extend to sides of skull
   c. **Occipital bone:** curves to form base of skull
   d. **Temporal bones:** located on sides of skull
   e. **Sphenoid bone:** helps form base and sides of skull, as well as part of the orbits

   Which of these bones contribute to forming the face? frontal, zygomatic, maxilla, manible

   Which could best be associated with wearing glasses? nasal bone.

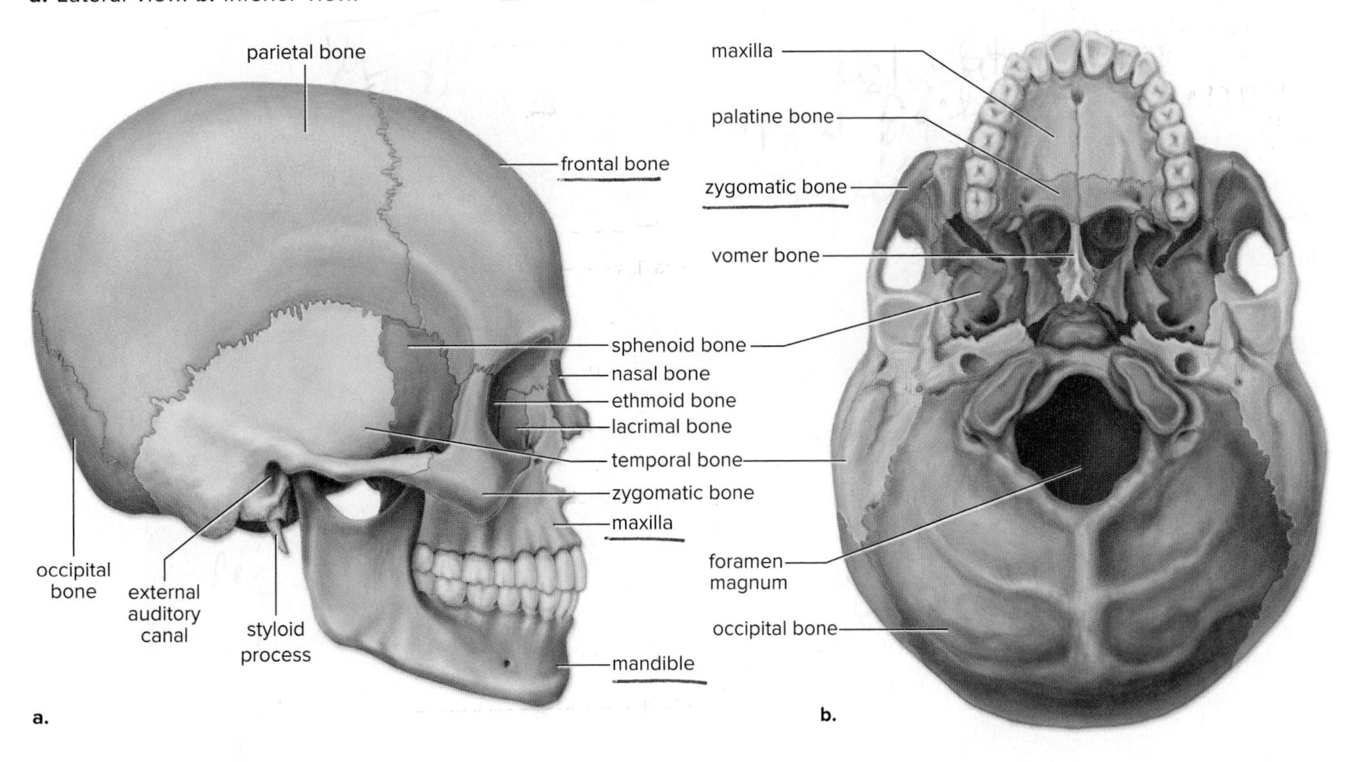

Figure 12.4 Skull.
a. Lateral view. **b.** Inferior view.

2. Facial bones (Fig. 12.5)
The most prominent of the facial bones are the mandible, the maxillae (sing., maxilla), the zygomatic bones, and the nasal bones. As shown previously, certain of the cranial bones contribute to the face. The frontal bone forms the forehead and thas supraorbital ridges where the eyebrows are located. The temporal bone and the wings of the sphenoid bone account for the flattened areas we call the temples. Note the following:

a. **Mandible:** the lower jaw
b. **Maxillae:** the upper jaw and anterior portion of the hard palate
c. **Palatine bones:** posterior portion of hard palate and floor of nasal cavity
d. **Zygomatic bones:** cheekbones
e. **Nasal bones:** bridge of nose

Which of these is movable and allows you to chew your food? Mandible and Maxillae.

Figure 12.5 Bones of the face.
© image 100/Corbis RF

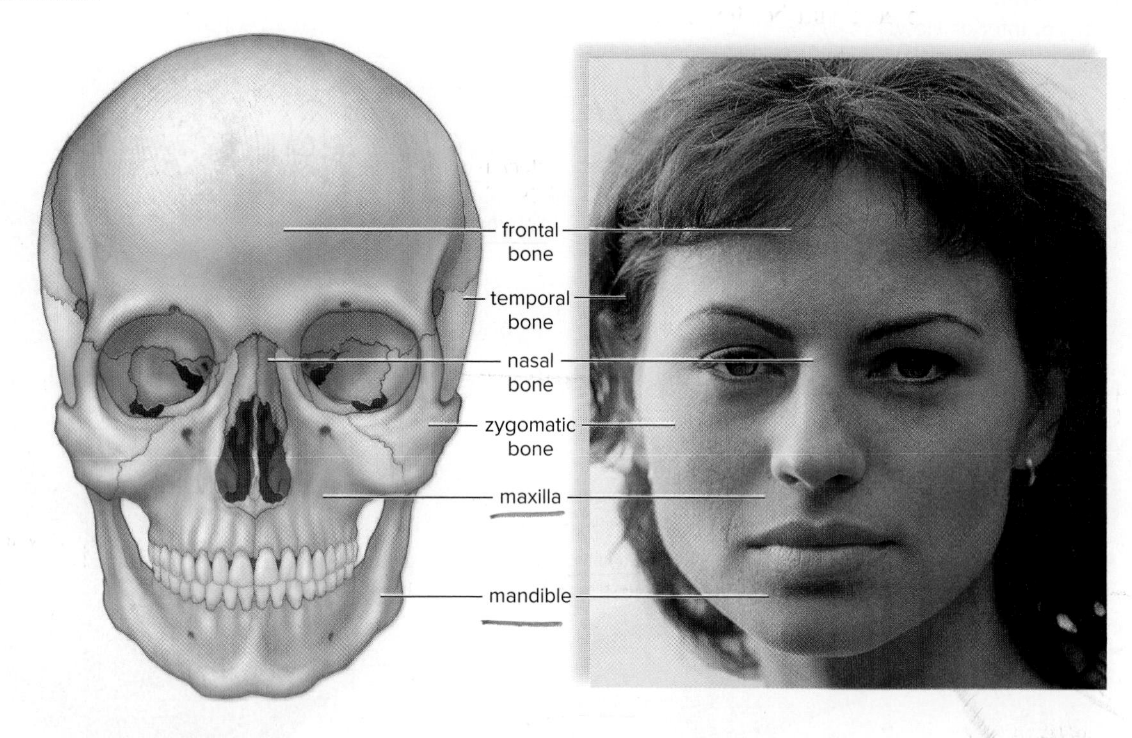

3. The **vertebral column** (Fig. 12.6) provides support and houses the **spinal cord.** It is composed of many vertebrae separated from one another by intervertebral disks. The vertebral column customarily is divided into five series:

   a. Seven **cervical vertebrae** (forming the neck region)
   b. Twelve **thoracic vertebrae** (with which the ribs articulate)
   c. Five **lumbar vertebrae** (in the abdominal region)
   d. Five fused sacral vertebrae, called the **sacrum**
   e. Four fused caudal vertebrae forming the **coccyx** in humans

   Which of the vertebrae can be associated with the chest? thoracic

   Which of the vertebrae could be the cause of your aching back? lumbar

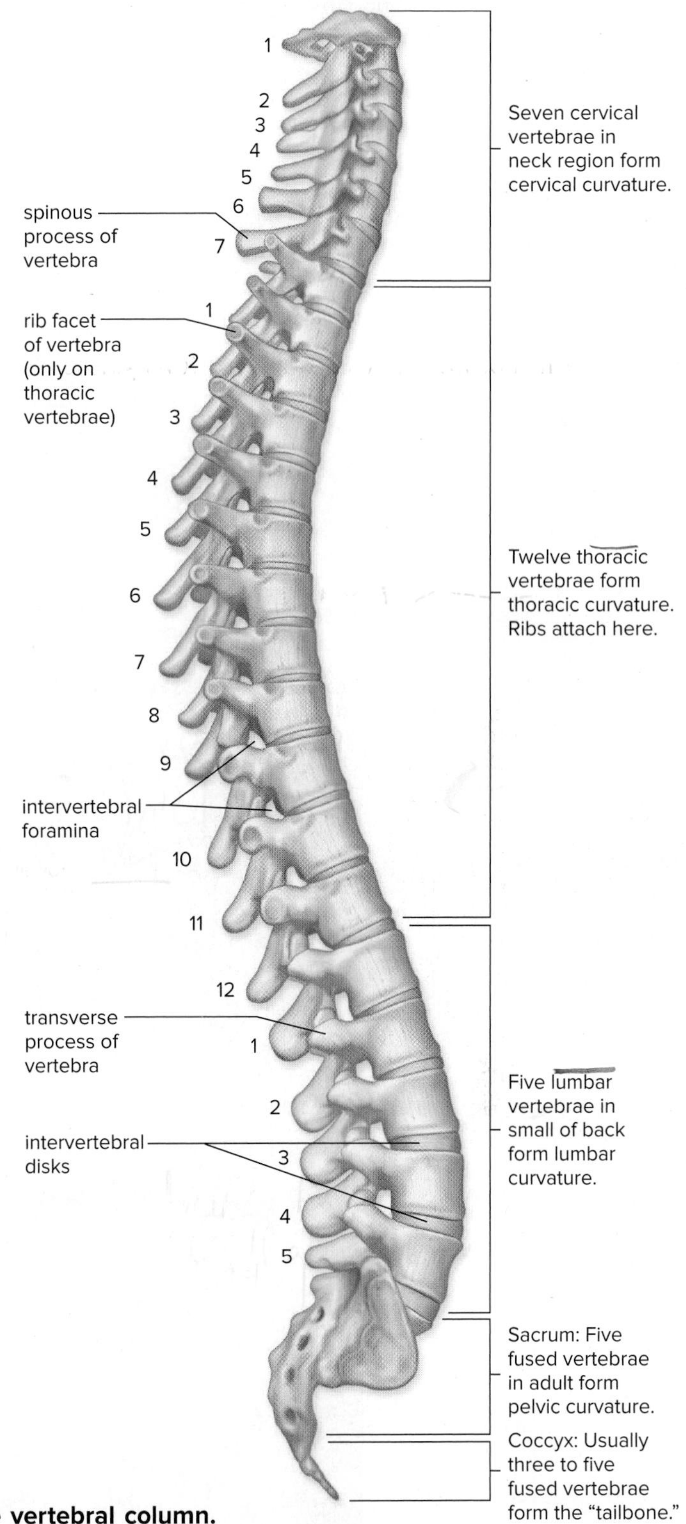

Figure 12.6 The vertebral column.

4. The twelve pairs of **ribs** and their associated muscles form a bony case that supports the thoracic cavity wall (Fig. 12.7). The ribs connect posteriorly with the thoracic vertebrae, and some are also attached by cartilage directly or indirectly to the sternum. Those ribs without any anterior attachment are called **floating ribs.**

Which of the ribs help form the protective part of the rib cage? True ribs

Figure 12.7 The ribcage.
a. Art. **b.** Photograph.
(b) © Eric Wise

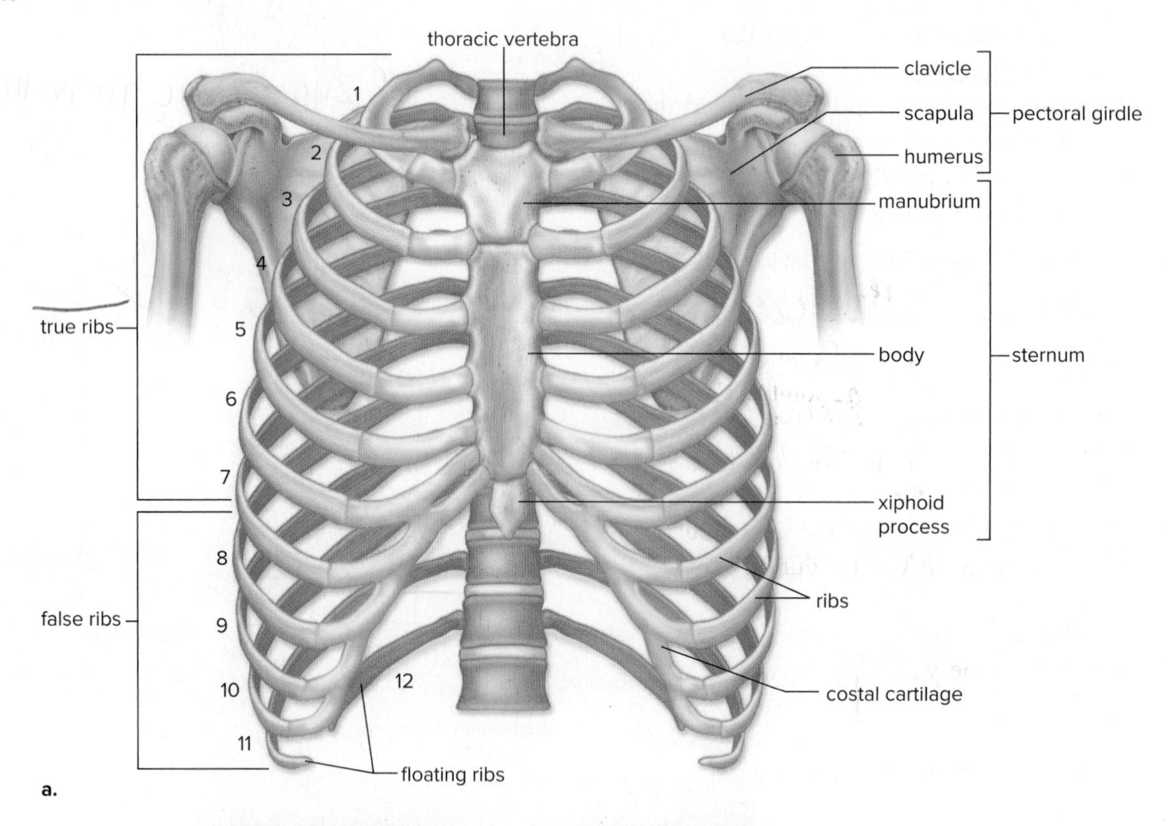

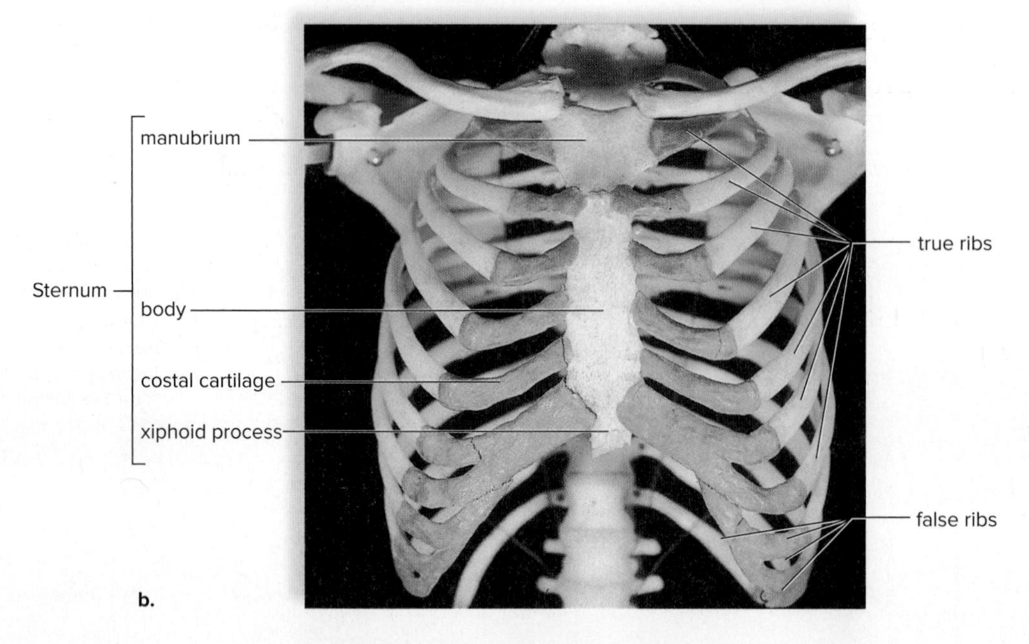

Observation: Appendicular Skeleton

Examine a human skeleton, and with the help of Figure 12.8, identify the following bones:

1. The **pectoral girdles,** which support the upper limbs, are composed of the **clavicle** (collarbone) and **scapula** (shoulder bone).

   Why is the shoulder apt to become dislocated? ____________________

2. The upper limb (arm plus the forearm) is composed of the following:
   - **a. Humerus:** the large long bone of the arm
   - **b. Radius:** the long bone of the forearm, with a pivot joint at the elbow that allows rotational motion
   - **c. Ulna:** the other long bone of the forearm, with a hinge joint at the elbow that allows motion in only one plane. Take hold of your elbow, and twist the forearm to show that the radius rotates over the ulna but the ulna doesn't move during this action.
   - **d. Carpals:** a group of small bones forming the wrist
   - **e. Metacarpals:** slender bones forming the palm
   - **f. Phalanges:** the bones of the fingers

   Which of these bones would you use to pick up a teacup? Phalanges and Carpals

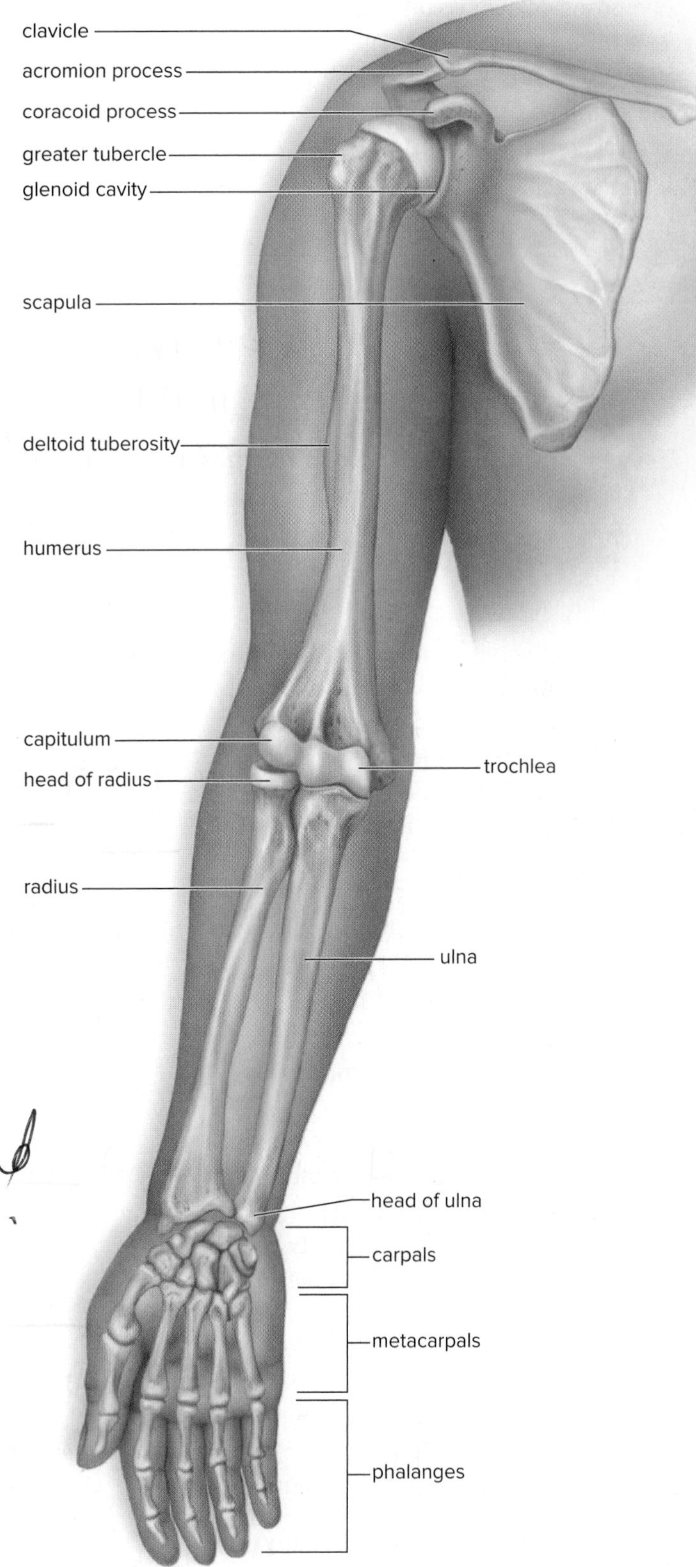

Figure 12.8 The bones of the pectoral girdle and upper limb.

3. The **pelvic girdle** forms the basal support for the lower limbs and is composed of two **coxal** (hip) **bones** (Fig. 12.9). Each coxal bone consists of the ilium, which is superior to the other two: the pubis and the ischium. (The pubis is ventral to the ischium.)

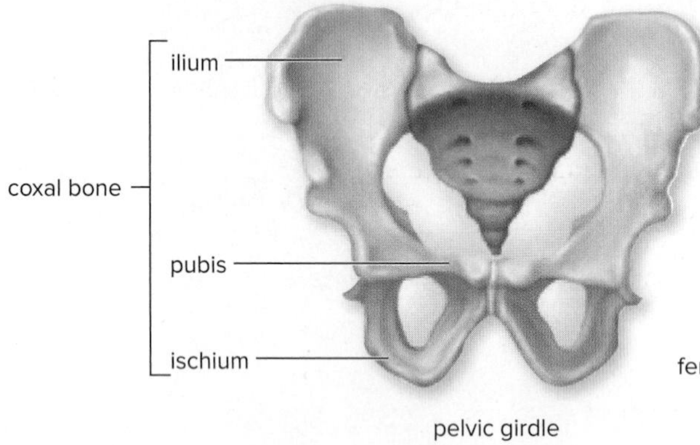

The female pelvis is much broader and shallower than that of the male. The angle between the pubic bones looks like a U in females and a V in males. How is it advantageous for the female pelvis to be broader and more shallow than that of a male? ________________

4. The lower limb (the thigh plus the leg) is composed of a series of loosely articulated bones, including the following:
   a. **Femur:** the long bone of the thigh
   b. **Patella:** kneecap
   c. **Tibia:** the larger of the two long bones of the leg; feel for the bump on the inside of the ankle
   d. **Fibula:** the smaller of the two long bones of the leg; feel for the bump on the outside of the ankle
   e. **Tarsals:** a group of small bones forming the ankle
   f. **Metatarsals:** slender anterior bones of the foot
   g. **Phalanges:** the bones of the toes

   Which of these bones would you use to kick a soccer ball? ________________

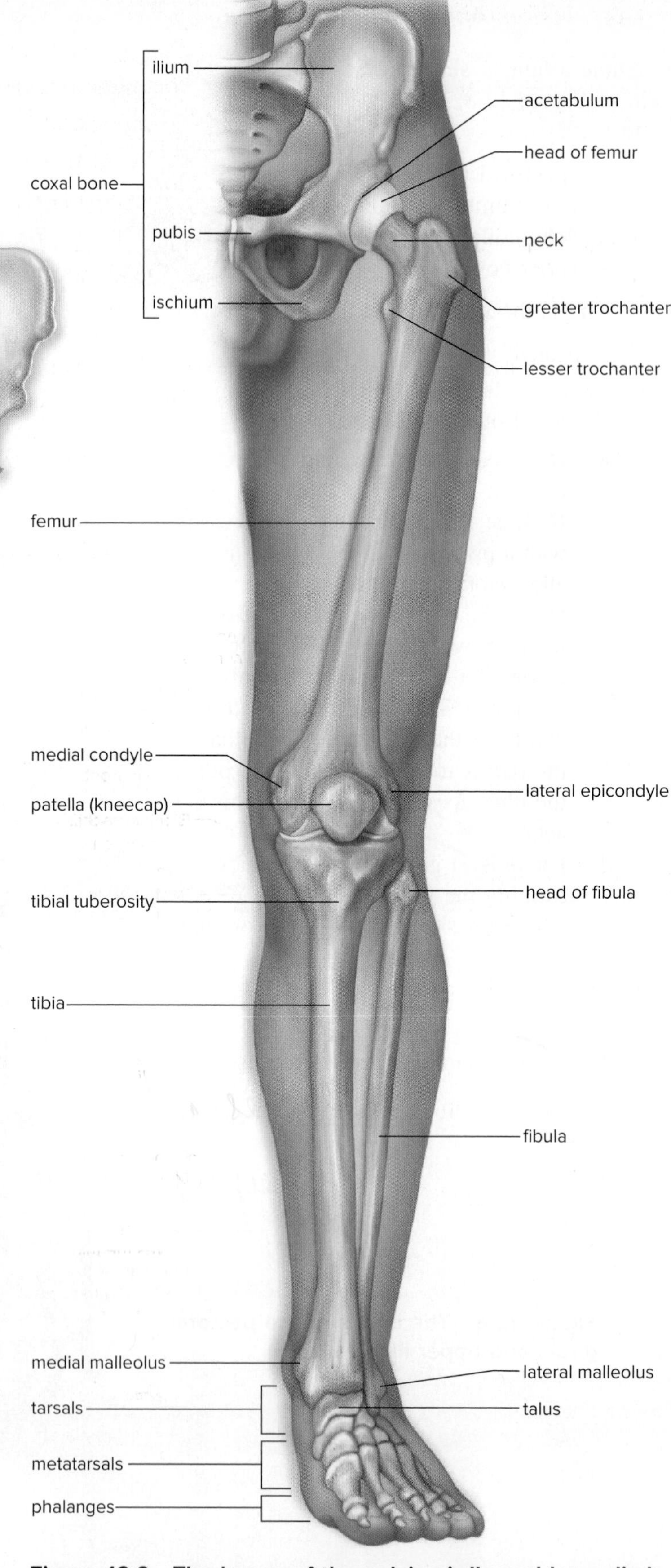

Figure 12.9 The bones of the pelvic girdle and lower limb.

12.3 The Skeletal Muscles

This laboratory is also concerned with skeletal muscles—those muscles that make up the bulk of the human body. With the help of Figure 12.10, identify the major muscles of the body.

Naming Muscles

Muscles are named for various characteristics, as shown in the following list:

1. **Size:** The gluteus maximus is the largest muscle, and it forms the buttocks.
2. **Shape:** The deltoid is shaped like a Greek letter delta, or triangle.
3. **Direction of fibers:** The rectus abdominis is a longitudinal muscle of the abdomen (*rectus* means "straight").
4. **Location:** The frontalis overlies the frontal bone.
5. **Number of attachments:** The biceps brachii has two attachments, or origins.
6. **Action:** The extensor digitorum extends the fingers, or digits.

Figure 12.10 Human musculature.
Superficial skeletal muscles in (**a**) anterior and (**b**) posterior view.

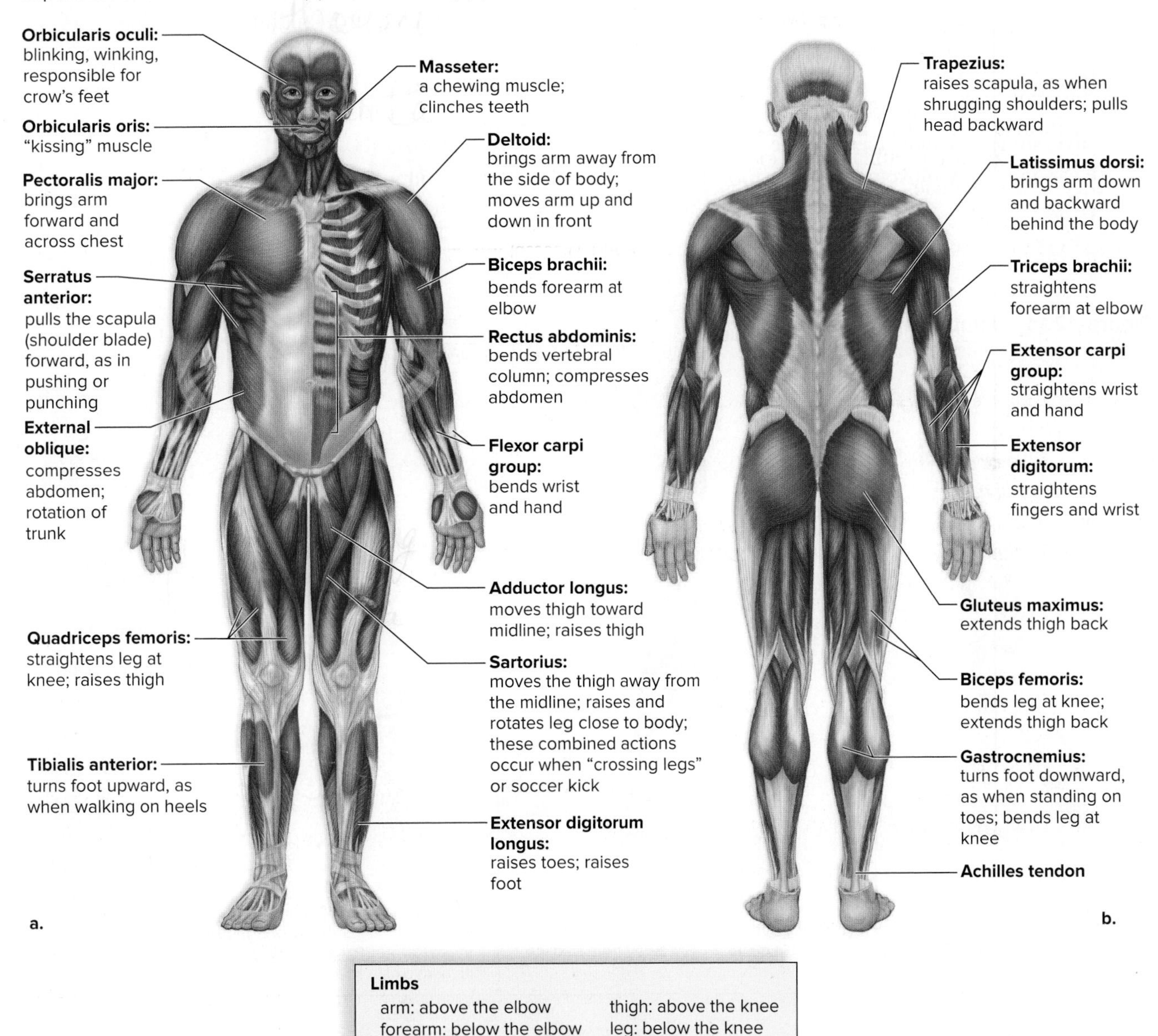

Limbs

arm: above the elbow	thigh: above the knee
forearm: below the elbow	leg: below the knee

Match these muscles to these functions from Fig. 12.10.

Doing a pelvic tilt and curving the spine ___________

Swinging movements of arms during walking and swimming ___________

Pulls forearm toward you when rowing ___________

Helps maintain the trunk in an erect posture ___________

Extends and separates the fingers ___________

Wrinkles the skin of the forehead ___________

Joint Movements

Figure 12.11 demonstrates the following types of joint movements:

Flexion	Moving jointed body parts toward each other
Extension	Moving jointed body parts away from each other
Adduction	Moving a part toward a vertical plane running through the longitudinal midline of the body
Abduction	Moving a part away from a vertical plane running through the longitudinal midline of the body
Rotation	Moving a body part around its own axis; **circumduction** is moving a body part in a wide circle.
Inversion	A movement of the foot in which the sole is turned inward
Eversion	A movement of the foot in which the sole is turned outward

Figure 12.11 Joint movements.
a. Flexion and extension. **b.** Adduction and abduction. **c.** Rotation and circumduction. **d.** Inversion and eversion.

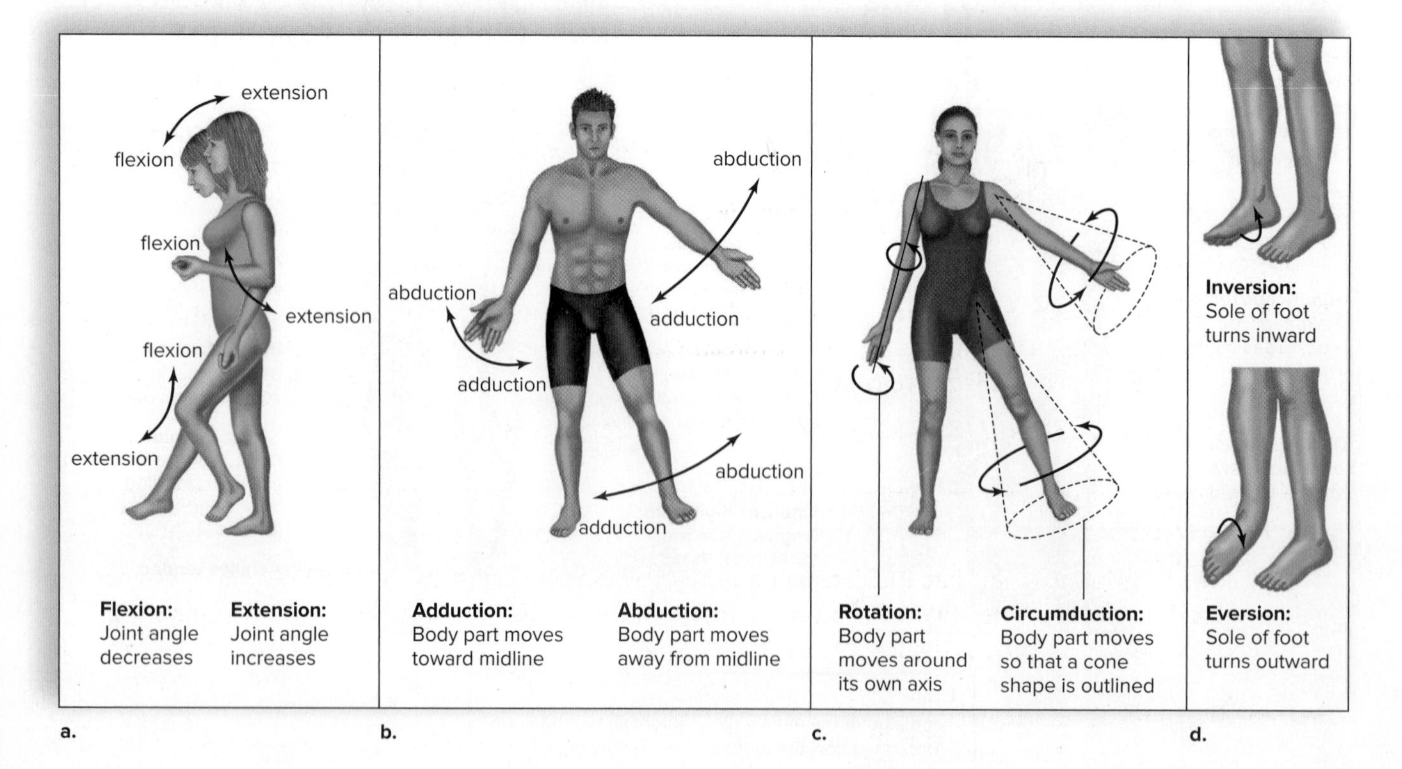

Antagonistic Pairs

Skeletal muscles are attached to the skeleton, and their contraction causes the movement of bones at a joint. Muscles shorten when they contract, so they can only pull; they cannot push. Therefore, muscles work in **antagonistic pairs.** Usually, contraction of one member of the pair causes a bone to move in one direction, and contraction of the other member of the pair causes the same bone to move in an opposite direction.

Are the muscles shown in Figure 12.1 antagonistic?______ How so? ______________________________

____

Observation: Antagonistic Pairs

Locate the following antagonistic pairs in Figure 12.10. In each case, state their opposing actions by inserting one of these functions: *flexes, extends, adducts,* or *abducts.*

1. The biceps brachii flex the forearm.
   The triceps brachii extends the forearm.
2. The sartorius abducts the thigh.
   The adductor longus adducts the thigh.
3. The quadriceps femoris extends the leg.
   The biceps femoris flexes the leg.

Isometric and Isotonic Contractions

A muscle contains many muscle fibers. When a muscle contracts, usually some fibers undergo isotonic contraction, and others undergo isometric contraction. When the tension of muscle fibers is sufficient to lift a load, many fibers change length as they lift the load. The muscle contraction is said to be **isotonic** (same tension). In contrast, when the tension of muscle fibers is used only to support rather than lift a load, the muscle contraction is said to be **isometric** (same length). The length of many fibers remains the same, but their tension still changes.

Experimental Procedure: Isometric and Isotonic Contractions

Note: The upper limb is composed of the arm plus the forearm.

Isotonic Contraction

1. Start with your left forearm resting on a table. Watch the anterior surface of your left arm while you slowly bend your elbow and bring your left forearm toward the arm. An isotonic contraction of the biceps brachii produces this movement.
2. If muscle contraction produces movement, is this an isometric or isotonic contraction? ________________

   ____

Isometric Contraction

1. Place the palm of your left hand underneath a tabletop. Push up against the table while you have your right hand cupped over the anterior surface of your left arm so that you can feel the muscle there undergo an isometric contraction.

2. Is the biceps brachii or the triceps brachii located on the anterior surface of the arm?

____

3. What change did you notice in the firmness of this muscle as it contracted? _______________

____

4. Did your forearm move as you pushed up against the table? _______________

5. Given your answer to question 4, did this muscle's fibers shorten as you pushed up against the tabletop? _______________

12.4 Mechanism of Skeletal Muscle Fiber Contraction

A whole skeletal muscle is made up of many cells, usually called muscle fibers. Muscle fibers can be examined using a light microscope and using an electron microscope.

Skeletal Muscle Fibers

Muscle fibers are striated—that is, they have alternating light and dark bands. These striations can be observed in a light micrograph of muscle fibers and in an electron micrograph in longitudinal section.

Observation: Skeletal Muscle Tissue

1. Examine a prepared slide of skeletal muscle using a light microscope. Identify the long, multinucleated fibers arranged in a parallel fashion. How do you know the muscle fibers are striated? _______________

____

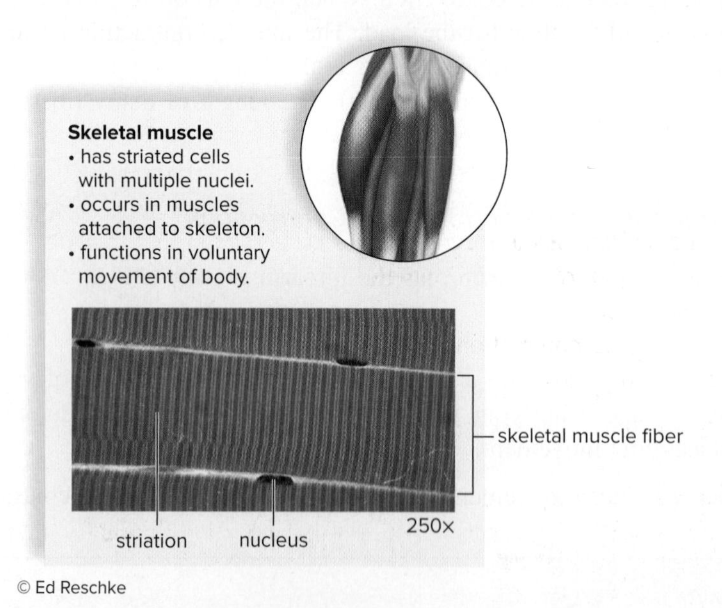

2. When a skeletal muscle fiber is examined using an electron microscope, it is possible to see that the fibers contain myofibrils which in turn contain actin and myosin filaments (Fig. 12.12). It is the placement of these filaments that cause muscle fibers to be striated.

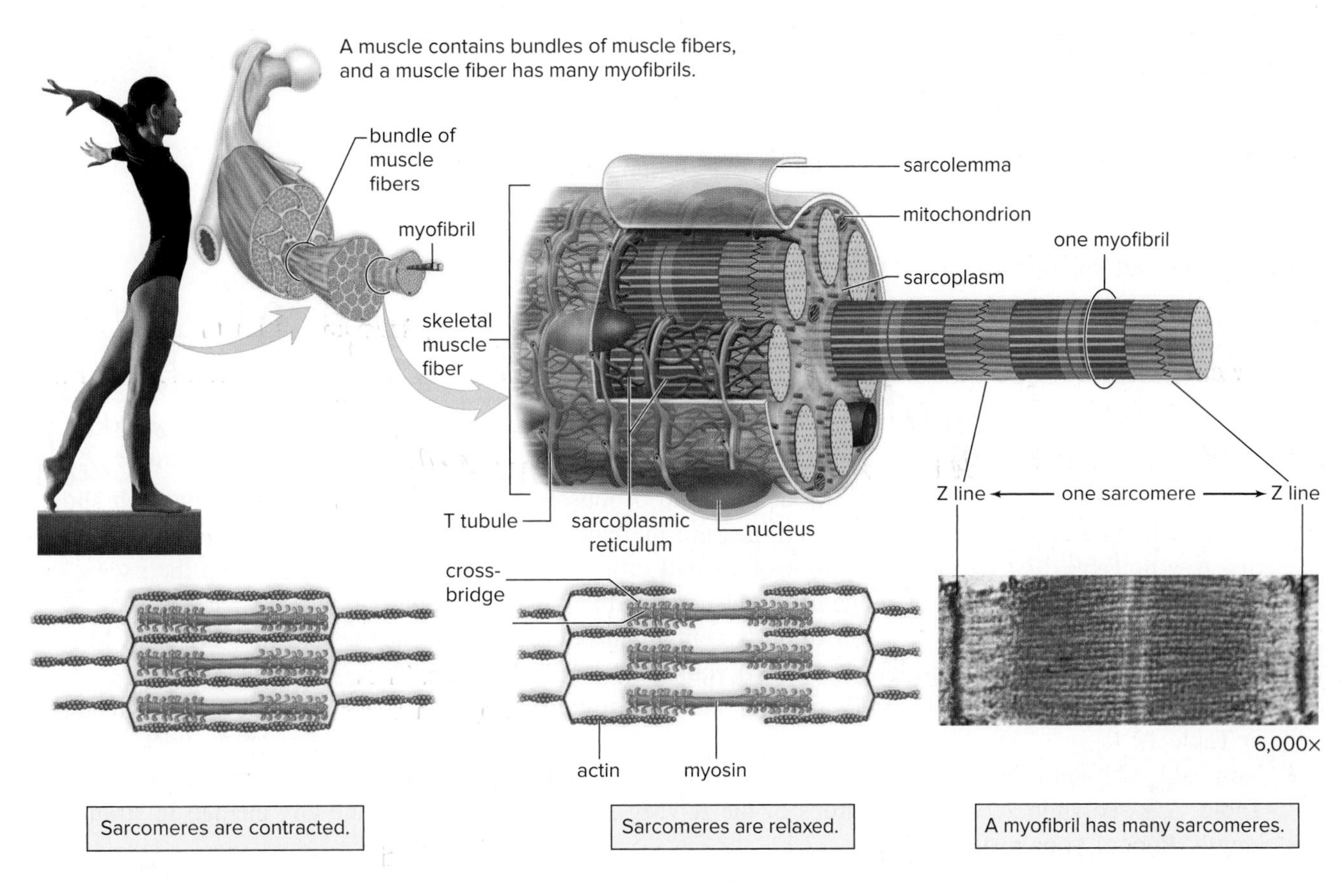

Figure 12.12 Electron microscopy of a skeletal muscle fiber.

(gymnast) © Fuse/Getty RF; (sarcomere): © Biology Media/Science Source

Skeletal Muscle Fiber Contraction

Electron microscopy helped investigators determine what causes the striated appearance of a muscle fiber and also its contraction. A muscle fiber contains hundreds, even thousands, of contractile portions called **myofibrils** divided into units called **sarcomeres.** Each sarcomere contains **myosin filaments** and **actin filaments** (Fig. 12.12). The sarcomeres contract when myosin cross-bridges attach to and pull the actin filaments to the center of the sarcomeres.

We are going to study muscle contraction with this data in mind:

1. Myosin is an enzyme that can break down ATP as an energy source for muscle contraction.
2. Enzymes often require cofactors, which can be coenzymes or certain ions, such as the K^+ and Mg^{2+} required by myosin.
3. When ATP, K^+, and Mg^{2+} are present, myosin cross-bridges attach to actin filaments and pull them to the center of sarcomeres within myofibrils. In this way, the actin filaments slide past myosin filaments and the sarcomeres, and a muscle fiber contracts.

Experimental Procedure: Muscle Fiber Contraction

During this procedure you are going to expose muscle fibers to different solutions:

1. Glycerol (A viscous liquid that keeps the muscle tissue moist.)
2. K^+ and Mg^{2+} salt solution alone
3. ATP solution alone
4. ATP and salt solution

Hypothesize which of these (1–4) will produce contraction, and explain. when ATP, and salt solution combine they cause myosin cross-bridges attach to actin filaments and slide past to myosin filaments and the sarcomere and muscle fiber contracts.

The Procedure

1. Label two slides, slide 1 and slide 2. Mount a strand of muscle fibers in a drop of glycerol on each slide. Place each slide on a millimeter ruler, and measure the length of the strand. Record these lengths in the first row in Table 12.1. If there is more than a small drop of glycerol on the slides, soak up the excess on a piece of lens paper held at the edge of the glycerol farthest from the fiber strand.
2. To slide 1, add a few drops of a salt solution containing potassium and magnesium ions (K^+ and Mg^{2+}), and note any change in strand length. Record your results in Table 12.1.
3. To slide 2, add a few drops of ATP solution, and note any change in strand length. Record your results in Table 12.1.
4. Now add ATP solution to slide 1. Note any change in strand length, and record your results in Table 12.1. To slide 2, add a few drops of the K^+/Mg^{2+} salt solution, and note any change in strand length. Record your results in Table 12.1.

Table 12.1 Glycerinated Muscle Contraction

Solution	Length of Muscle Fiber	
	Slide 1	Slide 2
1. Glycerol alone	mm	mm
2. K^+/Mg^{2+} salt solution alone	mm	—
3. ATP alone	—	mm
4. Both ATP and salt solution	mm	mm

Conclusion: Muscle Fiber Contraction

- Was your hypothesis stated above supported? ____________ Why or why not? ____________________

____

____

- To demonstrate that you understand the requirements for contraction, state the function of each of the substances listed in Table 12.2.

Table 12.2 Summary of Muscle Fiber Contraction	
Substance	**Function**
Myosin	
Actin	
K^+/Mg^{2+} salt solution	
ATP	

Application for Daily Living

Bone Marrow Transplants

We are accustomed to thinking of a transplant as a procedure that replaces a nonfunctioning organ with one that works properly. A red bone marrow transplant doesn't quite work like that. Instead, red bone marrow is injected into the bloodstream and the recipient is receiving the cells that occur in the red bone marrow. The red bone marrow contains the precious stem cells capable of producing all the various cells in the blood. And they aren't ordinarily in the blood! However, these cells usually find their way home—the bones that ordinarily contain red bone marrow in adults, such as the sternum, breast bone, skull, hips, ribs, and spine.

The health reasons for needing a red bone marrow transplant are numerous, but chief among them are cancer patients whose own bone marrow was destroyed by treatment of cancer. As with any transplant, a careful match between donor and recipient is required.

Laboratory Review 12

_______________ **1.** Is compact bone located in the diaphysis or in the epiphyses?

_______________ **2.** Does compact bone or spongy bone contain red bone marrow?

_______________ **3.** What are bone cells called?

_______________ **4.** What are the vertebrae in the neck region called?

_______________ **5.** Name the strongest bone in the lower limb.

_______________ **6.** What bones are part of a pectoral girdle?

_______________ **7.** What type of joint movement occurs when a muscle moves a limb toward the midline of the body?

_______________ **8.** What type of joint movement occurs when a muscle moves a body part around its own axis?

_______________ **9.** Skeletal muscle is voluntary, and its appearance is _________ because of the placement of actin and myosin filaments.

_______________ **10.** Glycerinated muscle requires the addition of what molecule to supply the energy for muscle contraction?

_______________ **11.** Actin and myosin are what type of biological molecule?

_______________ **12.** Does the quadriceps femoris flex or extend the leg?

_______________ **13.** Does the biceps brachii flex or extend the forearm?

_______________ **14.** What muscle forms the buttocks?

_______________ **15.** Name the muscle group antagonistic to the quadriceps femoris group.

Thought Questions

16. What bones protect the thoracic cavity?

17. When you see glycerinated muscle shorten, what is happening microscopically?

18. Release of a neurotransmitter at a neuromuscular junction causes a muscle fiber to contract. Label the neurotransmitter in this figure.

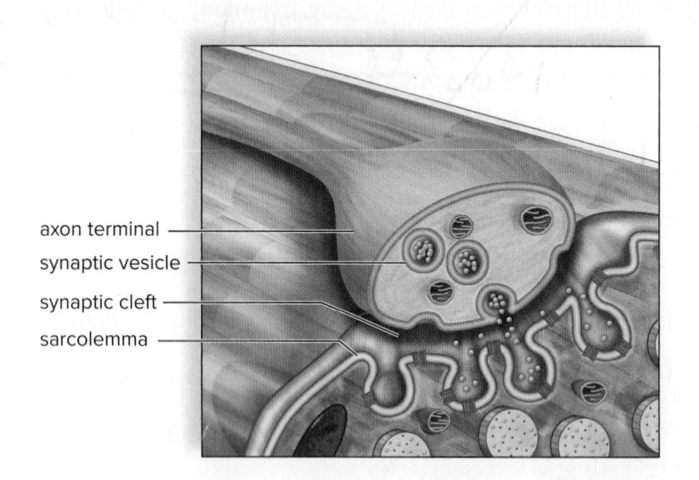

LABORATORY

13

Nervous System and Senses

Learning Outcomes

13.1 Central Nervous System

- Identify the parts of the brain studied, and state the functions of each part.
- Give examples to show that the parts of the brain work together.
- Describe the anatomy of the spinal cord and tell how the cord functions as a relay station.

Prelab Question: Distinguish between the cerebral cortex and the cerebrum.

13.2 Peripheral Nervous System

- Distinguish between cranial nerves and spinal nerves on the basis of location and function.
- Describe the anatomy and physiology of a spinal reflex arc.

Prelab Question: How does the brain become aware that you have removed your hand from a hot stove?

13.3 The Human Eye

- Identify the parts of the eye and state a function for each part.

Prelab Question: What part of the eye contains the sensory receptors for sight?

13.4 The Human Ear

- Identify the parts of the ear and state a function for each part.

Prelab Question: What part of the ear contains the sensory receptors for hearing?

13.5 Sensory Receptors in Human Skin

- Describe the anatomy of the human skin and explain the distribution and function of sensory receptors.
- Relate the abundance of touch receptors to the ability to distinguish between two different touch points.

Prelab Question: What part of the skin contains sensory receptors?

13.6 Human Chemoreceptors

- Relate the ability to distinguish foods to the senses of smell and taste.

Prelab Question: Taste is dependent on what types of chemical stimuli?

***Application for Daily Living:* LASIK Surgery**

Introduction

The nervous system has two major divisions: the central nervous system (CNS) consisting of the brain and spinal cord and the peripheral nervous system (PNS), which contains cranial nerves and spinal nerves (Fig. 13.1). Sensory receptors detect changes in environmental stimuli, and nerve impulses move along sensory nerve fibers to the brain and the spinal cord. The brain and spinal cord sum up the data before sending impulses via motor nerve fibers to effectors (muscles and glands) so a response to stimuli is possible. Nervous tissue consists of neurons; whereas the brain and spinal cord contain all parts of neurons, nerves contain only axons.

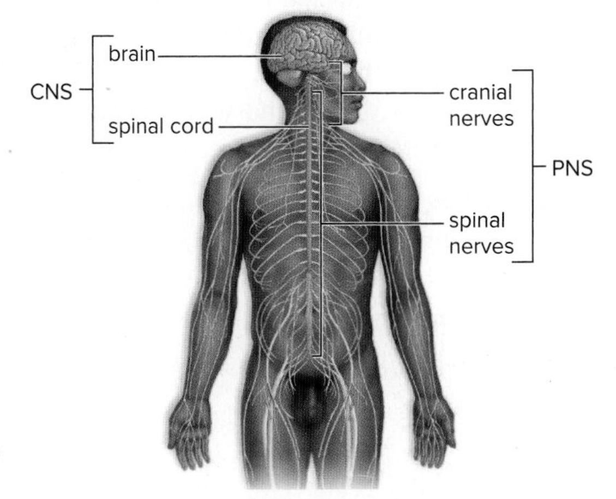

Figure 13.1 The nervous system.
The central nervous system (CNS) is in the midline of the body, and the peripheral nervous system (PNS) is outside the CNS.

13.1 Central Nervous System

The brain is the enlarged anterior end of the spinal cord; it contains centers that receive input from and can command other regions of the nervous system.

> **Latex gloves** Wear protective latex gloves when handling preserved animal organs. Use protective eyewear and exercise caution when using sharp instruments during this laboratory. Wash hands thoroughly upon completion of this laboratory.

Preserved Sheep Brain

The sheep brain (Fig. 13.2) is often used to study the brain. It is easily available and large enough that individual parts can be identified.

Observation: Preserved Sheep Brain

Examine the exterior and a midsagittal (longitudinal) section of a preserved sheep brain or a model of the human brain, and with the help of Figure 13.2, identify the following.

1. **Ventricles:** interconnecting spaces that produce and serve as a reservoir for cerebrospinal fluid, which cushions the brain. Toward the anterior, note the lateral ventricle (on one longitudinal section) and similarly a lateral ventricle (on the other longitudinal section). Trace the second ventricle to the third and then the fourth ventricles.
2. **Cerebrum:** most developed area of the brain; responsible for higher mental capabilities. The cerebrum is divided into the right and left **cerebral hemispheres,** joined by the **corpus callosum,** a broad sheet of white matter. The outer portion of the cerebrum is highly convoluted and divided into the following surface lobes (see Fig. 13.4):
   a. **Frontal lobe:** controls motor functions and permits voluntary muscle control; it also is responsible for abilities to think, problem solve, speak, and smell.
   b. **Parietal lobe:** receives information from sensory receptors located in the skin and also the taste receptors in the mouth. A groove called the **central sulcus** separates the frontal lobe from the parietal lobe.
   c. **Occipital lobe:** interprets visual input and combines visual images with other sensory experiences. The optic nerves split and enter opposite sides of the brain at the optic chiasma, located in the diencephalon.
   d. **Temporal lobe:** has sensory areas for hearing and smelling. The olfactory bulb contains nerve fibers that communicate with the olfactory cells in the nasal passages and take nerve impulses to the temporal lobe.
3. **Diencephalon:** portion of the brain where the third ventricle is located. The hypothalamus and thalamus are also located here.
   a. **Thalamus:** two connected lobes located in the roof of the third ventricle. The thalamus is the highest portion of the brain to receive sensory impulses before the cerebrum. It is believed to control which received impulses are passed on to the cerebrum. For this reason, the thalamus sometimes is called the "gatekeeper to the cerebrum."
   b. **Hypothalamus:** forms the floor of the third ventricle and contains control centers for appetite, body temperature, and water balance. Its primary function is homeostasis. The hypothalamus also has centers for pleasure, reproductive behavior, hostility, and pain.
4. **Cerebellum:** located just posterior to the cerebrum as you observe the brain dorsally, the cerebellum's two lobes make it appear rather like a butterfly. In cross section, the cerebellum has an internal pattern that looks like a tree. The cerebellum coordinates equilibrium and motor activity to produce smooth movements.

Figure 13.2 The sheep brain.

(a-c) © Dr. J. Timothy Cannon

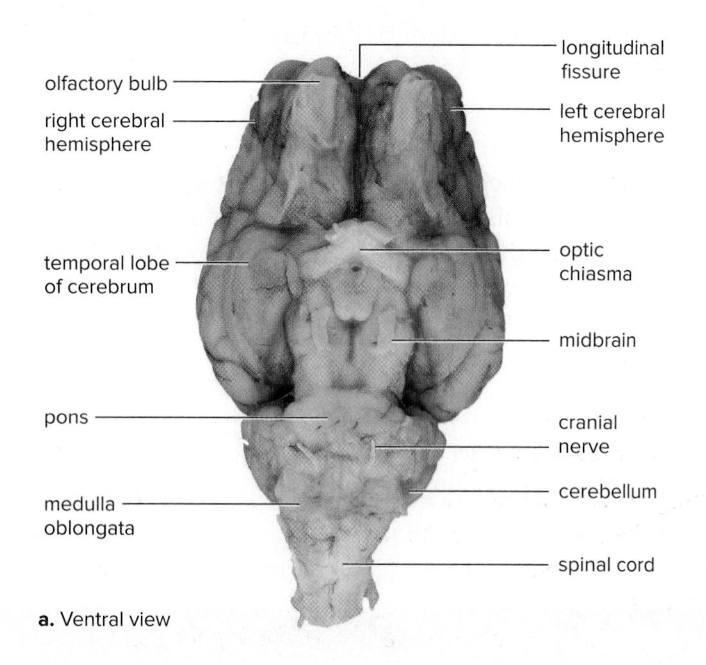

a. Ventral view

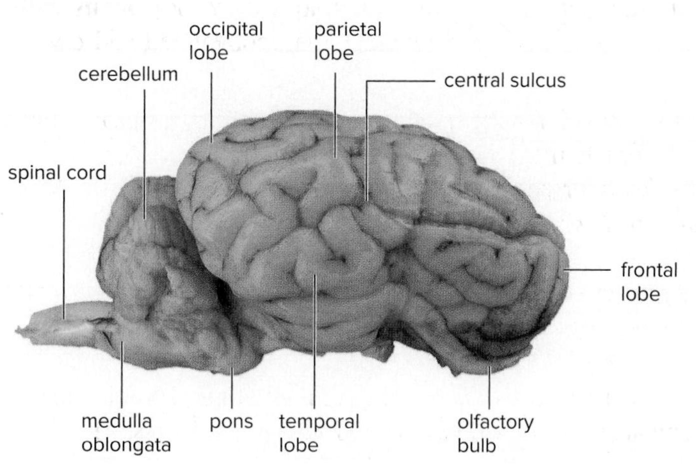

b. Lateral view

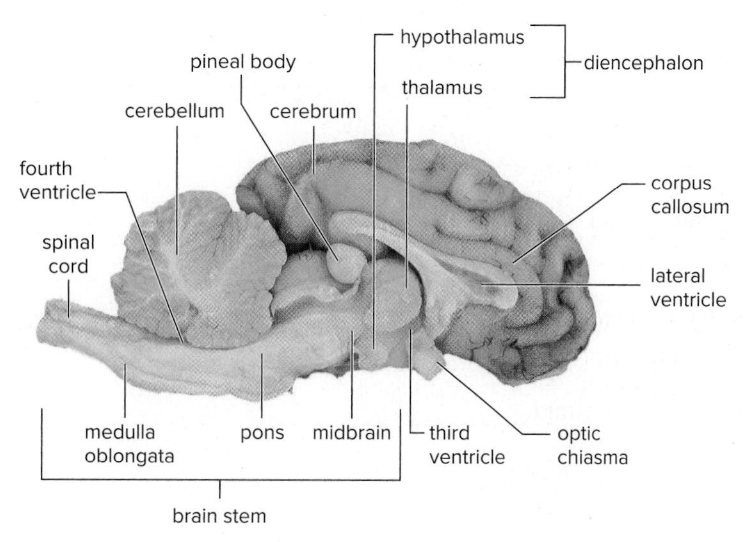

c. Longitudinal section

5. **Brain stem:** Part of the brain that connects with the spinal cord. Because it includes the pons and medulla oblongata, it contains centers for the functioning of internal organs; because of its location, it serves as a relay station for nerve impulses passing from the cord to the brain. Therefore, it helps keep the rest of the brain alert and functioning.
   a. **Midbrain:** anterior to the pons, the midbrain serves as a relay station for sensory input and motor output. It also contains a reflex center for eye muscles.
   b. **Pons:** the ventral, bulblike enlargement on the brain stem. It serves as a passageway for nerve impulses running between the medulla and the higher brain regions.
   c. **Medulla oblongata** (or simply **medulla**): the most posterior portion of the brain stem. It controls internal organs; for example, cardiac and breathing control centers are present in the medulla. Nerve impulses pass from the spinal cord through the medulla to higher brain regions.

The Human Brain

Label Figure 13.3 where indicated. Based on your knowledge of the sheep brain, complete Table 13.1 by stating the major functions of each part of the brain listed.

Table 13.1 Summary of Brain Functions

Part	Major Functions
Cerebrum	
Cerebellum	
Diencephalon Thalamus	
Hypothalamus	
Brain stem Midbrain	
Pons	
Medulla oblongata	

Which parts of the brain would work together to achieve the following?

1. Good eye–hand coordination ______________________________
2. Concentrating on homework when TV is playing ______________________________
3. Avoiding dark alleys while walking home at night ______________________________
4. Keeping the blood pressure constant ______________________________

Figure 13.3 The human brain (longitudinal section).
The cerebrum is larger in humans than in sheep. Label this diagram where indicated.

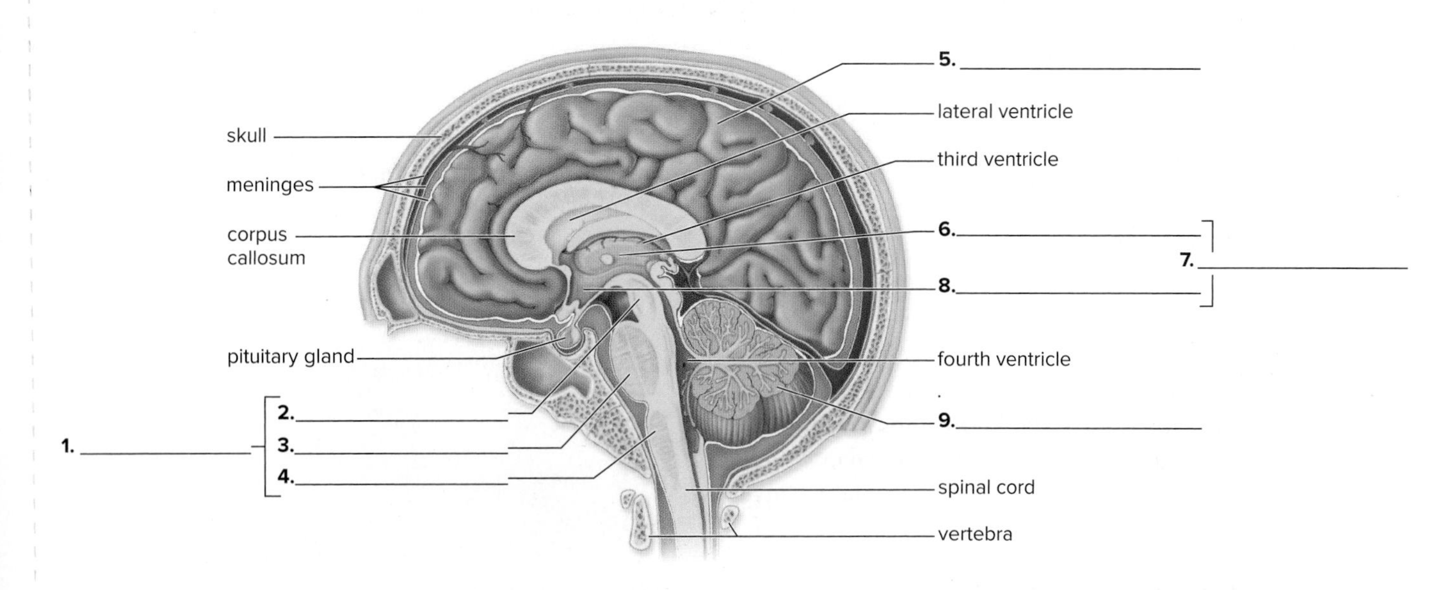

Cerebral Lobes

As stated previously, the outer portion of the cerebrum is highly convoluted and divided into lobes as illustrated in Figure 13.4. The various sense organs send nerve impulses to a particular lobe where the nerve impulses are integrated to give us our senses of vision, hearing, smell, taste, and touch. Although not stated in Figure 13.4, the frontal lobe helps us remember smells of some significance to us. *Write the name of a sense next to the appropriate lobe in Figure 13.4.*

Figure 13.4 The cerebral lobes.
Each lobe has centers for integrating nerve impulses received from a particular type of sense organ. Our five senses result from this activity of the brain.

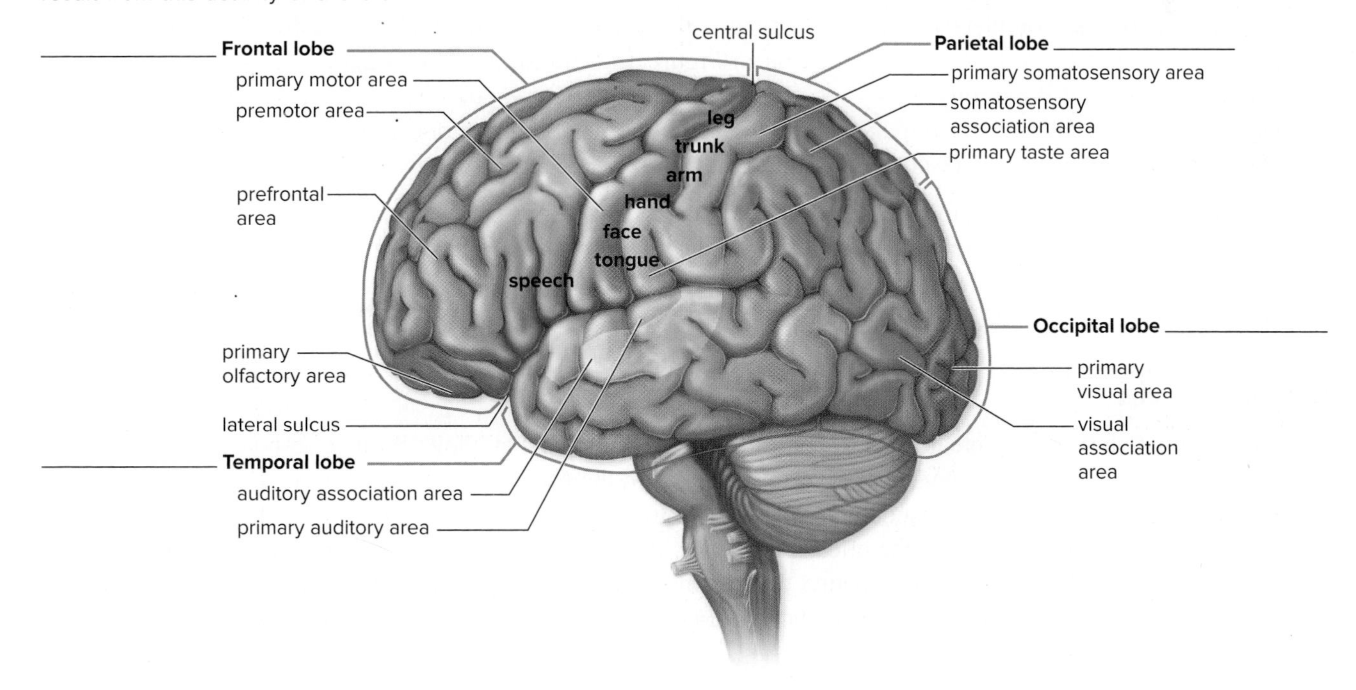

The Spinal Cord

The spinal cord is a part of the central nervous system. It lies in the middorsal region of the body and is protected by the vertebral column.

Observation: The Spinal Cord

1. Examine a prepared slide of a cross section of the spinal cord under the lowest magnification possible.
2. Identify the following with the help of Figure 13.5:
   a. **Gray matter:** a central, butterfly-shaped area composed of masses of short nerve fibers, interneurons, and motor neuron cell bodies.
   b. **White matter:** masses of long fibers that lie outside the gray matter and carry impulses up and down the spinal cord. In living animals, white matter appears white because an insulating myelin sheath surrounds long fibers.

Figure 13.5 The spinal cord.
Photomicrograph of spinal cord cross section.

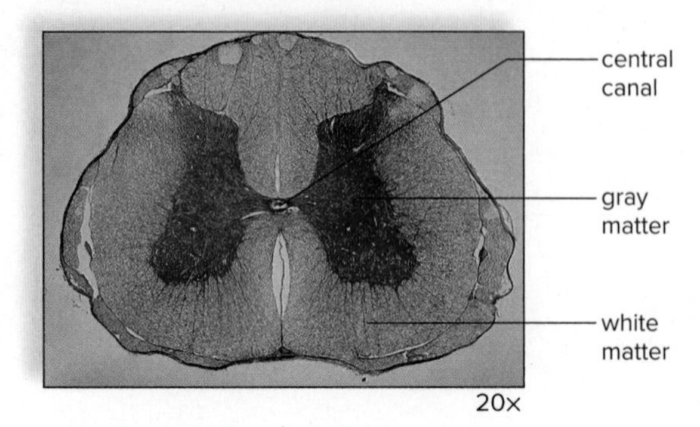

13.2 Peripheral Nervous System

The peripheral nervous system contains the cranial nerves and the spinal nerves. Twelve pairs of cranial nerves project from the inferior surface of the brain. The cranial nerves are largely concerned with nervous communication between the head, neck, and facial regions of the body and the brain. The 31 pairs of spinal nerves emerge from either side of the spinal cord (Fig. 13.6).

Spinal Nerves

Each spinal nerve contains long fibers of sensory neurons and long fibers of motor neurons. In Figure 13.6, identify the following:

1. **Sensory neuron:** takes nerve impulses from a sensory receptor to the spinal cord. The cell body of a sensory neuron is in the dorsal root ganglion.
2. **Interneuron:** lies completely within the spinal cord. Some interneurons have long fibers and take nerve impulses to and from the brain. The neuron in Figure 13.6 transmits nerve impulses from the sensory neuron to the motor neuron.
3. **Motor neuron:** takes nerve impulses from the spinal cord to an effector—in this case, a muscle. Muscle contraction is one type of response to stimuli.

Suppose you were walking barefoot and stepped on a prickly sandbur. Describe the pathway of information, starting with the pain receptor in your foot, that would allow you to both feel and respond to this unwelcome stimulus. ______________________________

Spinal Reflexes

A **reflex** is an involuntary and predictable response to a given stimulus that allows a quick response to environmental stimuli without communicating with the brain. In the spinal reflexes that follow, stretch receptors detect the tap, and sensory neurons conduct nerve impulses to interneurons in the spinal cord. The interneurons send a message via motor neurons to the effectors, muscles in the leg or foot. These reflexes are involuntary because the brain is not involved in formulating the response. Consciousness of the stimulus lags behind the response because information must be sent up the spinal cord to the brain before you can become aware of the tap.

Figure 13.6 Spinal nerves and spinal cord.
The arrows mark the path of nerve impulses from a sensory receptor to an effector. When a spinal reflex occurs, a sensory receptor is stimulated and generates nerve impulses that pass along the sensory neuron, the interneuron, and motor neuron causing the effector to respond.

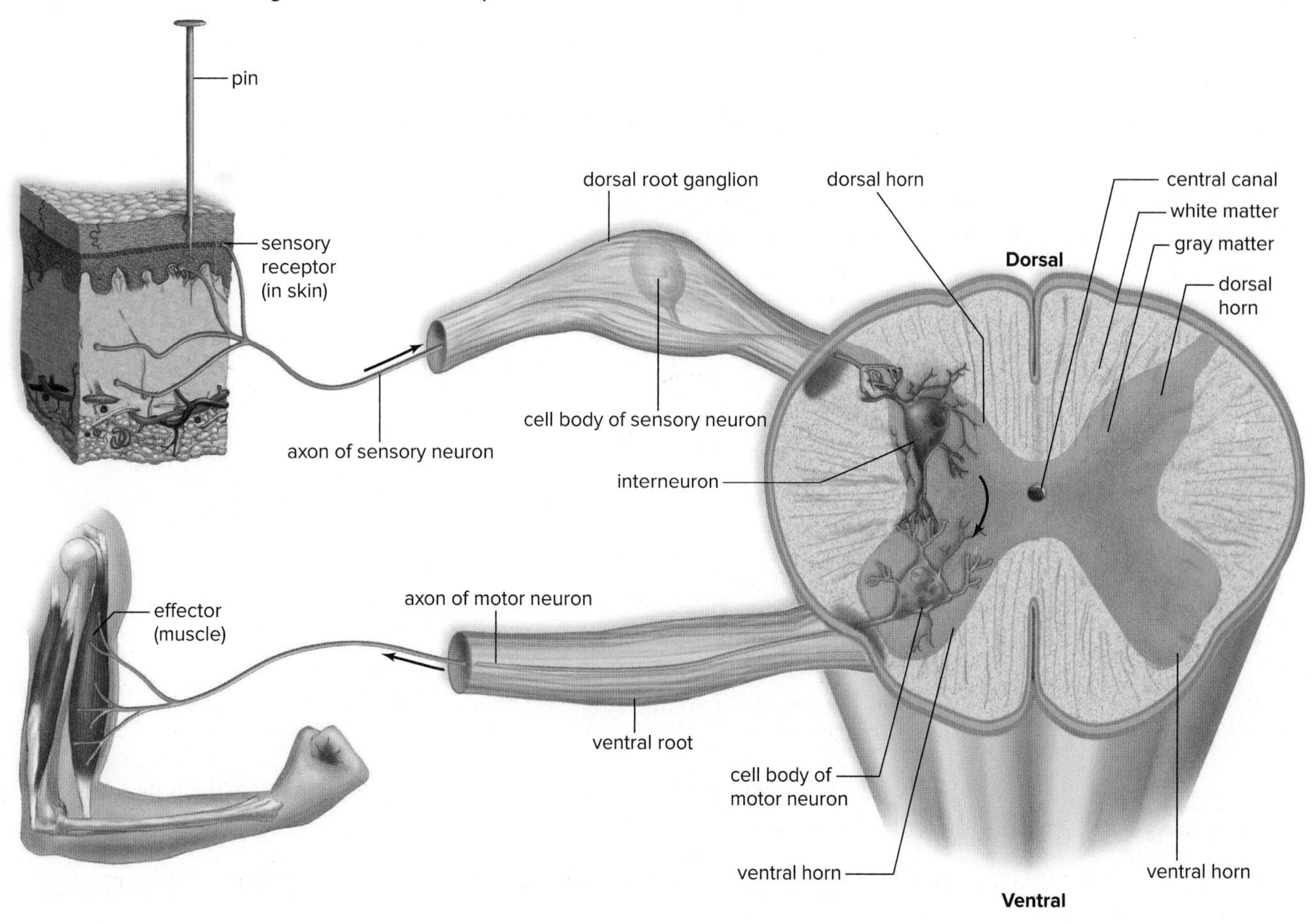

Experimental Procedure: Spinal Reflex

One easily tested tendon reflex involves the **patellar tendon.** When this tendon is tapped with a reflex hammer (Fig. 13.7) or, in this experiment, with a meter stick, the attached muscle contracts and the leg rises. The stretch receptor generates nerve impulses which are transmitted along sensory neurons to the spinal cord. Nerve impulses from the cord then pass along motor neurons and stimulate the muscle, causing it to contract. Receptors in other tendons, such as the Achilles tendon, respond similarly.

Knee-Jerk (Patellar) Reflex

1. Have the subject sit on a table so that his or her legs hang freely.
2. Sharply tap one of the patellar tendons just below the patella (kneecap) with a meter stick.
3. In this relaxed state, does the leg flex (move toward the buttocks) or extend (move away from the buttocks)? ______________________

Figure 13.7 Knee-jerk reflex.
The quick response when the patellar tendon is stimulated by tapping with a reflex hammer indicates that a reflex has occurred. © P.H. Gerbier/SPL/Science Source

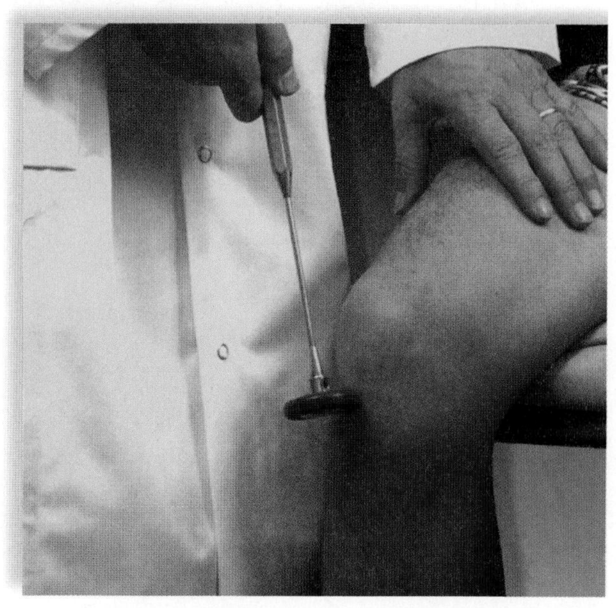

Knee-jerk (patellar) reflex

13.3 The Human Eye

First you will observe an eye model and then perform two experimental procedures to test your own eyes.

Observation: The Human Eye

The eye is a special sense organ for detecting light rays in the environment. Examine a human eye model, and identify the structures listed in Table 13.2 and depicted in Figure 13.8. As you identify the structures state their functions.

Table 13.2 Parts of the Human Eye

Part	Location	Function
Sclera	Outer layer of eye	Protects and supports eyeball
Cornea	Transparent portion of sclera	Refracts light rays
Choroid	Middle layer of eye	Absorbs stray light rays
Retina	Inner layer of eye	Contains receptors for sight
Rod cells (black and white vision)		
Cone cells (color vision)		
Fovea centralis	Special region of retina for cones	Makes acute vision possible
Lens	Between compartments	Refracts and focuses light rays
Ciliary body	Extension from choroid	Holds lens in place; functions in accommodation
Iris	Most anterior portion of choroid	Regulates light entrance
Pupil	Opening in middle of iris	Admits light
Aqueous and vitreous humors	Fluid media of eye	Transmit and refract light rays
Optic nerve	Extension from posterior of eye	Transmits impulses to occipital lobe of brain

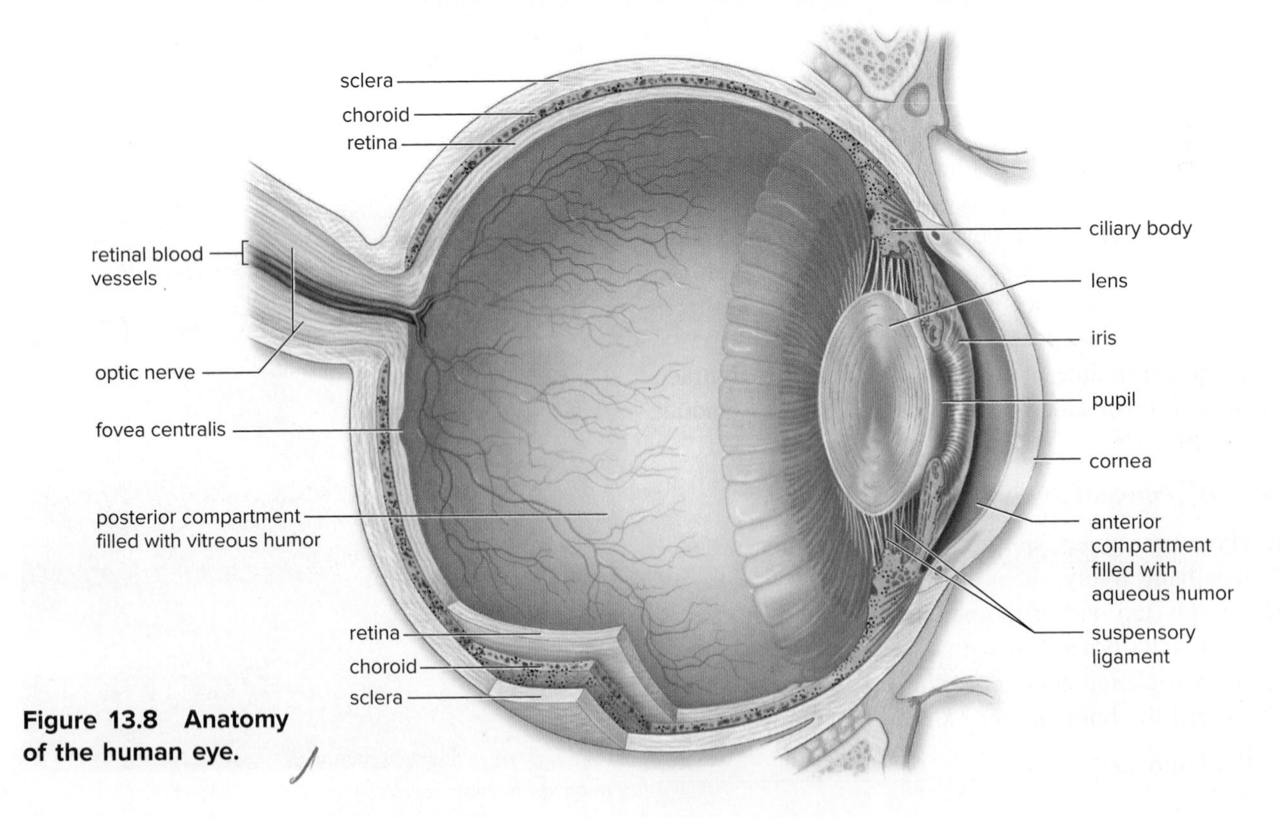

Figure 13.8 Anatomy of the human eye.

The Blind Spot of the Eye

The **blind spot** occurs where the optic nerve fibers exit the retina. No vision is possible at this location because of the absence of rod cells and cone cells.

Experimental Procedure: Blind Spot of the Eye

This Experimental Procedure requires a laboratory partner. Figure 13.9 shows a small circle and a cross several centimeters apart.

Figure 13.9 Blind spot.
This dark circle (or cross) will disappear at one location because there are no rod cells or cone cells at each eye's blind spot, where vision does not occur.

Left Eye

1. Hold Figure 13.9 approximately 30 cm from your eyes. The cross should be directly in front of your left eye. If you wear glasses, keep them on.
2. Close your right eye.
3. Stare only at the cross with your left eye. You should also be able to see the circle in the same field of vision. Slowly move the paper toward you until the circle disappears.
4. Repeat the procedure as many times as needed to find the blind spot.
5. Then slowly move the paper closer to your eyes until the circle reappears. Because only your left eye is open, you have found the blind spot of your left eye.
6. With your partner's help, measure the distance from your eye to the paper when the circle first disappeared. Left eye: _________ cm

Right Eye

1. Hold Figure 13.9 approximately 30 cm from your eyes. The circle should be directly in front of your right eye. If you wear glasses, keep them on.
2. Close your left eye.
3. Stare only at the circle with your right eye. You should also be able to see the cross in the same field of vision. Slowly move the paper toward you until the cross disappears.
4. Repeat the procedure as many times as needed to find the blind spot.
5. Then slowly move the paper closer to your eyes until the cross reappears. Because only your right eye is open, you have found the blind spot of your right eye.
6. With your partner's help, measure the distance from your eye to the paper when the cross first disappeared. Right eye: _________ cm

In this exercise, you created an artificial situation in which you became aware of how your perception of the world can be constrained by the eye's anatomy. Although the eye detects patterns of light and color, it is the brain that determines what we visually perceive. The brain can fill in data that is missing based in part on past experiences.

Accommodation of the Eye

When the eye accommodates to see objects at different distances, the shape of the lens changes. The lens shape is controlled by the ciliary muscles attached to it. When you are looking at a distant object, the lens is in a flattened state. When you are looking at a closer object, the lens becomes more rounded. The elasticity of the lens determines how well the eye can accommodate. Lens elasticity decreases with increasing age, a condition called **presbyopia.** Presbyopia is the reason many older people need bifocals to see near objects.

Experimental Procedure: Accommodation of the Eye

This Experimental Procedure requires a laboratory partner. It tests accommodation of either your left or right eye.

Figure 13.10 Accommodation.
When testing the ability of your eyes to accommodate to see a near object, always keep the pencil in this position.

1. Hold a pencil upright by the eraser and at arm's length in front of whichever of your eyes you are testing (Fig. 13.10).
2. Close the opposite eye.
3. Move the pencil from arm's length toward your eye.
4. Focus on the end of the pencil.
5. Move the pencil toward you until the end is out of focus. Measure the distance (in centimeters) between the pencil and your eye: ________ cm
6. At what distance can your eye no longer accommodate for distance? ________ cm
7. If you wear glasses, repeat this experiment without your glasses, and note the accommodation distance of your eye without glasses: ____________ cm. (Contact lens wearers need not make these determinations, and they should write the words *contact lens* in this blank.)
8. The "younger" lens can easily accommodate for closer distances. The nearest point at which the end of the pencil can be clearly seen is called the **near point.** The more elastic the lens, the "younger" the eye (Table 13.3). How "old" is the eye you tested? ________

Table 13.3 Near Point and Age Correlation

Age (years)	10	20	30	40	50	60
Near point (cm)	9	10	13	18	50	83

13.4 The Human Ear

The human ear, whose parts are listed and depicted in Table 13.4 and Figure 13.11, serves two functions: hearing and balance.

Observation: The Human Ear

Examine a human ear model, and find the structures depicted in Figure 13.11 and listed in Table 13.4.

Table 13.4 Parts of the Human Ear

Part	Medium	Function	Mechanoreceptor
Outer ear	Air		
Pinna		Collects sound waves	—
Auditory canal		Filters air	—
Middle ear	Air		
Tympanic membrane and ossicles		Amplify sound waves	—
Auditory tube		Equalizes air pressure	—
Inner ear	Fluid		
Semicircular canals		Rotational equilibrium	Stereocilia embedded in cupula
Vestibule (contains utricle and saccule)		Gravitational equilibrium	Stereocilia embedded in otolithic membrane
Cochlea (spiral organ)		Hearing	Stereocilia embedded in tectorial membrane

Figure 13.11 Anatomy of the human ear.
The outer ear extends from the pinna to the tympanic membrane. The middle ear extends from the tympanic membrane to the oval window. The inner ear encompasses the semicircular canals, the vestibule, and the cochlea.

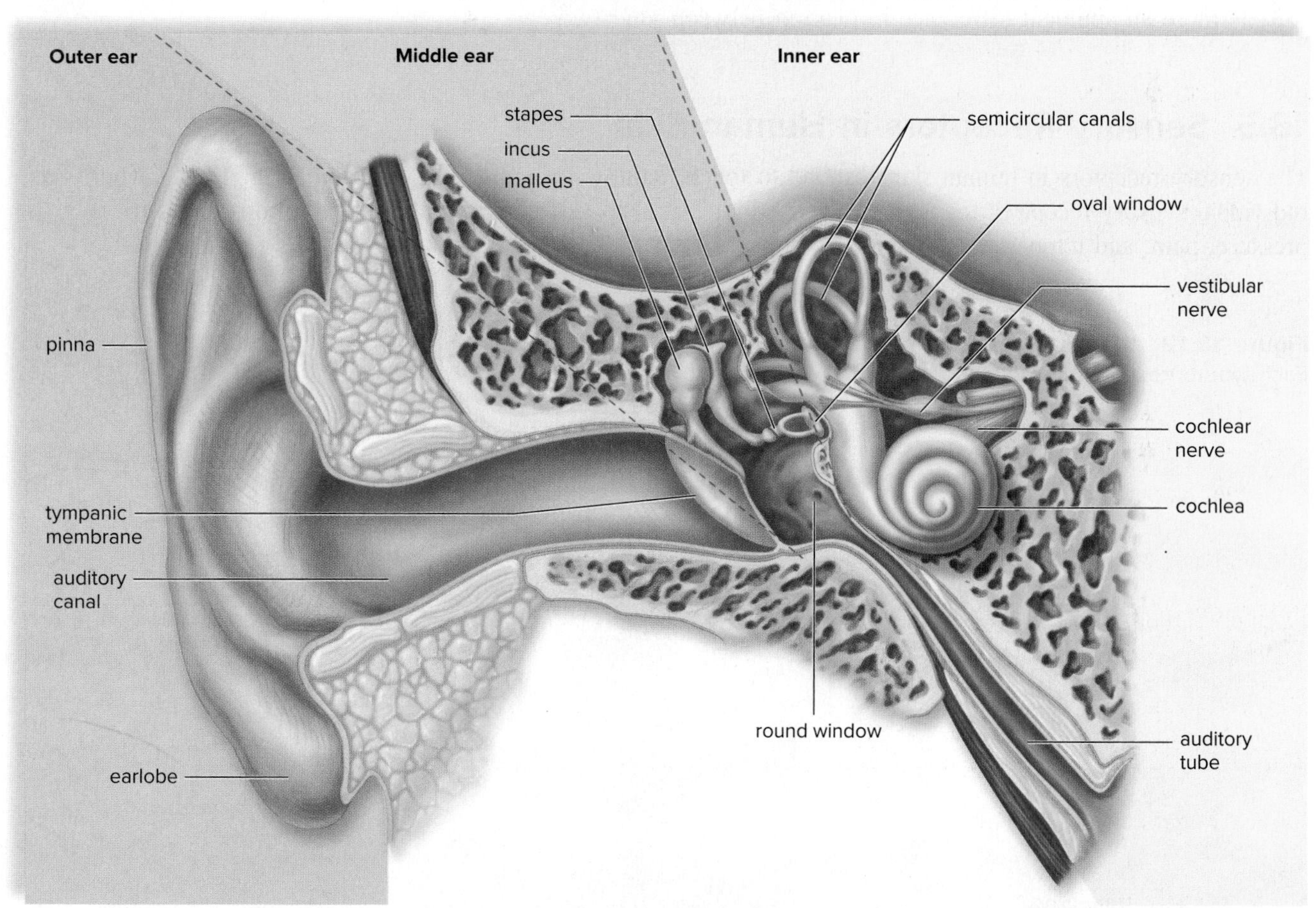

Physiology of the Human Ear When you hear, sound waves are picked up by the **tympanic membrane** and amplified by the **malleus, incus,** and **stapes.** This creates pressure waves in the canals of the **cochlea** that lead to stimulation of **hair cells,** the receptors for hearing. Hair cells in the utricle and saccule of the vestibule and in semicircular canals are receptors for equilibrium (i.e., balance). Nerve impulses from the sensory receptors in the ears travel by way of the cochlear nerve and the vestibular nerve to the brain and eventually are interpreted by the ________________ lobe of the brain.

Experimental Procedure: Locating Sound

Humans locate the direction of sound according to how fast it is detected by either or both ears. A difference in the hearing ability of the two ears can lead to a mistaken judgment about the direction of sound. You and a laboratory partner should perform this Experimental Procedure on each other. Enter the data for your ears, not your partner's ears, in the spaces provided.

1. Ask the subject to be seated, with eyes closed. Then strike a tuning fork or rap two spoons together at the five locations listed in number 2. Use a random order.
2. Ask the subject to give the exact location of the sound in relation to his or her head. Record the subject's perceptions when the sound is
   - **a.** Directly below and behind the head ______________________________
   - **b.** Directly behind the head ______________________________
   - **c.** Directly above the head ______________________________
   - **d.** Directly in front of the face ______________________________
   - **e.** To the side of the head ______________________________
3. Is there an apparent difference in hearing between your two ears? ______________________________

13.5 Sensory Receptors in Human Skin

The sensory receptors in human skin respond to touch, pain, temperature, and pressure (Fig. 13.12). There are individual sensory receptors for each of these stimuli, as well as free nerve endings able to respond to pressure, pain, and temperature.

Figure 13.12 Sensory receptors in the skin.
Each type of receptor shown responds primarily to a particular stimulus.

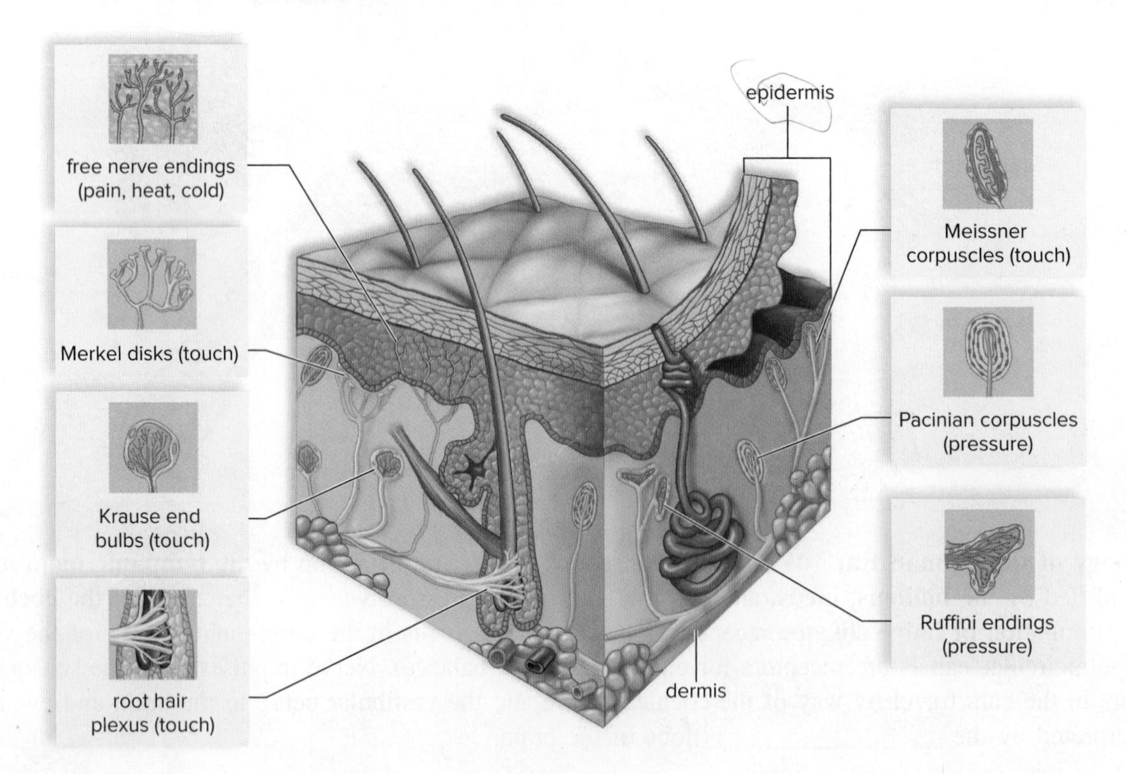

Sense of Touch

The dermis of the skin contains touch receptors, whose concentration differs in various parts of the body.

Experimental Procedure: Sense of Touch

You will need a laboratory partner to perform this Experimental Procedure. Enter your data, not the data of your partner, in the spaces provided.

1. Ask the subject to be seated, with eyes closed.
2. Then test the subject's ability to discriminate between the two points of a hairpin or a pair of scissors at the four locations listed in number 5.
3. Hold the points of the hairpin or scissors on the given skin area, with both of the points simultaneously and gently touching the subject.
4. Ask the subject whether the experience involves one or two touch sensations.
5. Record the shortest distance between the hairpin or scissor points for a two-point discrimination.

   a. Forearm: ________ mm

   b. Back of the neck: ________ mm

   c. Index finger: ________ mm

   d. Back of the hand: ________ mm

6. Which of these areas apparently contains the greatest density of touch receptors? ________________

   Why is this useful? ________________________________

7. Do you have a sense of touch at every point in your skin? ________ Explain. ________________

   ____

8. What specific part of the brain processes nerve impulses from touch and pain receptors? See Figure 13.4.

   ____

Sense of Heat and Cold

Temperature receptors respond to a change in temperature.

Experimental Procedure: Sense of Heat and Cold

1. Obtain three 1,000 mL beakers, and fill one with ice water, one with tap water at room temperature, and one with warm water (45°–50°C).
2. Immerse your left hand in the ice-water beaker and your right hand in the warm-water beaker for 30 seconds.
3. Then place both hands in the beaker with room-temperature tap water.
4. Record the sensation in the right and left hands.

   a. Right hand: Num________________________________

   b. Left hand: Tingling.________________________________

5. Explain your results. Feels Cold.________________________________

   ____

13.6 Human Chemoreceptors

The taste receptors, called ____________________, located in the mouth, and the smell receptors, called ____________________, located in the nasal cavities, are the chemoreceptors that respond to molecules in the air and water. Nerve impulses from taste receptors go to the ____________________ lobe of the brain while those from smell receptors go to the ____________________ lobe of the brain.

Experimental Procedure: Sense of Taste and Smell

You will need a laboratory partner to perform the following procedures. It will not be necessary for all tests to be performed on both partners. You should take turns being either the subject or the experimenter. Dispose of used cotton swabs in a hazardous waste container or as directed by your instructor.

1. Students work in groups. Each group has one experimenter and several subjects.
2. The experimenter should obtain a LifeSavers candy from the various flavors available, without letting the subject know what flavor it is.
3. The subject closes both eyes and holds his or her nose.
4. The experimenter gives the LifeSavers candy to the subject, who places it on his or her tongue.
5. The subject, while still holding his or her nose, guesses the flavor of the candy. The experimenter records the guess in Table 13.5.
6. The subject releases his or her nose and guesses the flavor again. The experimenter records the guess and the actual flavor in Table 13.5.

Table 13.5 Taste and Smell Experiment

Subject	Actual Flavor	Flavor While Holding Nose	Flavor After Releasing Nose
1			
2			
3			
4			
5			

Conclusions: Sense of Taste and Smell

- From your results, how would you say that smell affects the taste of LifeSavers candy?
  ____
- What do you conclude about the effect of smell on your sense of taste?
  ____

Application for Daily Living

LASIK Surgery

The focusing ability of the lens to place an image on the retina so we can see is assisted by the cornea. The cornea refracts or bends the light rays, and then the lens takes on the chore from then on. This accounts for why LASIK eye surgery, which reshapes the cornea, works. The traditional LASIK vision correction involves two steps: (1) First, the surgeon has to make a flap in the outer surface of the eye to expose the underlying cornea. (2) Then, the cornea is reshaped.

Anyone considering undergoing LASIK eye surgery should be aware that various side effects have been seen from dry eyes to severe glare when driving at night. Therefore, the procedure should be discussed with a physician or optician first. They can recommend a reliable clinic and surgeon and will also be able to advise whether there is any reason why LASIK eye surgery might not work for you.

Laboratory Review 13

______________ 1. Is the sciatic nerve part of the central nervous system or the peripheral nervous system?

______________ 2. What part of the brain is divided into right and left hemispheres?

______________ 3. What portion of the brain looks like a tree in cross section?

______________ 4. What is the most posterior portion of the brain stem?

______________ 5. What cerebral lobe is associated with the sense of vision?

______________ 6. What structures protect the spinal cord?

______________ 7. What makes the white matter of the spinal cord appear white?

______________ 8. What type of neuron is found completely within the central nervous system?

______________ 9. What type of neuron is responsible for transmitting nerve impulses from the spinal cord to an effector?

______________ 10. What part of the eye refracts and focuses light rays?

______________ 11. What part of the ear contains the sensory receptors for hearing?

______________ 12. What is the anatomical name for the eardrum that picks up sound waves?

______________ 13. What layer of the skin contains Meissner and Pacinian corpuscles?

______________ 14. What are the names of the chemoreceptors that respond to molecules in air and water?

Thought Questions

15. Explain why people who develop cataracts (cloudy eye lenses) require surgery to implant new lenses.

16. When you stared at the dark cross with your left eye, why did the dark circle "disappear" from view as you moved the illustration toward you?

17. In a drag race, drivers must wait until the green light is illuminated before they can move their vehicle. Explain why a time delay exists, based on the information presented in this exercise.

LABORATORY

14

Reproduction and Development

Learning Outcomes

14.1 Male and Female Reproductive Systems

- Trace the path of sperm in the male and the egg in the female.
- Locate on a model and state a function for each major organ in the male and female reproductive systems.
- Identify a slide of the testis and the ovary and the major structures in these organs.
- Describe the specialization of a sperm and of an egg for their functions.

Prelab Question: What is the most significant biological contribution of a male and a female to an offspring?

14.2 Embryonic Development

- Identify the cellular stages of development with reference to slides of early sea star development.
- Identify the tissue stages of development with reference to slides of frog development.
- Associate the germ layers with the development of various organs.
- Identify which organs develop first in a vertebrate embryo (e.g., frog, chick, and human).
- When presented with a sequence of human embryos, point out and discuss aspects of their increasing complexity.

Prelab Question: Which human organs develop first during development?

14.3 Extraembryonic Membranes, the Placenta, and the Umbilical Cord

- Distinguish between and give a function for the extraembryonic membranes, the placenta, and the umbilical cord.
- Trace the development of the extraembryonic membranes during embryonic development, and state a function for each membrane.

Prelab Question: Which extraembryonic membrane becomes the placenta?

14.4 Fetal Development

- Trace the main events of human fetal development.

Prelab Question: Account for why the respiratory system is not functional until the sixth month or later.

***Application for Daily Living:* Cord Around Baby's Neck**

Introduction

The sperm and egg contribute chromosomes to the offspring. The testes in males and the ovaries in females produce the gametes and also the hormones that maintain the organs and the characteristics we associate with a male and female. In humans, embryonic development begins when the sperm fertilizes the egg and a zygote is formed. Cell division produces a ball of cells that arrange themselves into three layers, called the **germ layers.** It is possible to associate the development of particular organs with a specific germ layer. Embryonic development comes to a close when all the basic organs have formed. Refinements and an increase in size occur during fetal development.

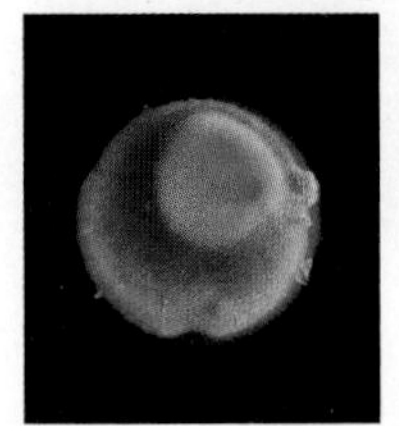
zygote

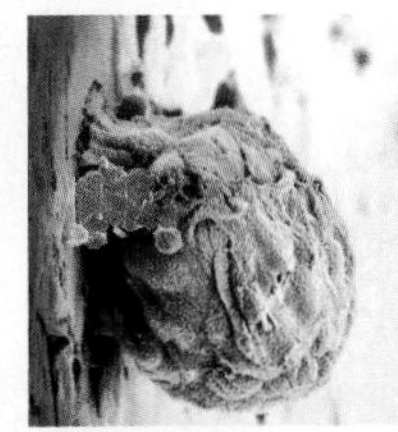
embryo at 1 week; implants in uterine wall

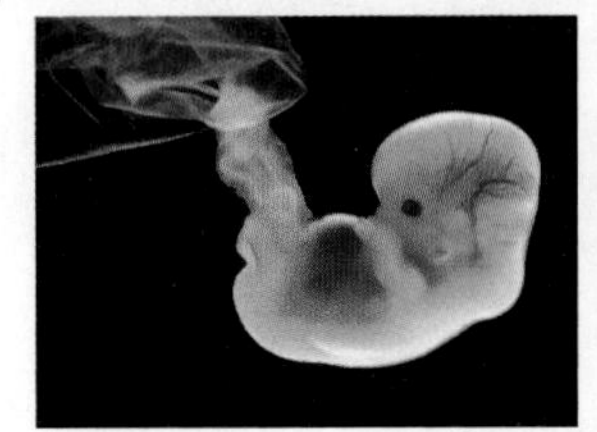
embryo at 8 weeks

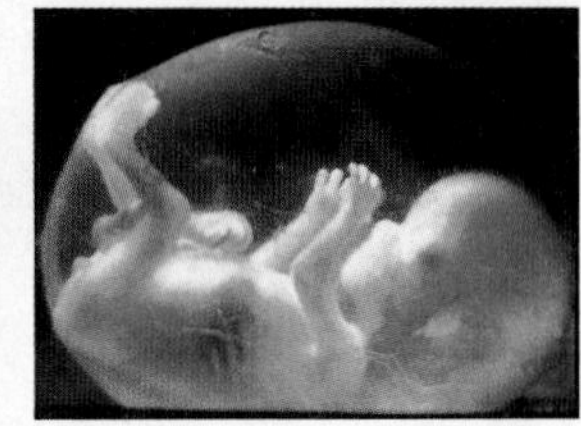
fetus at 3 months

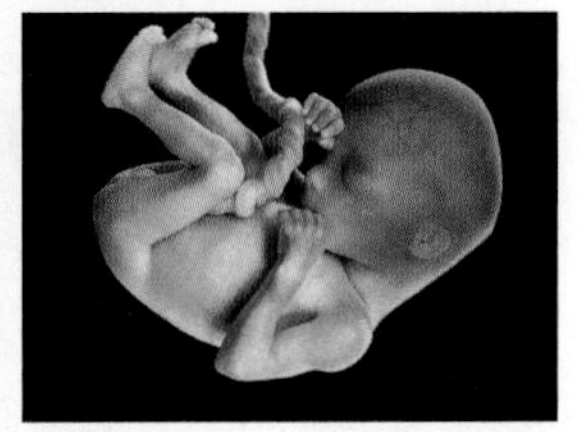
fetus at 5 months

(zygote) © Anatomical Travelogue/Science Source; (embryo, 1 week) © Bettman/Corbis; (embryo, 8 weeks) © Neil Harding/Getty Images; (fetus, 3 months) © Petit Format/Science Source; (fetus, 5 months) © John Watney/Science Source

14.1 Male and Female Reproductive Systems

Only the urinary and reproductive systems have a different anatomy and physiology in males and females.

Observation: Male Reproductive System Model

The functions of the male reproductive system are to (1) produce sperm and the male sex hormones within the testes, (2) transport the sperm in ducts until they exit through the penis, (3) nourish and provide a medium for sperm in which they can survive, and (4) deliver sperm to the vagina of the female. Identify the following organs in a model (Fig. 14.1):

1. **Testes.** The primary sex organs of males where the sperm and male sex hormones (e.g., testosterone) are produced. The testes fulfill the first function noted previously for the male reproductive system.
2. **Epididymis.** A tightly coiled, threadlike tube where sperm are stored until they are mature and capable of fertilizing an egg.
3. **Vas deferens.** A muscular tube that conducts sperm to the ejaculatory duct.
4. **Ejaculatory duct.** When the vas deferens and a duct from the seminal vesicle meet, they form a short ejaculatory duct, which passes through the prostate gland and empties into the urethra.
5. **Urethra.** The male urethra is located in the penis. The urethra either conducts semen or it conducts urine to the exterior and does not pass them both at the same time. What function of the male reproductive system previously listed is fulfilled by the ducts (2–5) just described? ______________________________
6. **Seminal vesicles,** the **prostate gland,** and the **bulbourethral glands.** These glands do not produce hormones, but they do produce secretions that account for why semen (the substance that exits the penis during ejaculation) is a thick fluid. The secretions of these glands are favorable to the health of the sperm. What function noted previously for the male reproductive system is fulfilled by these glands? ______________________________

   The prostate gland frequently enlarges in older men and blocks the urethra. Under these conditions, a male is apt to have difficulty urinating. Explain. ______________________________

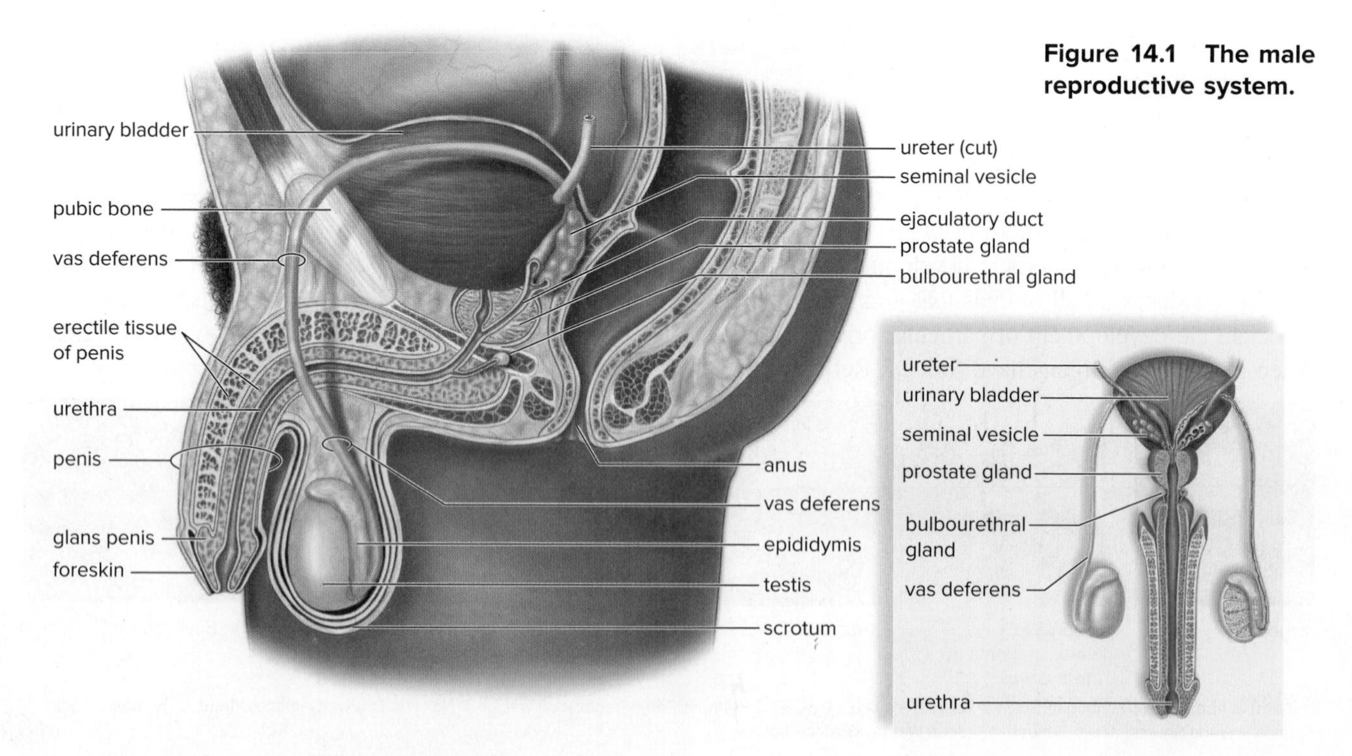

Figure 14.1 The male reproductive system.

Observation: Microscopic Examination of the Testis and Sperm

Testes

1. Obtain a slide of the testis. As shown in Figure 14.2*a,* a testis (sing.) contains many seminiferous tubules where spermatogenesis (production of sperm) occurs. Note under low power the circular nature of a tubule.
2. Switch to high power, and observe one tubule in particular. Between the tubules, try to identify interstitial cells, which produce testosterone. Testosterone enters the body by way of the blood and not by way of ducts. Explain why a vasectomy, a cutting of the vas deferens, causes a male to be sterile but has no effect on his masculinity. ____

____
3. **Spermatogenesis** occurs in the seminiferous tubules (Fig. 14.2*b*). Identify:

   Sertoli cells nuclei. The cytoplasm of a Sertoli cell surrounds a cell undergoing spermatogenesis.

   Germ cells. First spermatogonia, next spermatocytes, and finally spermatids occur, but they are hard to identify. Males produce four viable spermatids for each spermatogenesis. Each spermatid becomes a sperm.

   Sperm. Sperm tails look like thin, fine, dark lines in the lumen of the tubule. The sperm are fully developed but immature. They will complete their maturation in the epididymis. Why is it important for sperm cells to have a tail? ____

   ____

Sperm

Obtain a prepared slide of sperm and compare what you see to Figure 14.2*c*. An acrosome contains enzymes that digest a path through cells that surround an egg and the outer covering of an egg so that one sperm can enter. State another way sperm are specialized. ____

The ejaculated semen of a normal male contains several hundred million sperm, each a product of spermatogenesis, but only one sperm normally enters an egg. Speculate why so many sperm are needed for fertilization to occur. ____

____

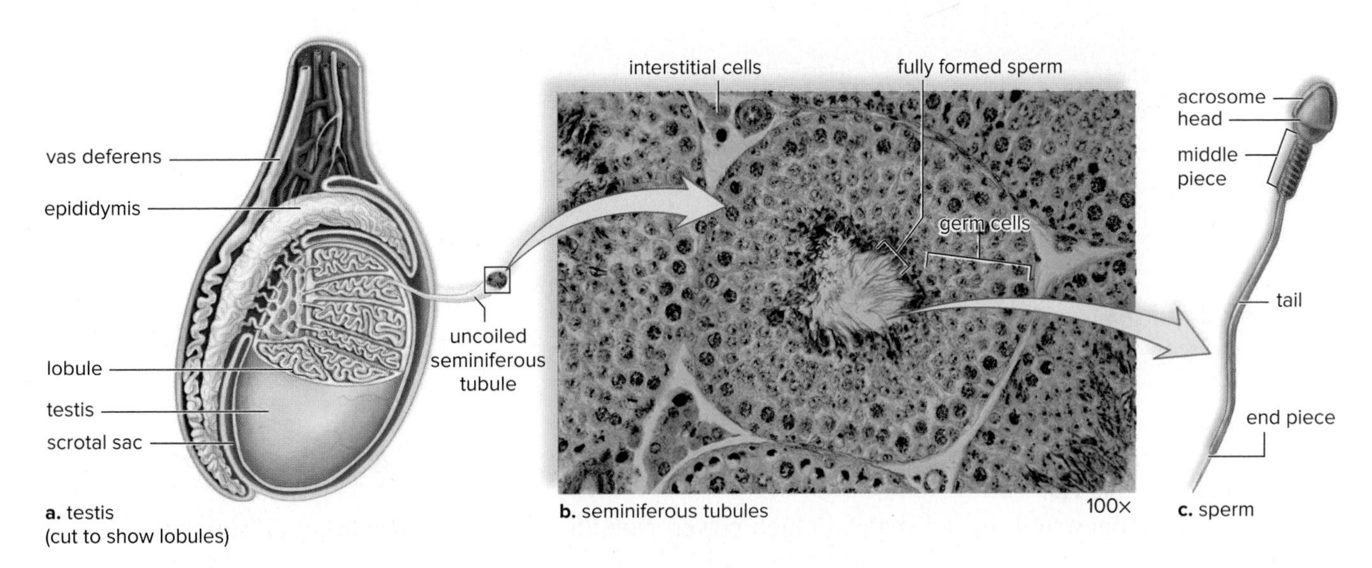

a. testis (cut to show lobules) **b.** seminiferous tubules **c.** sperm

Figure 14.2 Testis and sperm.

Observation: Female Reproductive System Model

The function of the female reproductive system (Fig. 14.3) is to (1) produce eggs and sex hormones within the ovaries; (2) transport eggs from the ovaries to the uterus; (3) receive the sperm in the vagina, also the birth canal; and (4) protect the fertilized egg (zygote) until it matures and is born. Identify the following organs in a model (see Fig. 14.3):

1. **Ovaries.** The primary sex organs of females where eggs and female sex hormones (e.g., estrogen and progesterone) are produced.
2. **Oviducts.** At the end nearest the ovaries, the oviducts have fimbriae that sweep over an ovary so that a released egg enters the oviduct. Cilia that line the oviduct and muscular contraction of its walls propel the egg toward the uterus. Fertilization, if it occurs, happens in an oviduct.
3. **Uterus.** A thick-walled, muscular organ about the size and shape of an inverted pear, where development of the embryo and fetus occur. The cervix surrounds the opening to the uterus.
4. **Vagina.** The uterus empties into the vagina, which makes a 45° angle with the small of the back.

Figure 14.3 The female reproductive system.

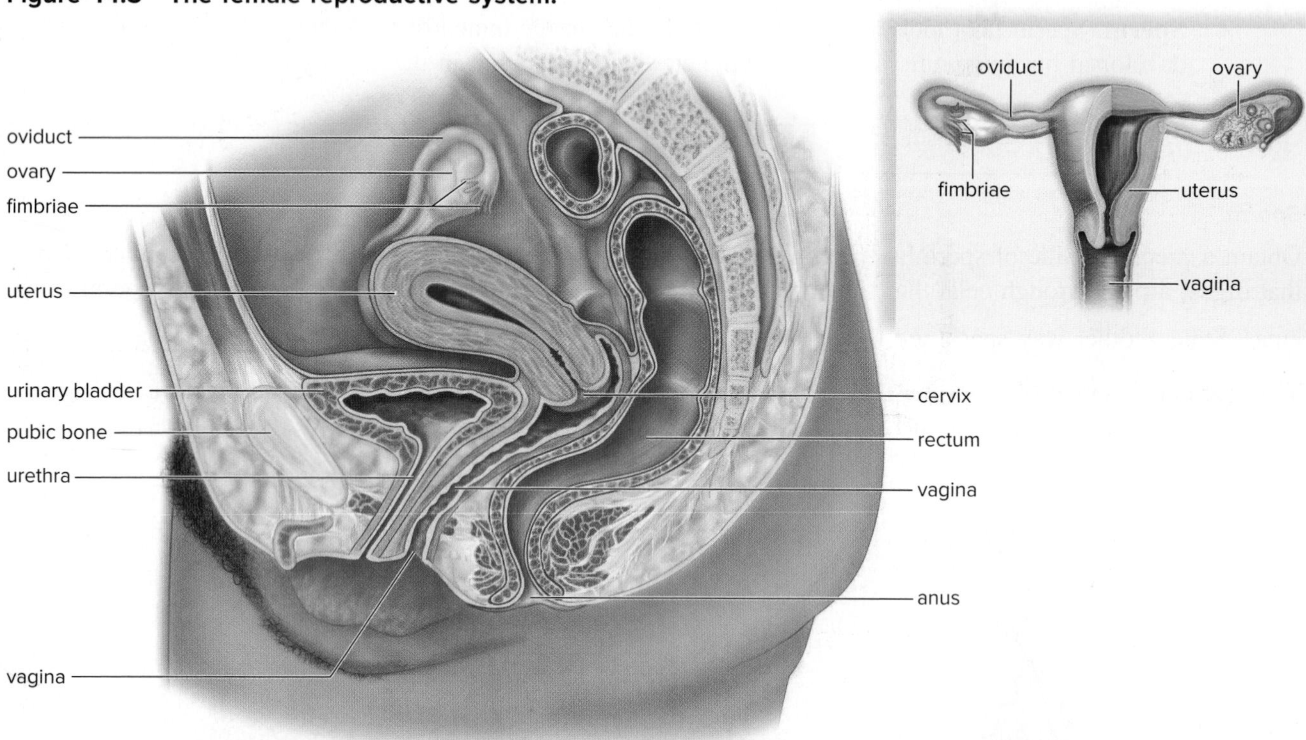

Observation: Microscopic Examination of the Ovary

Ovary

1. Examine a prepared slide of an ovary, and refer to Figure 14.4 for help in identifying the structures.
2. Locate a **primary follicle,** which appears as a circle of small cells surrounding a somewhat larger cell, the **primary oocyte.**
3. Find a **secondary follicle,** and switch to high power. Note the **secondary oocyte** (egg), surrounded by numerous cells, to one side of the liquid-filled follicle. Why is it important for females to produce an egg that has plentiful cytoplasm? ____

4. Look for a large, fluid-filled **vesicular** (Graafian) **follicle,** which contains a mature secondary oocyte to one side. This follicle will be next to the outer surface of the ovary because it is the type of follicle that releases the egg during ovulation.
5. Look for the remains of the **corpus luteum,** which will look like scar tissue. The corpus luteum develops after the vesicular follicle has released its egg, and then later it deteriorates. Not all slides will contain a corpus luteum and a vesicular follicle because they may not have been present when the slide was made.
6. **Oogenesis**
   The primary follicle and the secondary follicle are the site of oogenesis (production of an egg). Each oogenesis event in females produces a single egg. In contrast to males, females produce one egg a month.

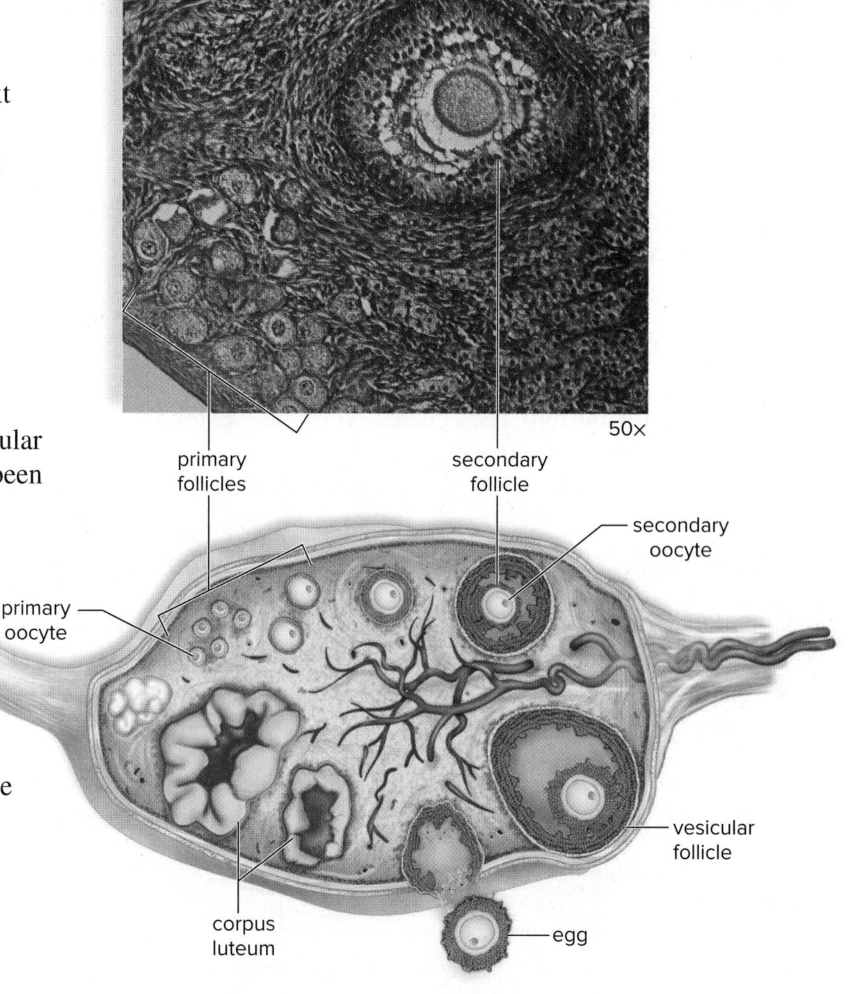

Figure 14.4 Ovary and oocytes.

Summary of Reproductive Systems

- Complete Table 14.1 to describe the differences between the male and female human reproductive systems. A **gonad** is an organ that produces gametes, the sperm and egg.

Table 14.1 Comparison of Human Male and Female Reproductive Systems

	Male	Female
Gonad		
Duct from gonad		
Duct from gonad leads to		
Copulatory organ		

14.2 Embryonic Development

We will divide embryonic development into three stages: cellular, tissue layer, and organ development. In human beings, it takes 2 months to complete embryonic development. It is impossible for us to view the stages of embryonic development in a human being, so we will use the sea star, frog, and chick as our observational material.

Cellular Stages of Development

The cellular stages of development include the following:

- **Zygote formation:** A single sperm fertilizes an egg and the result is a zygote, the first cell of the new individual.
- **Morula formation:** The zygote divides into a number of smaller cells until there is a cluster of 16–32 cells called a morula.
- **Blastula formation:** The morula becomes a blastula, a hollow ball of cells.

Observation: Cellular Stages of Development in the Sea Star

The cellular stages of development are remarkably similar in all animals. Therefore, we can view slides of sea star development to study the cellular stages of human development (Fig. 14.5). A sea star is an invertebrate that develops in the ocean and, therefore, will develop easily in the laboratory where it can be observed.

Obtain slides or view a model of sea star development and note the following:

1. **Zygote.** Both plants and animals begin life as a single cell, a zygote. A zygote contains chromosomes from each parent. Explain. ______________________________

   ____
2. **Cleavage.** View slides showing various numbers of cells due to the process of cleavage, cell division without growth until the morula stage. Is the morula about the same size as the zygote? ______________

   Explain. ______________________________

   ____
3. **Blastula.** The cavity of a blastula is called the blastocoel. *Label blastocoel in Figure 14.5.* The formation of a hollow cavity is important to the next stage of development.

Figure 14.5 Starfish development.
All animals, including starfish and humans, go through the same cellular stages from cleavage to blastula.

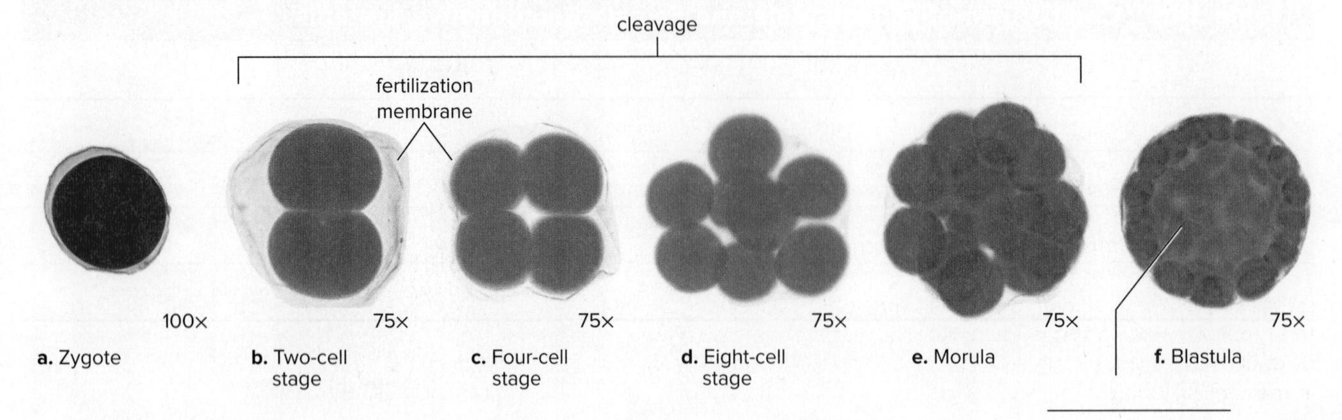

Observation: Cellular Stages of Development in Humans

In Figure 14.6, or in a model of human development, observe the same stages of development already observed in sea star slides. Also, observe that fertilization in humans occurs in an oviduct following ovulation. As the embryo undergoes cleavage, it travels in the oviduct to the uterus.

If the embryo splits at the 2-cell stage, the result is identical twins. (Fraternal twins arise when two separate eggs are fertilized.) How might you account for the development of identical triplets?

____

____

The blastula in humans is called a blastocyst. The blastocyst contains an **inner cell mass** that becomes the embryo, and the outer group of cells (the trophoblast) will become membranes that nourish and protect it. At about day 6, the blastocyst has reached the uterus and implants into the uterine wall, where it will receive nourishment from the mother's bloodstream.

What's the main difference between the cellular stages in a sea star and in a human?

____

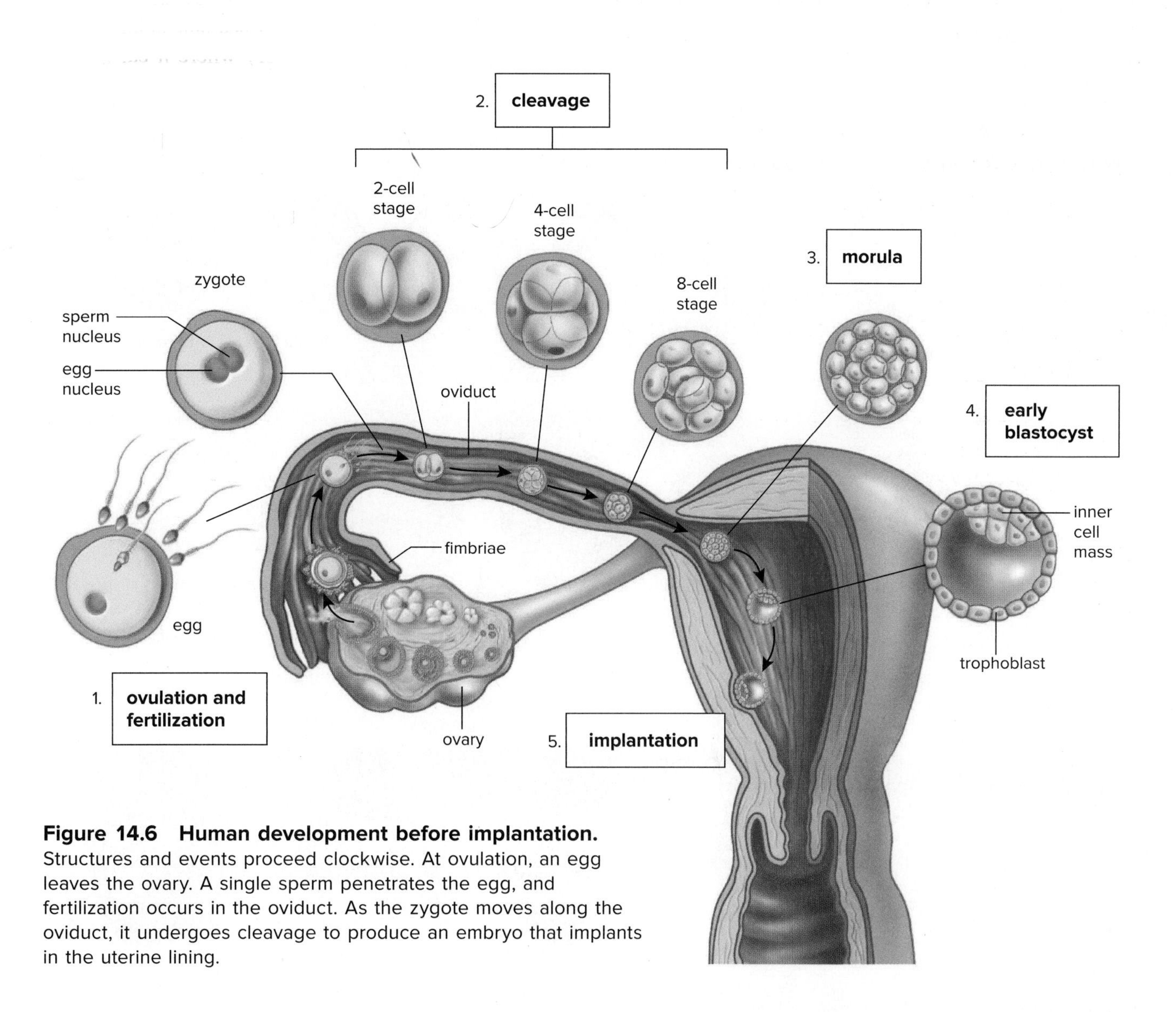

Figure 14.6 Human development before implantation. Structures and events proceed clockwise. At ovulation, an egg leaves the ovary. A single sperm penetrates the egg, and fertilization occurs in the oviduct. As the zygote moves along the oviduct, it undergoes cleavage to produce an embryo that implants in the uterine lining.

Tissue Stages of Development

The tissue stages of development include the following:

- **Early gastrula stage.** This stage begins when certain cells begin to push or invaginate into the blastocoel, creating a double layer of cells. The outer layer is called the ectoderm, and the inner layer is called the endoderm.
- **Late gastrula stage.** Gastrulation is not complete until there are three layers of cells. The third layer called mesoderm occurs between the other two layers already mentioned.

Observation: Tissue Stages of Development in a Frog

It is traditional to view frog gastrulation. A frog is a vertebrate, and so its development is expected to be closer to that of a human than is a sea star. In Figure 14.7, note that the yellow (vegetal pole) cells are heavily laden with yolk, and the blue (animal pole) cells are the ones that invaginate into the blastocoel forming the early gastrula.

1. **Early gastrula stage.** Obtain a cross section of a frog gastrula. Most likely, your slide is the equivalent of Figure 14.7*b*, number 3, in which case you will see two cavities, the old blastocoel and newly forming *archenteron,* which forms once the animal pole cells have invaginated. The archenteron will become the digestive tract.
2. **Late gastrula stage.** In Figure 14.7, note that a third layer of cells, the mesoderm, is colored red and that it develops between the ectoderm and endoderm.

Figure 14.7 Drawings of frog developmental stages.
a. During cleavage, the number of cells increases but overall size remains the same. **b.** During gastrulation, three tissue layers form. **c.** During neurulation, the notochord and neural tube form.

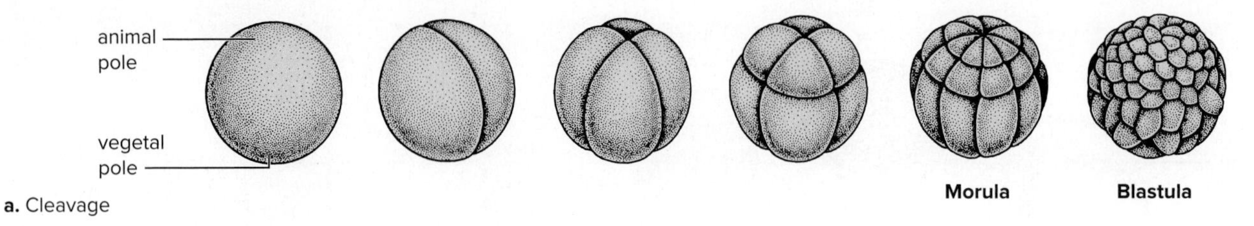

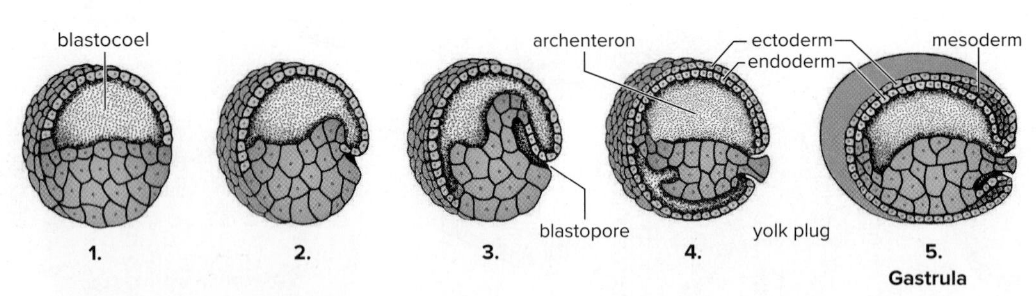

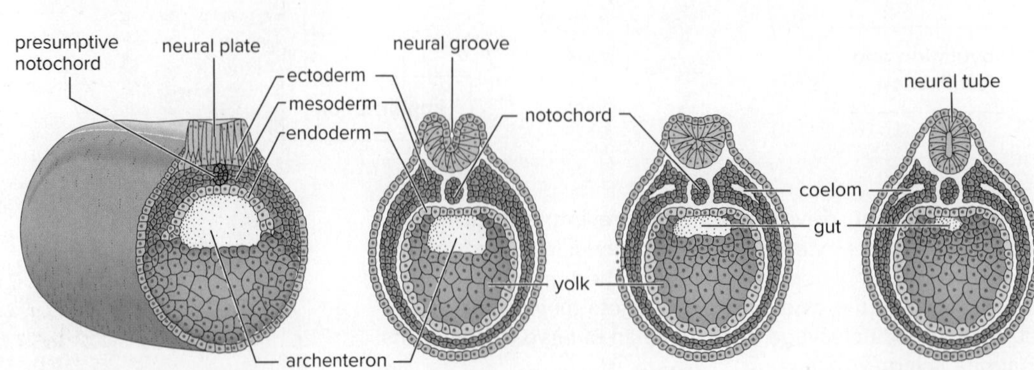

Observation: Tissue Stages of Development in a Human

In a model of human development, observe the same stages of development already observed in frog slides. After implantation, gastrulation in humans turns the inner cell mass into the **embryonic disk.** Figure 14.8 shows the embryonic disk, which has the three layers of cells we have been discussing: the ectoderm, mesoderm, and endoderm. Figure 14.8 also shows the significance of these layers, often called the **germ layers.** The future organs of an individual can be traced back to one of the germ layers.

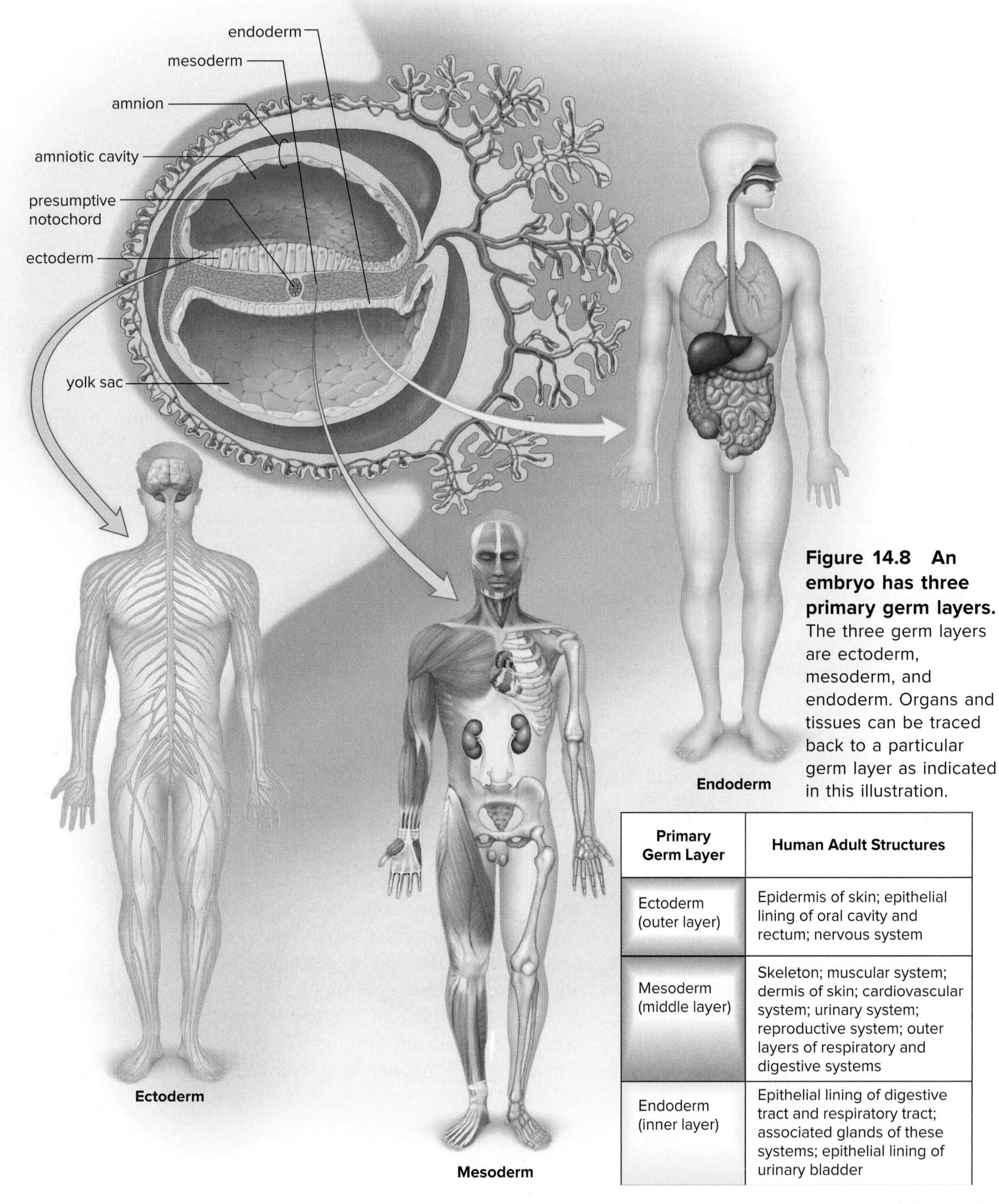

Figure 14.8 An embryo has three primary germ layers. The three germ layers are ectoderm, mesoderm, and endoderm. Organs and tissues can be traced back to a particular germ layer as indicated in this illustration.

Primary Germ Layer	Human Adult Structures
Ectoderm (outer layer)	Epidermis of skin; epithelial lining of oral cavity and rectum; nervous system
Mesoderm (middle layer)	Skeleton; muscular system; dermis of skin; cardiovascular system; urinary system; reproductive system; outer layers of respiratory and digestive systems
Endoderm (inner layer)	Epithelial lining of digestive tract and respiratory tract; associated glands of these systems; epithelial lining of urinary bladder

Organ Stages of Development

As soon as all three embryonic tissue layers (ectoderm, endoderm and mesoderm) are established, the organ level of development begins. It continues until development is complete. The first organs to develop are the

- Digestive tract. You have already observed the start of the archenteron during gastrulation.
- Spinal cord and brain
- Heart

Observation: Development of the Spinal Cord and Brain

One of the first systems to form is the nervous system. Why might it be beneficial for the nervous system to begin development first? ____

____

1. Obtain a cross section of a frog neurula stage, and match it to one of the drawings in Figure 14.7*c*.

   Which drawing seems to best match your slide? ______________________________

   Your instructor will confirm your match for you.

2. A neural tube develops from ectoderm (Fig. 14.7*c*). Can you see how? ______________________

   When neural folds rise up and fuse, the neural tube has formed. The neural tube, which runs the length of the embryo, is the first sign of the central nervous system. The nerve cord, also called the spinal cord, and the brain both develop from the neural tube.

Notice how the neural tube develops above the notochord, a dorsal supporting rod that later becomes the vertebral column. Why would you expect the neural tube, which becomes the spinal cord, to develop in the same vicinity as the notochord, which becomes the vertebral column? ______________________

____

Observation: Development of the Heart

A chick embryo offers an opportunity to view a beating heart in an embryo. Your instructor may show you various stages. In particular you will want to observe the 48-hour chick embryo.

Observing Live Chick Embryos

Use the following procedure for selecting and opening the eggs of live chick embryos:

1. Choose an egg of the proper age to remove from the incubator, and put a penciled × on the uppermost side. The embryo is just below the shell.
2. Add warmed chicken Ringer solution to a finger bowl until the bowl is about half full. (Chicken Ringer solution is an isotonic salt solution for chick tissue that maintains the living state.) The chicken Ringer solution should not cover the yolk of the egg.
3. On the edge of the dish, gently crack the egg on the side opposite the ×.
4. With your thumbs placed over the ×, hold the egg in the chicken Ringer solution while you pry it open from below and allow its contents to enter the solution. If you open the egg too slowly or too quickly, the shell may damage the delicate membranes surrounding the embryo.

Observation: Forty-Eight-Hour Chick Embryo

1. Follow the standard procedure for selecting and opening an egg containing a 48-hour chick embryo.
2. The embryo has turned so that the head region is lying on its side. Refer to Figure 14.9, and identify the following:
   a. **Shape of the embryo,** which has started to bend. The head is now almost touching the heart.
   b. **Heart,** contracting and circulating blood. Can you make out a ventricle, an atrium, and the aortic arches in the region below the head? Later, only one aortic arch will remain.
   c. **Vitelline arteries** and **veins,** which extend over the yolk. The vitelline veins carry nutrients from the yolk sac to the embryo.
   d. **Brain** with several distinct regions.
   e. **Eye,** which has a developing lens.
   f. **Margin (edge) of the amnion,** which can be seen above the vitelline arteries (see next section for amnion).
   g. **Somites,** blocks of developing muscle tissue that differentiate from mesoderm, which now number 24 pairs.
   h. **Caudal fold** of the amnion. The embryo will be completely enveloped when the head fold and caudal fold meet the margin of the amnion.
   i. **Neural tube,** which runs the length of the embryo, is the first sign of the central nervous system.

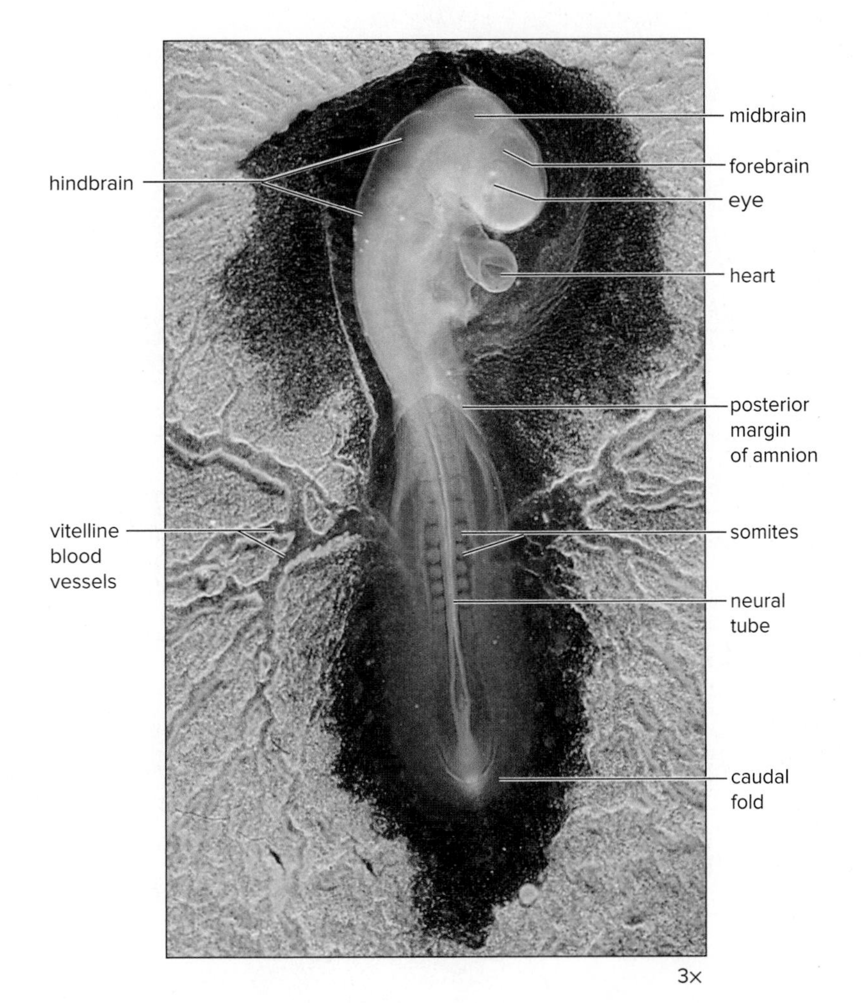

Figure 14.9 Forty-eight-hour chick embryo.
The most prominent organs are labeled.

Observation: Development of Human Organs

Study models or other study aids available that show the development of the nervous system and the heart in human beings and/or show models of human embryos of different ages. Also view Figure 14.10, which depicts the external appearance of the embryo from the fourth to the seventh week of development.

During the embryonic period of development, the growing baby is susceptible to environmental influences, including the following:

- Drugs, such as alcohol; certain prescriptions; and recreational drugs. These can cause birth defects.
- Infections such as rubella, also called German measles, and other viral infections.
- Nutritional deficiencies.
- X-rays or radiation therapy.

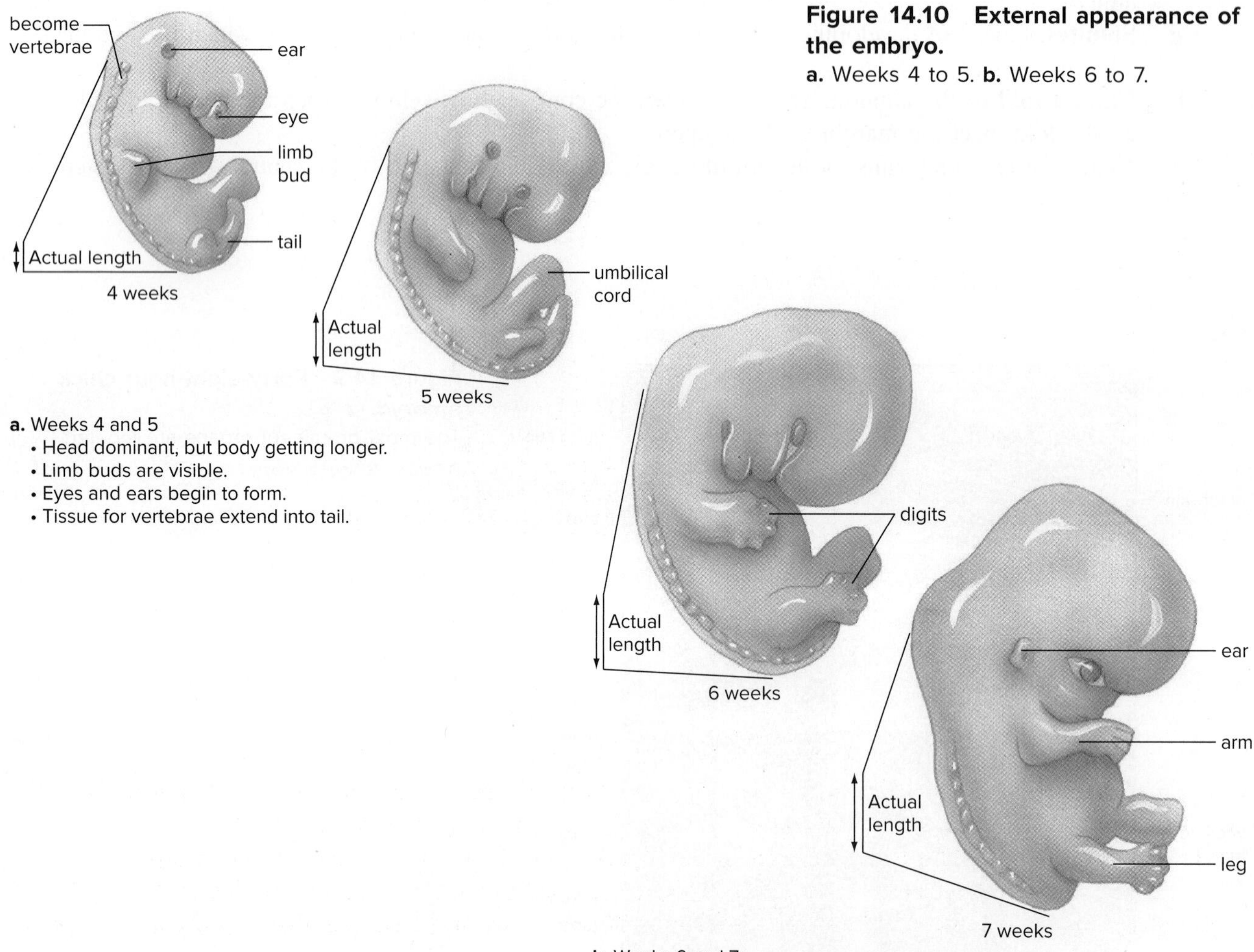

Figure 14.10 External appearance of the embryo.
a. Weeks 4 to 5. **b.** Weeks 6 to 7.

14.3 Extraembryonic Membranes, the Placenta, and the Umbilical Cord

- The **extraembryonic membranes** take their name from the observation that they are not part of the embryo proper. They are outside the embryo, and therefore they are "extra."
- The **placenta** is the structure that provides the embryo with nutrient molecules and oxygen and takes away its waste molecules, such as carbon dioxide. The fetal half of the placenta is the chorionic villi, which contain fetal capillaries. The maternal half of the placenta is capillaries in the uterine wall.
- The **umbilical cord** is a tubular structure that contains two of the extraembryonic membranes (the allantois and the yolk sac) and also the umbilical blood vessels. The umbilical blood vessels bring fetal blood to and from the placenta. When a baby is born and begins to breathe on its own, the umbilical cord is cut and the remnants become the navel.

In this drawing, *label the umbilical cord,* which contains the umbilical blood vessels. Also *label the placenta,* which contains the maternal blood vessels.

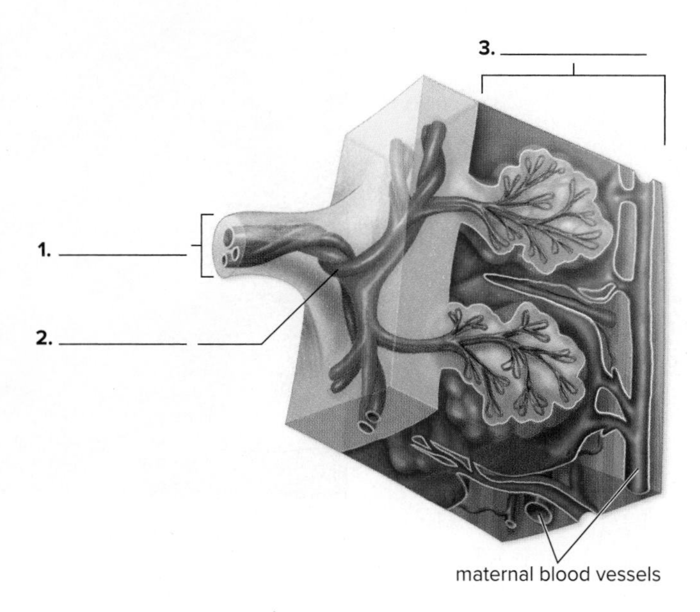

Observation: The Extraembryonic Membranes

In a model, and in Figure 14.11, trace the development of the extraembryonic membranes. Also, note the development of the placenta and the umbilical cord. The extraembryonic membranes are the

- **Chorion.** The chorion begins to form at the blastocyst stage of development. The outer layer of cells surrounding the inner cell mass of the blastocyst becomes the chorion. Notice in Figures 14.8 and 14.11, the treelike **chorionic villi** are a part of the chorion that will become the placenta.
- **Amnion.** Forms the amniotic cavity, which envelops the fetus and contains the amniotic fluid that cushions and protects the fetus (Fig. 14.12). All animals, whether the sea star, the frog, the chick, or a human, develop in an aqueous environment. Birth of a human is imminent when "the water breaks," the loss of the amniotic fluid.
- **Allantois.** The allantois extends into the umbilical cord. It accumulates the small amount of urine produced by the fetal kidneys and later gives rise to the urinary bladder. Its blood vessels become the umbilical blood vessels.
- **Yolk sac.** The yolk sac is the first embryonic membrane to appear. In the chick, the yolk sac does contain yolk, food for the developing embryo. In humans, the yolk sac contains plentiful blood vessels and is the first site of blood cell formation.

Figure 14.11 Development of extraembryonic membranes.
a. At first, no organs are present in the embryo, only tissues. The amniotic cavity is above the embryonic disk, and the yolk sac is below. The chorionic villi are present. **b, c.** The allantois and yolk sac, two more extraembryonic membranes, are positioned inside the body stalk as it becomes the umbilical cord. **d.** At 35+ days, all membranes are present, and the umbilical cord takes blood vessels between the embryo and the chorion (placenta).

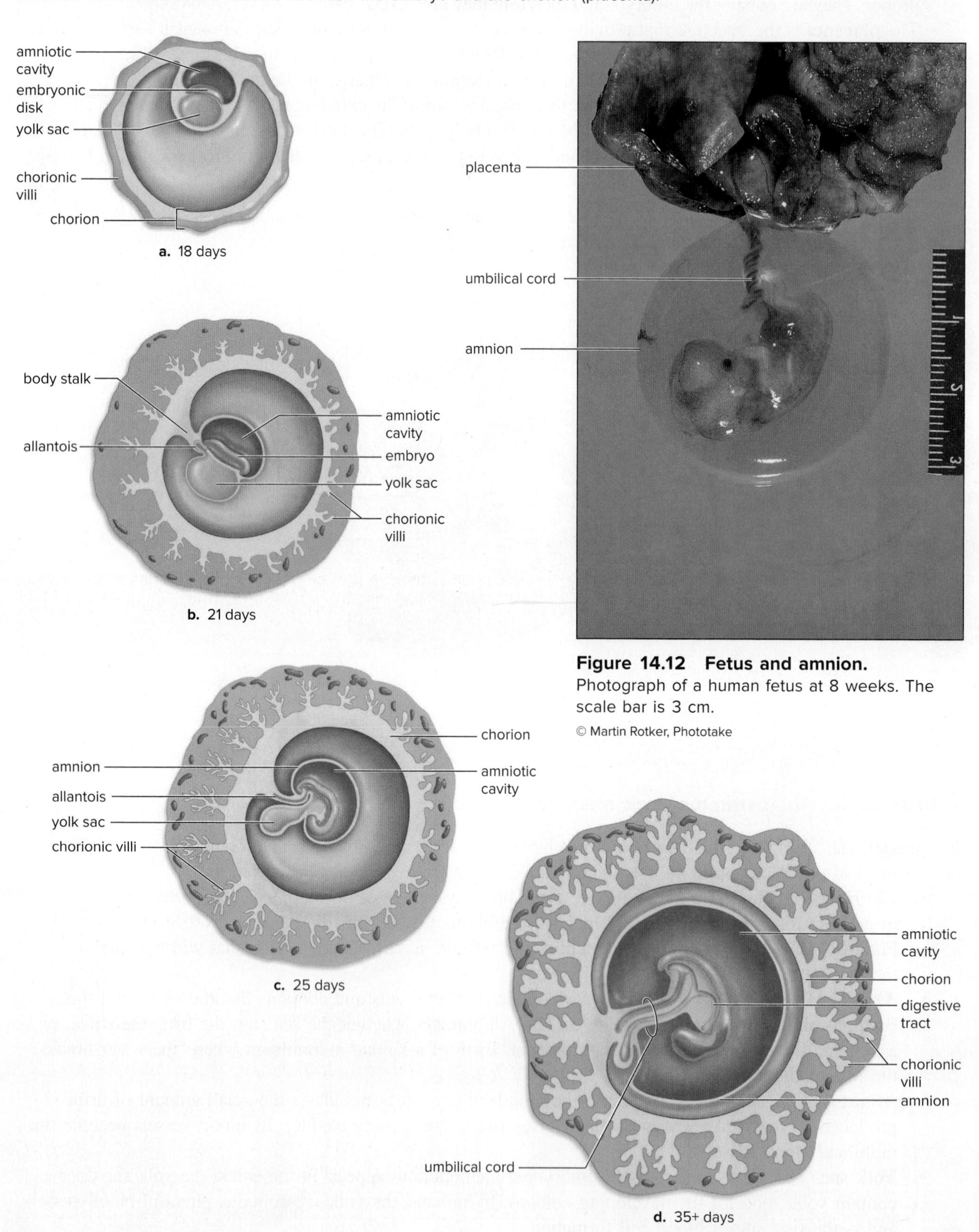

Figure 14.12 Fetus and amnion.
Photograph of a human fetus at 8 weeks. The scale bar is 3 cm.

14.4 Fetal Development

During fetal development (last seven months), the skeleton becomes ossified (bony), reproductive organs form, arms and legs develop fully, and the fetus enlarges in size and gains weight.

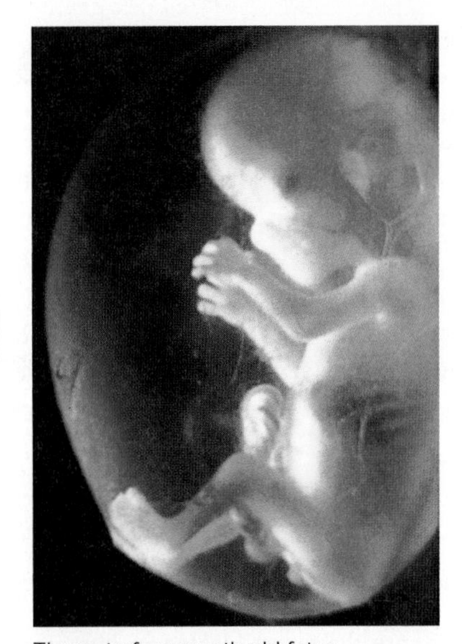

Three- to four-month-old fetus

Seven- to eight-month-old fetus

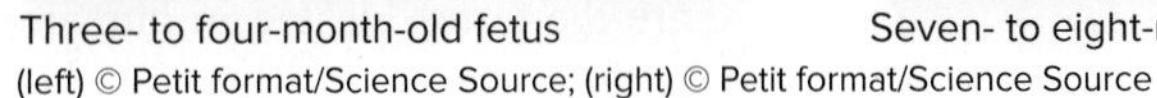

(left) © Petit format/Science Source; (right) © Petit format/Science Source

Observation: Fetal Development

1. Using Table 14.2 as a guide, examine models of fetal development.
2. In Table 14.2, note the following.
   a. **External genitals:** About the third month, it is possible to tell male from female if an ultrasound is done.
   b. **Quickening:** Fetal movement is felt during the fourth or fifth months.
   c. **Vernix caseosa:** Beginning with the fifth month, the skin is covered with a cheesy coating called vernix caseosa.
   d. **Lanugo:** During the sixth and seventh months, the body is covered with fine, downy hair termed lanugo.
3. All organ systems become fully functional during fetal development.

Table 14.2 Fetal Development

Month	Events for Mother	Events for Baby
Third month	Uterus is the size of a grapefruit.	Possible to distinguish sex. Fingernails appear.
Fourth month	Fetal movement is felt by those who have been previously pregnant. Heartbeat is heard by stethoscope.	Bony skeleton visible. Hair begins to appear. 150 mm (6 in.), 170 g (6 oz.).
Fifth month	Fetal movement is felt by those who have not been previously pregnant. Uterus reaches up to level of umbilicus and pregnancy is obvious.	Protective cheesy coating, called vernix caseosa, begins to be deposited. Heartbeat can be heard.
Sixth month	Doctor can tell where baby's head, back, and limbs are. Breasts have enlarged, nipples and areolae are darkly pigmented, and colostrum is produced.	Body is covered with fine hair called lanugo. Skin is wrinkled and reddish.
Seventh month	Uterus reaches halfway between umbilicus and rib cage.	Testes descend into scrotum. Eyes are open. 300 mm (12 in.), 1,350 g (3 lb).
Eighth month	Weight gain is averaging about a pound a week. Difficulty in standing and walking because center of gravity is thrown forward.	Body hair begins to disappear. Subcutaneous fat begins to be deposited.
Ninth month	Uterus is up to rib cage, causing shortness of breath and heartburn. Sleeping becomes difficult.	Ready for birth. 530 mm (20½ in.), 3,400 g (7½ lb).

Application for Daily Living

Cord Around Baby's Neck

Babies are sometimes born with the umbilical cord wrapped around the neck. There isn't much room in the uterus, and this is a common place for the cord to be located. The umbilical cord around the neck is not as dangerous as it sounds because the baby gets its oxygen by way of the placenta and the umbilical blood vessels and not the lungs. If the umbilical cord becomes overly stretched, oxygen may not be delivered as it should. This causes the fetal heart to slow down. First, the mother can change her position and/or be given oxygen, but if the baby's heart slows to below 100 beats for any length of time, then a cesarean delivery may be in order.

Laboratory Review 14

_______________ **1.** What do male sperm and female eggs contribute to offspring?

_______________ **2.** Where are the testes located in human males?

_______________ **3.** What is the function of the vas deferens?

_______________ **4.** What is the function of the prostate gland?

_______________ **5.** Where are the ovaries located?

_______________ **6.** What is the function of the uterus?

_______________ **7.** Where are sperm produced in the testes?

_______________ **8.** What structure in the ovary contains the developing oocyte (egg)?

_______________ **9.** Name the stage at which the embryo is a ball of cells.

_______________ **10.** Which extraembryonic membrane participates in the formation of the placenta?

_______________ **11.** What process directly follows the zygote stage?

_______________ **12.** What overall change in the skeleton occurs during fetal development?

_______________ **13.** Do the lungs function during fetal development?

_______________ **14.** What is the function of the placenta?

Thought Questions

15. A vasectomy is a procedure in which both of the vas deferens are severed. Why would such a procedure cause sterility?

16. Why is early development similar among all animals?

17. Explain why shortness of breath, frequent urination, and heartburn are common symptoms during the last month of pregnancy.

LABORATORY

15

Mitosis and Meiosis

Learning Outcomes

15.1 The Cell Cycle

- Name and describe the stages of the cell cycle.
- Identify the phases of mitosis in models and microscope slides. Explain how the chromosome number stays constant.
- Describe cytokinesis.

Prelab Question: During what stage of the cell cycle does chromosome duplication occur, and why is this critical to mitosis?

15.2 Meiosis

- Name and describe the phases of meiosis I and meiosis II with attention to the movement of chromosomes.
- Explain how the chromosome number is reduced.

Prelab Question: What is synapsis, and why is it critical to meiosis?

15.3 Mitosis Versus Meiosis

- Compare the results of mitosis to meiosis.
- Contrast the behavior of chromosomes during mitosis with the behavior of chromosomes during meiosis I and II.

Prelab Question: How are mitosis and meiosis II similar? How are they different?

15.4 Gametogenesis

- Contrast spermatogenesis with oogenesis using diagrams and models.

Prelab Question: What differences between spermatogenesis and oogenesis help explain why males produce so many more sperm than females produce eggs?

Application for Daily Living: **Mitosis and Cancer**

Introduction

Dividing cells experience nuclear division, cytoplasmic division, and a period between divisions called interphase. During **interphase,** the nucleus appears normal, and the cell is performing its usual cellular functions. Also, the cell is increasing all of its components, including such organelles as the mitochondria, ribosomes, and centrioles. DNA replication (making an exact copy of the DNA) occurs toward the end of interphase. Thereafter, the chromosomes, which contain DNA, are duplicated and contain two chromatids held together at a **centromere.** These chromatids are called **sister chromatids.**

When the nucleus divides during **mitosis,** the daughter nuclei receive the same number of chromosomes and genetic material as the parent cell. During cytokinesis, the cytoplasm divides and two daughter cells are produced. Mitosis in humans permits growth and repair of tissues. During sexual reproduction, another form of division called **meiosis** occurs. Meiosis is a part of **gametogenesis,** the production of gametes (sex cells called sperm in males and eggs in females). As a result of meiosis, the daughter cells have half the number of chromosomes as the parent cell. Also, the chromatids sometimes exchange genetic material during crossing-over; therefore, the daughter cells do not have the same number of chromosomes and are not genetically identical to the parent cell, following meiosis.

15.1 The Cell Cycle

As stated in the introduction, the period between cell divisions is known as interphase. Early investigators noted little visible activity between cell divisions, so they dismissed this period as a resting state. But when they discovered that DNA replication occurs (and the chromosomes become duplicated) during interphase, the **cell cycle** concept was proposed. Investigators have also discovered that cytoplasmic organelle duplication occurs during G_1 and synthesis of the proteins involved in regulating cell division occurs during G_2. Thus, the cell cycle can be broken down into four stages (Fig. 15.1). State the event of each stage on the line provided:

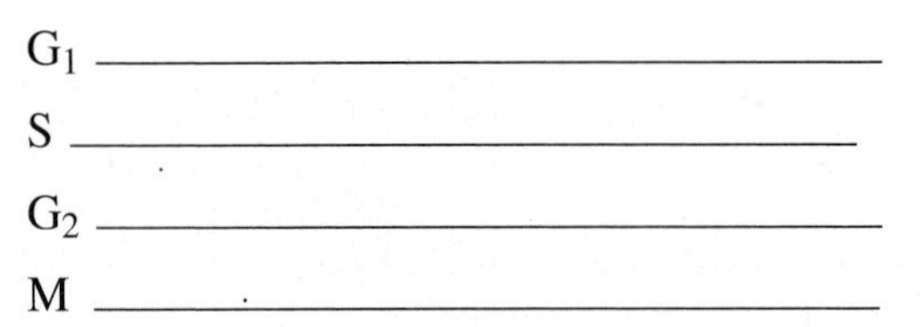

G_1 ______________________________

S ______________________________

G_2 ______________________________

M ______________________________

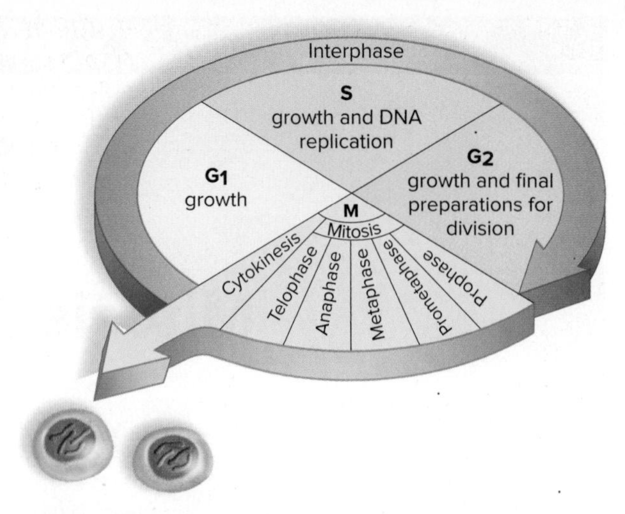

Figure 15.1 The cell cycle.
Immature cells repeatedly go through a cycle that consists of four stages: G1, S (for synthesis), G2, and M (for mitosis). Eventually, some daughter cells "break out" of the cell cycle and become specialized cells.

Why is the entire process called the "cell cycle"? ______________________________

____

Duration of the Cell Cycle

If mature human cells are speedily dividing, the length of the cell cycle is about 22–24 hours. Embryonic cells go through the cell cycle much more rapidly by omitting G_1 and G_2, so that the cell cycle includes only DNA replication and mitosis. Mature cells usually take this amount of time per phase: G_1 = 10 hours; S = 5–6 hours; G_2 = 3–4 hours; mitosis and cytokinesis = 2 hours.

Mitosis

Mitosis is nuclear division that results in two new nuclei, each having the same number of chromosomes as the original nucleus. The **parent cell** is the cell that divides, and the resulting cells are called **daughter cells.** If a parent cell has 46 chromosomes, how many chromosomes does each daughter cell have following mitosis?

____

When cell division is about to begin, chromatin starts to condense and compact to form visible, rodlike **sister chromatids** held together at the centromere (Fig. 15.2*a*). *Label the sister chromatids and the centromere in Figure 15.2*b. This illustration represents a **duplicated chromosome** as it would appear just before nuclear division occurs.

Spindle

Table 15.1 lists the structures that play a role during mitosis. The **spindle** is a structure that appears and brings about an orderly distribution of daughter chromosomes to the daughter cell nuclei. A spindle has fibers that stretch between two poles (ends) (Fig. 15.3). **Spindle fibers** are bundles of microtubules, protein cylinders found in the cytoplasm that can assemble and disassemble. The **centrosome,** which is the main microtubule-organizing center of the cell, divides before mitosis so that each pole of the spindle has a pair of centrosomes. Human cells contain two barrel-shaped organelles called **centrioles** in each centrosome and **asters,** arrays of short microtubules radiating from the poles. Thus, centrioles mark the location of the spindle poles.

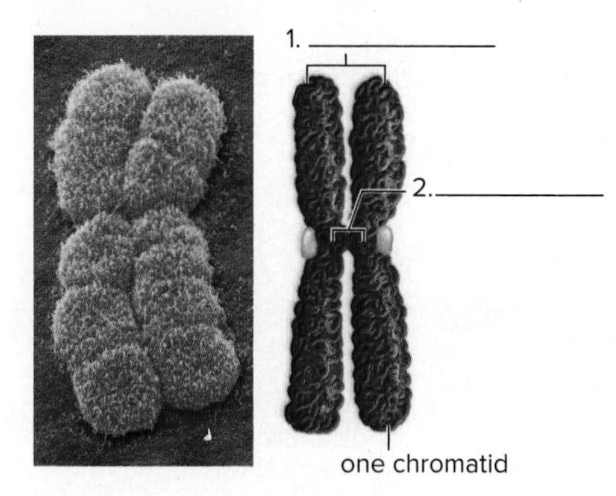

Figure 15.2 Duplicated chromosomes.
DNA replication results in a duplicated chromosome that consists of two sister chromatids held together at a centromere. **a.** Scanning electron micrograph of a duplicated chromosome. **b.** Drawing of a duplicated chromosome. During mitosis, each chromatid becomes a daughter chromosome.

Table 15.1 Structures Associated with Mitosis

Structure	Description
Nucleus	A large organelle containing the chromosomes and acting as a control center for the cells
Chromosome	Rod-shaped body in the nucleus seen during mitosis and meiosis that contains DNA and therefore the hereditary units, or genes
Nucleolus	An organelle found inside the nucleus; composed largely of RNA for ribosome formation
Spindle	Microtubule structure that brings about chromosome movement during nuclear division
Chromatids	The two identical parts of a chromosome following DNA replication; they become daughter chromosomes
Centromere	A constriction where duplicates (sister chromatids) of a chromosome are held together
Centrosome	The central microtubule-organizing center of cells; consists of granular material; in human cells, contains two centrioles
Centriole	A short, cylindrical organelle in human cells that contains microtubules. They mark the location of the spindle poles.
Aster	Short, radiating fibers produced by the centrioles; important during mitosis and meiosis

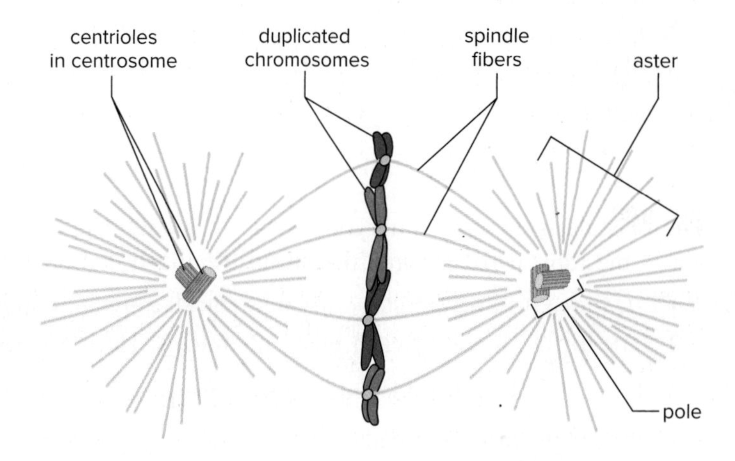

Figure 15.3 The spindle.
Spindle fibers (bundles of microtubules) help distribute the daughter chromosomes during mitosis.

Figure 15.4 Phases of mitosis in human cells.
This illustration extends across two pages. The colors signify that the chromosomes were inherited from different parents.
(all) © Ed Reschke

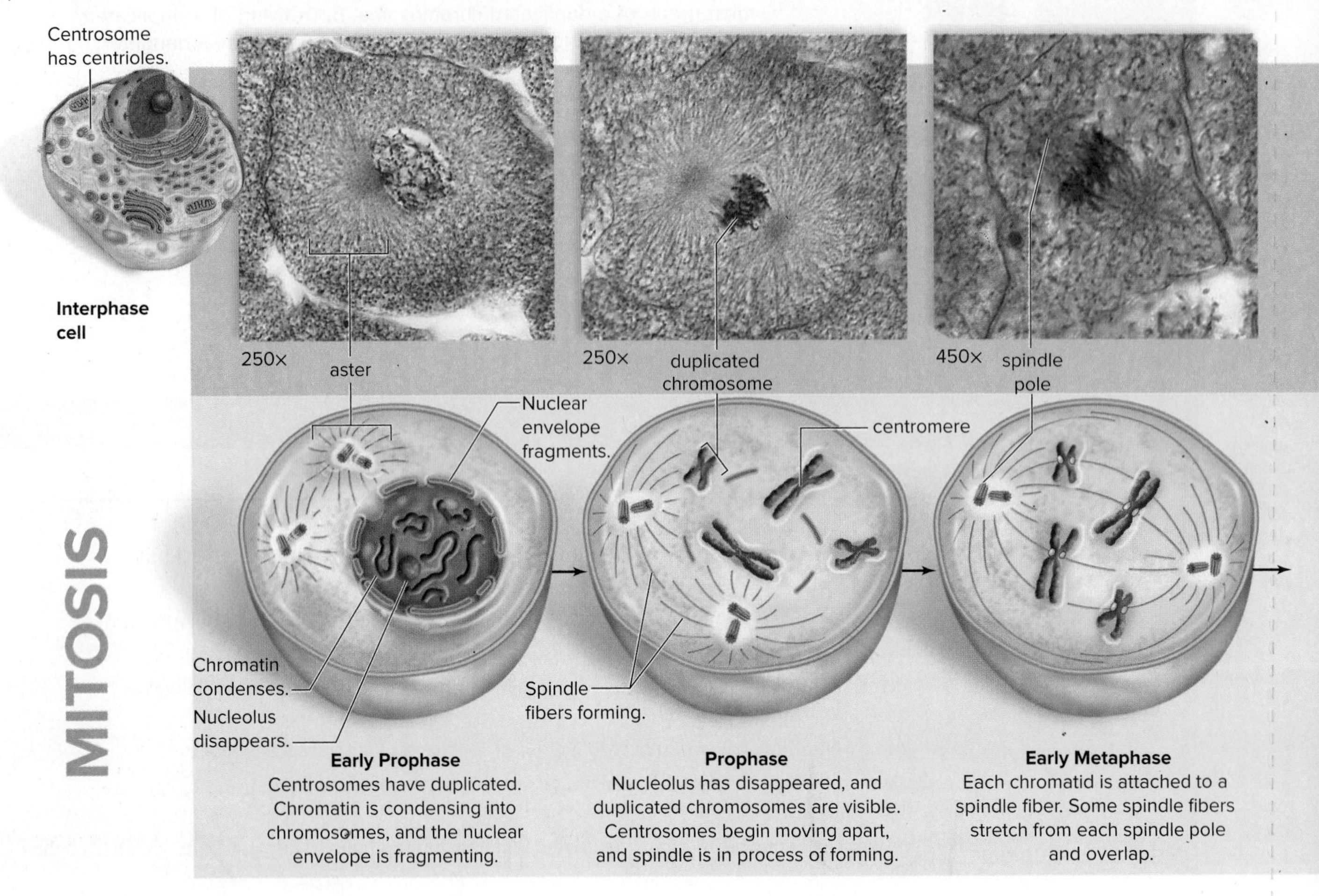

Early Prophase
Centrosomes have duplicated. Chromatin is condensing into chromosomes, and the nuclear envelope is fragmenting.

Prophase
Nucleolus has disappeared, and duplicated chromosomes are visible. Centrosomes begin moving apart, and spindle is in process of forming.

Early Metaphase
Each chromatid is attached to a spindle fiber. Some spindle fibers stretch from each spindle pole and overlap.

Observation: Mitosis

Mitosis Models

1. Using the descriptions given in Figure 15.4 as a guide, identify the phases of human cell mitosis in models of human cell mitosis.
2. Each species has its own chromosome number. Counting the number of centromeres tells you the number of chromosomes in the models. What is the number of chromosomes observed in each nucleus of the cells? ______________________________

Whitefish Blastula Slide

The blastula is an early embryonic stage in the development of humans. The **blastomeres** (blastula cells) that are in the top row of Figure 15.4 are in the illustrated phases of mitosis.

1. Examine a prepared slide of whitefish blastula cells undergoing mitotic cell division.
2. Try to find a cell in each phase of mitosis. Have a partner or your instructor check your identification.
3. Determine if you find more cells in prophase than the other phases. If so, what does that tell you about the length of time a cell stays in prophase? ______________________________

____

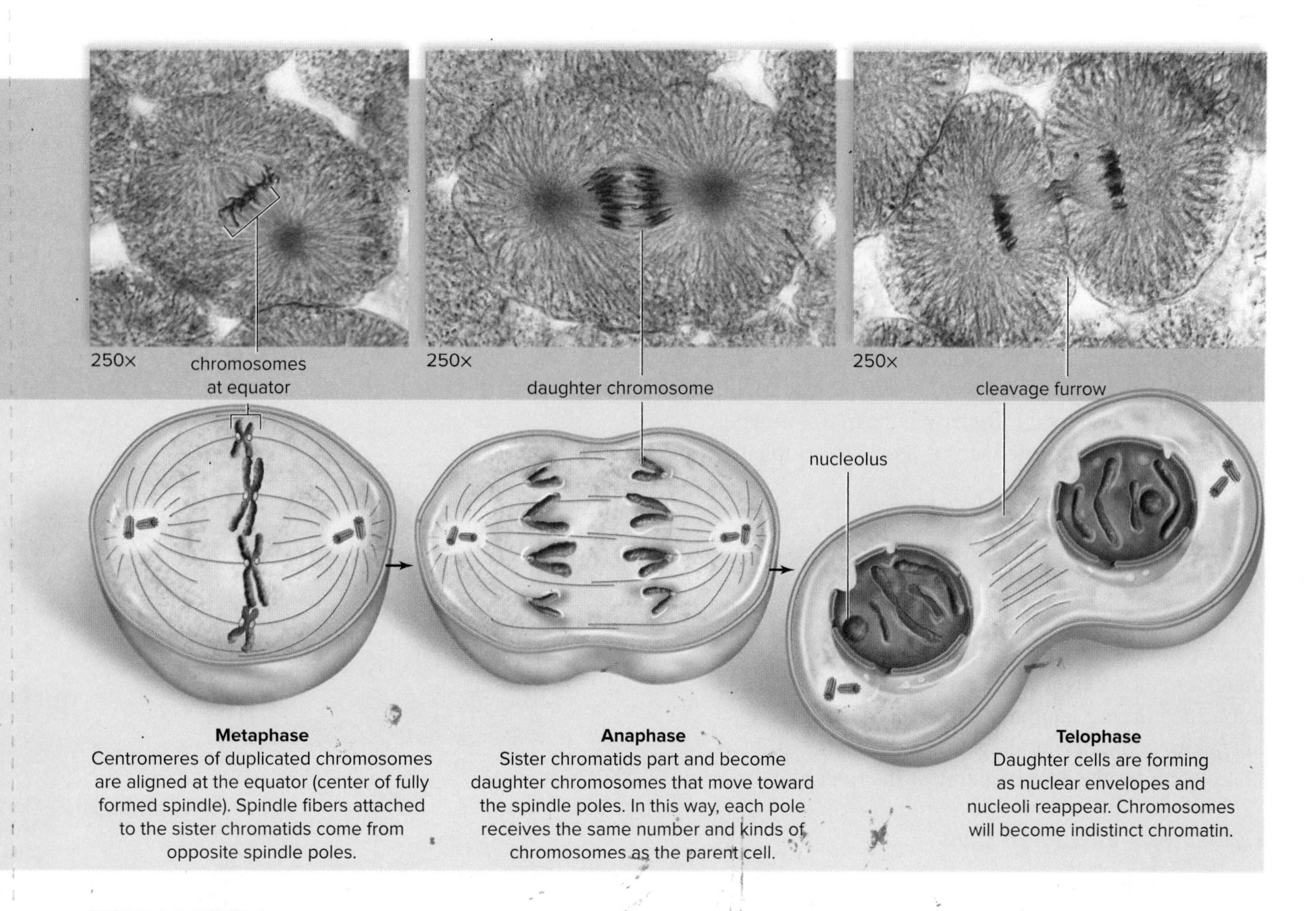

Metaphase
Centromeres of duplicated chromosomes are aligned at the equator (center of fully formed spindle). Spindle fibers attached to the sister chromatids come from opposite spindle poles.

Anaphase
Sister chromatids part and become daughter chromosomes that move toward the spindle poles. In this way, each pole receives the same number and kinds of chromosomes as the parent cell.

Telophase
Daughter cells are forming as nuclear envelopes and nucleoli reappear. Chromosomes will become indistinct chromatin.

Mitosis Phases

The phases of mitosis are **prophase, metaphase, anaphase,** and **telophase**—in that order (Fig. 15.4). Early stages of prophase and metaphase are also shown in this figure.

Prophase

During early prophase, the chromosomes continue to condense, the **nucleolus** disappears, and the nuclear envelope fragments. Still during prophase, the chromosomes have no apparent orientation within the cell. The already duplicated chromosomes are composed of two sister chromatids held together at a centromere. Counting the number of centromeres in diagrammatic drawings gives the number of chromosomes for the cell.

What is the chromosome number for the cells in Figure 15.4? ______________________________

The spindle begins to assemble as the centrosomes, each containing two centrioles, migrate to the poles.

Contrast the appearance of a cell in prophase with one in interphase. (*Hint:* See Fig. 15.1 and compare the nuclei and their contents.) ______________________________

____

Metaphase

The sister chromatids are now attached to the spindle, and the chromosomes are aligned at the equator of the spindle. The mitotic spindle occupies the region formerly occupied by the nucleus. Short microtubules radiate out in a starlike aster from the pair of centrioles located in each centrosome. The spindle consists of poles, asters, and fibers, bundles of parallel microtubules.

Contrast the appearance of a cell in metaphase with one in prophase. ______________________________

____

Anaphase

At the start of anaphase, the centromeres split, and the sister chromatids of each chromosome separate, giving rise to two daughter chromosomes. The daughter chromosomes begin to move toward opposite poles of the spindle. Each pole receives the diploid number of daughter chromosomes.

Contrast the appearance of a cell in anaphase with one in metaphase. ______________________________

____

Telophase

New nuclear envelopes form around the daughter chromosomes at the poles. Each daughter nucleus contains the same number and types of chromosomes as the parental cell. The chromosomes become more diffuse chromatin once again, and a nucleolus reappears in each daughter nucleus. Division of the cytoplasm by formation of a **cleavage furrow** is nearly complete.

Contrast the appearance of a cell in telophase with one in anaphase. ______________________________

____

Cytokinesis

Cytokinesis, division of the cytoplasm, usually accompanies mitosis. During cytokinesis, each daughter cell receives a share of the organelles that duplicated during interphase. Cytokinesis begins in anaphase, continues in telophase, and reaches completion by the start of the next interphase.

Cytokinesis in Human Cells

In human cells, a cleavage furrow, an indentation of the membrane between the daughter nuclei, begins as anaphase draws to a close. The cleavage furrow deepens as a band of actin filaments called the contractile ring slowly constricts the cell, forming two daughter cells (Fig. 15.5).

Were any of the cells of the whitefish blastula slide undergoing cytokinesis? __________

How do you know? ______________________

____

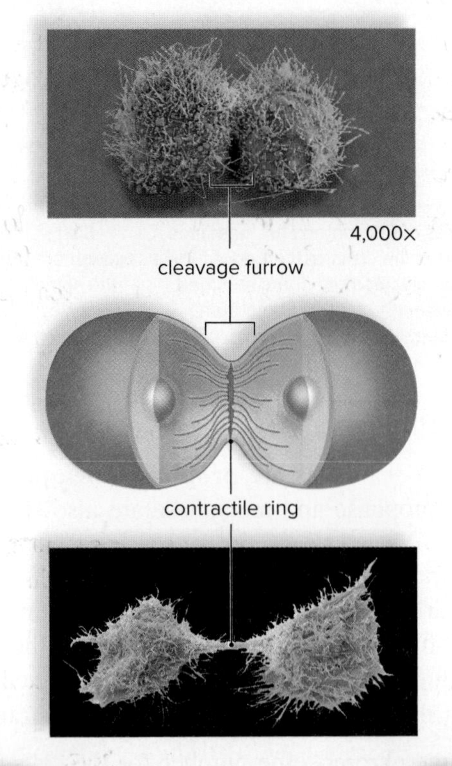

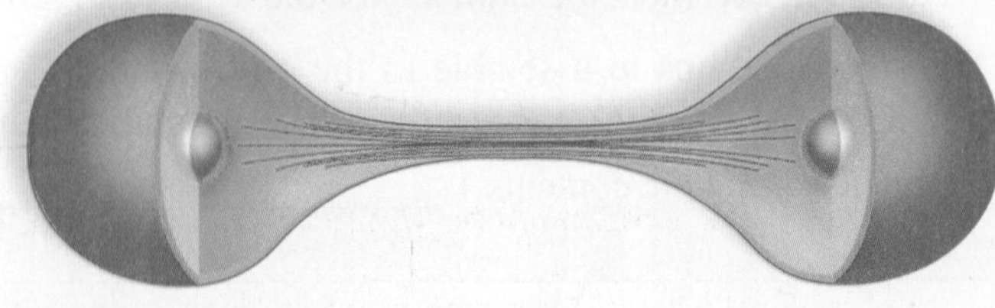

Figure 15.5 Cytokinesis in human cells. A single cell becomes two cells by a furrowing process. A contractile ring composed of actin filaments gradually gets smaller, and the cleavage furrow pinches the cell into two cells.

(top): Source: National Institutes of Health (NIH)/USHHS; (bottom): © Steve Gschmeissner/SPL/Getty RF

Summary of Mitotic Cell Division

1. The nuclei in the daughter cells have the ______________ number of chromosomes as the parent cell had.
2. Mitosis is cell division in which the chromosome number ______________

____

15.2 Meiosis

Meiosis is a form of nuclear division in which the chromosome number is reduced by half. The nucleus of the parent cell has the **diploid (2n) number** of chromosomes; the daughter nuclei, after meiosis is complete, have the **haploid (n) number** of chromosomes. In sexually reproducing species, meiosis must occur or the chromosome number would double with each generation.

A diploid cell nucleus contains **homologues,** also called homologous chromosomes. The homologues of each pair look alike and carry the genes for the same traits. Before meiosis begins, the chromosomes are already double stranded—they contain sister chromatids. Meiosis requires two divisions, called **meiosis I** and **meiosis II** (Fig. 15.6).

Meiosis I

During prophase of meiosis I, the spindle appears and the nuclear envelope and nucleolus disappear. Homologues line up next to one another during a process called synapsis. During **crossing-over,** the nonsister chromatids of a homologue pair exchange genetic material. At metaphase I, the homologue pairs line up at the equator of the spindle. During anaphase I, homologues separate and the chromosomes (still having two chromatids) move to each pole. In telophase I, the nuclear envelope and the nucleolus reappear as the spindle disappears. Each new nucleus contains one homologue from each pair of homologues.

Prophase I

Use Figure 15.6 to learn the major events of each phase of meiosis I. In prophase I the chromosomes are in a fragmented nuclear envelope, and the homologues undergo synapsis and nonsister chromatids undergo crossing-over. (The red-short and blue-short are homologues; red-long and blue-long are also homologues.)

Metaphase I

The homologues are at the equator, prepared to move apart toward the poles. Each homologue pair acts independently, and either homologue can be facing either pole. Notice that crossing-over is represented by an exchange of color between nonsister chromatids. Why do nonsister chromatids participate in crossing-over but sister chromatids do not? ____

____

(The difference in color between nonsister chromatids represents different genetic material.)

Anaphase I

The members of each homologue pair separate, and they move toward opposite poles. Now the nuclei will be haploid: they no longer have homologue pairs.

Telophase I

1. During telophase I, the chromosomes are at the poles. Disregard any crossing-over and state what combinations of chromosomes are at the poles? Fill in the following blanks with the words *red-long, red-short, blue-long,* and *blue-short:*

   Pole A: ____________________ and ____________________

   Pole B: ____________________ and ____________________

2. What other combinations would have been possible? (*Hint:* Alternate the colors at metaphase I.)

   Pole A: ____________________ and ____________________

   Pole B: ____________________ and ____________________

   All possible combinations of chromosomes can occur in daughter nuclei following meiosis I. Why would you not find two short chromosomes nor two long chromosomes at the poles of the spindle following telophase I? ____

Figure 15.6 Meiosis I and II in human cell drawings.
This illustration extends across two pages. Meiosis I begins on this page and continues on the next page; meiosis II begins on this page and continues on the next page. Note that only by coloring the homologues differently (red and blue) is it possible to show that the daughter cells following meiosis I vary genetically and to show the increased variation caused by crossing-over.

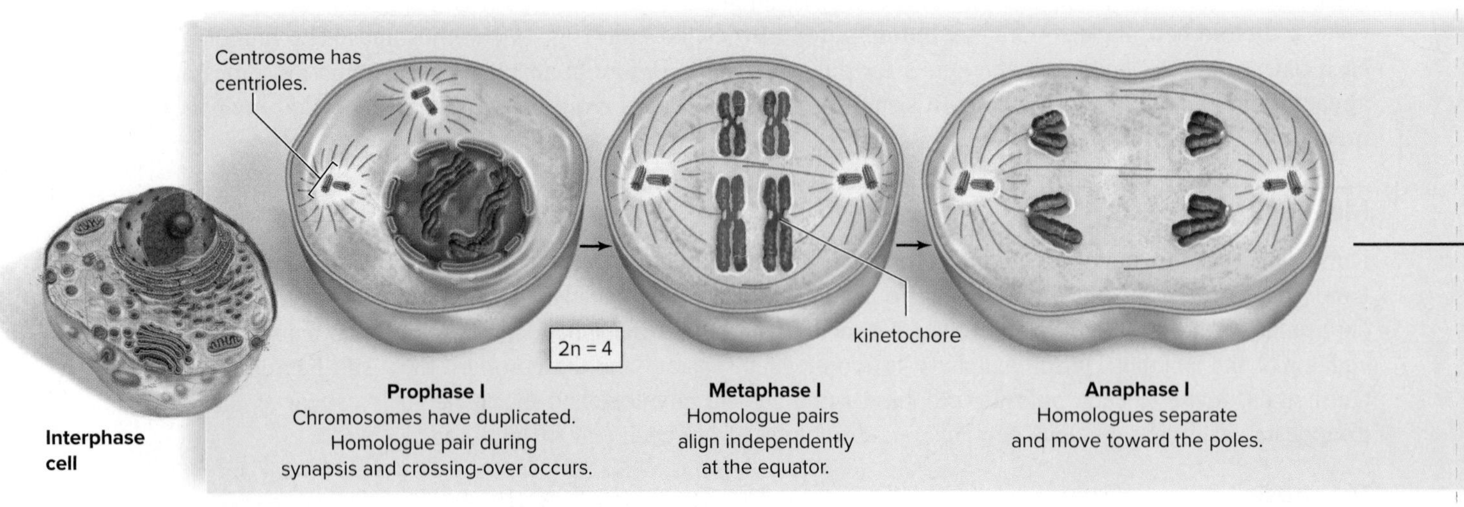

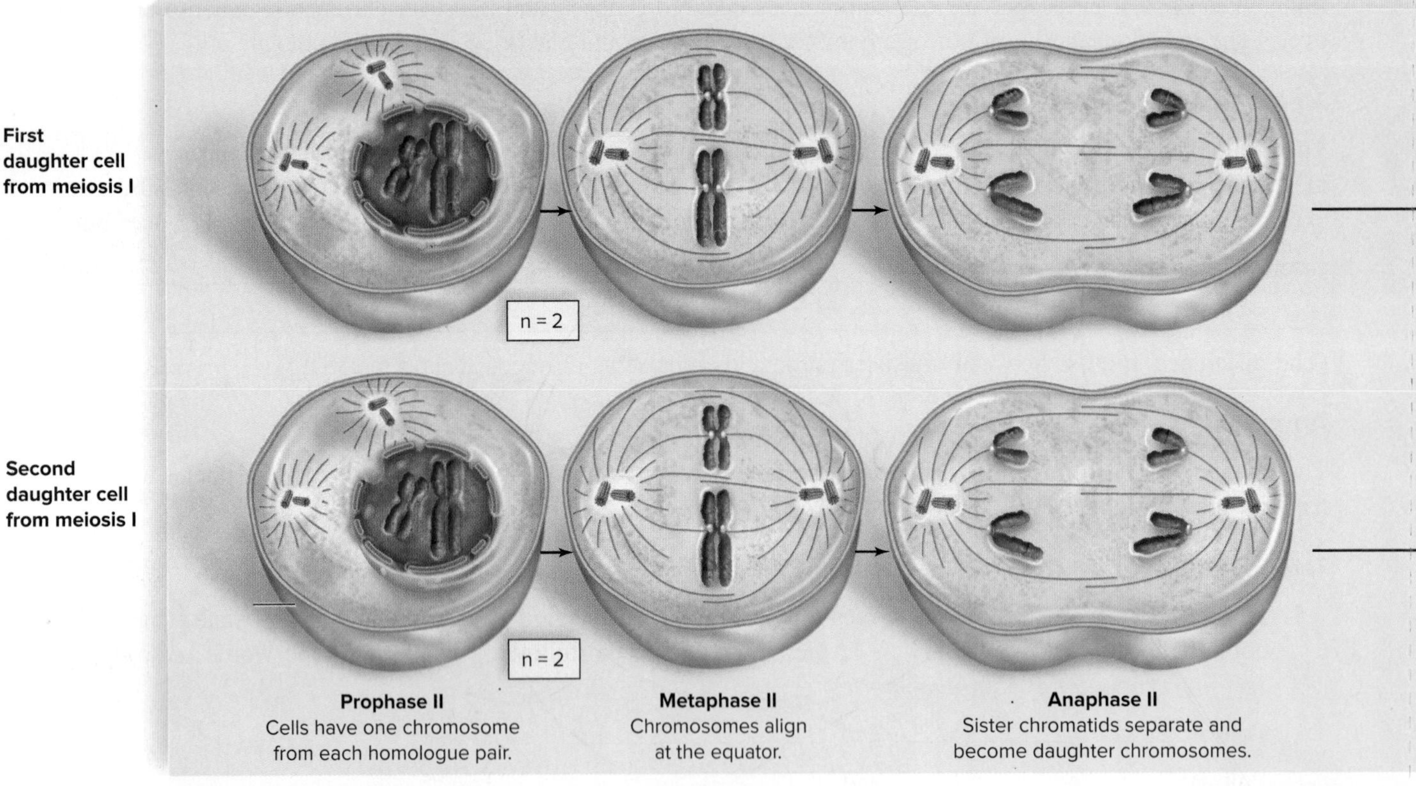

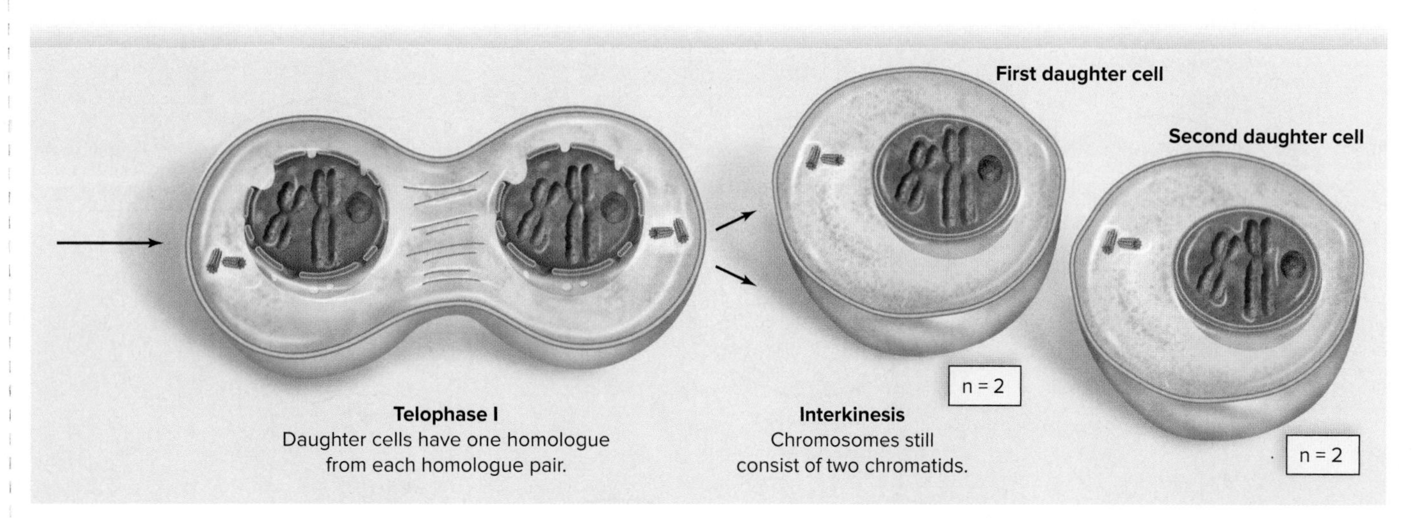

MEIOSIS I cont'd

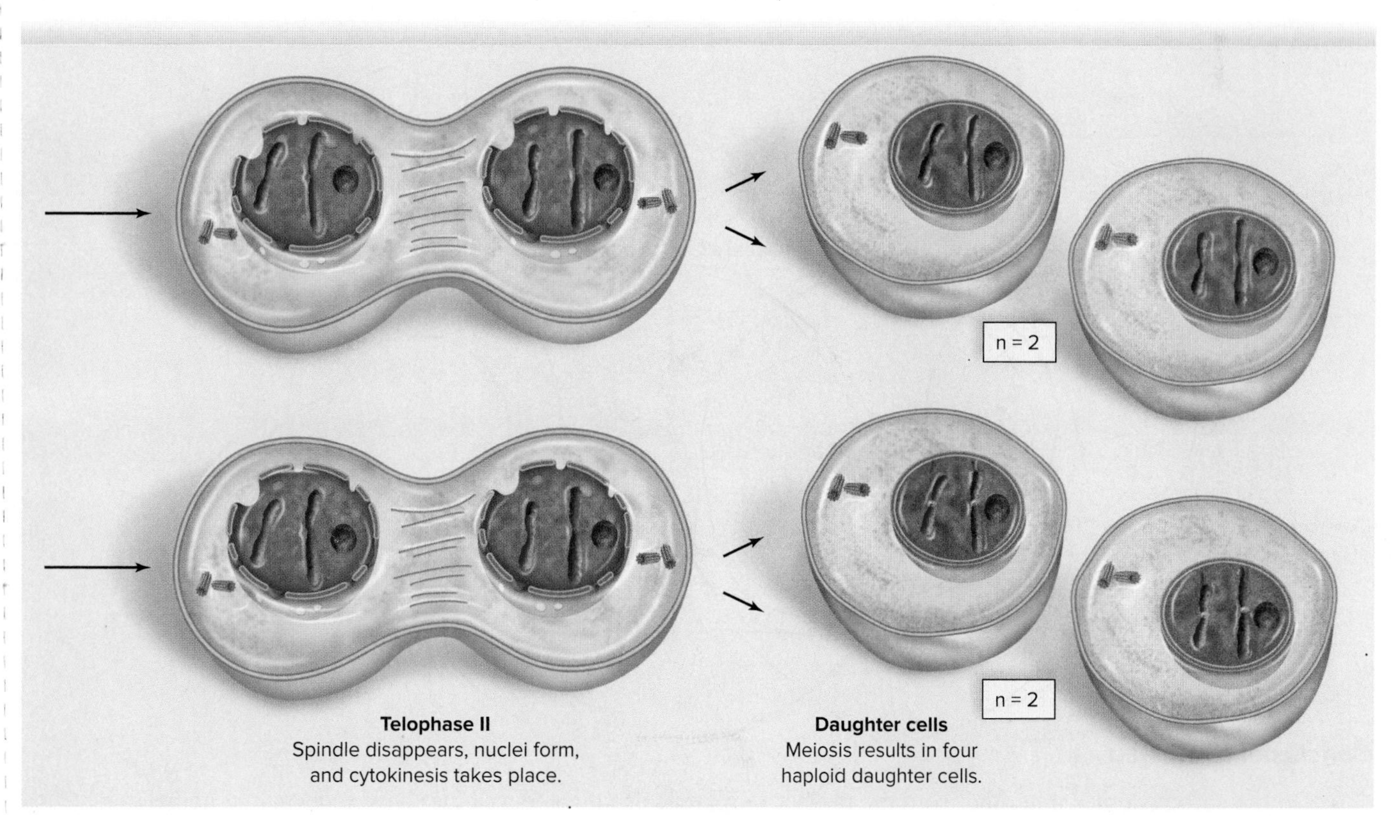

MEIOSIS II cont'd

Events of Meiosis I

Show the events of meiosis I by adding two pairs of simplified homologues (chromatids can be straight lines) to Figure 15.7. Remember to color members of the homologue pair differently (red/blue) and to draw the pairs with different lengths. Assume that nonsister chromatids have already experienced crossing-over when you begin.

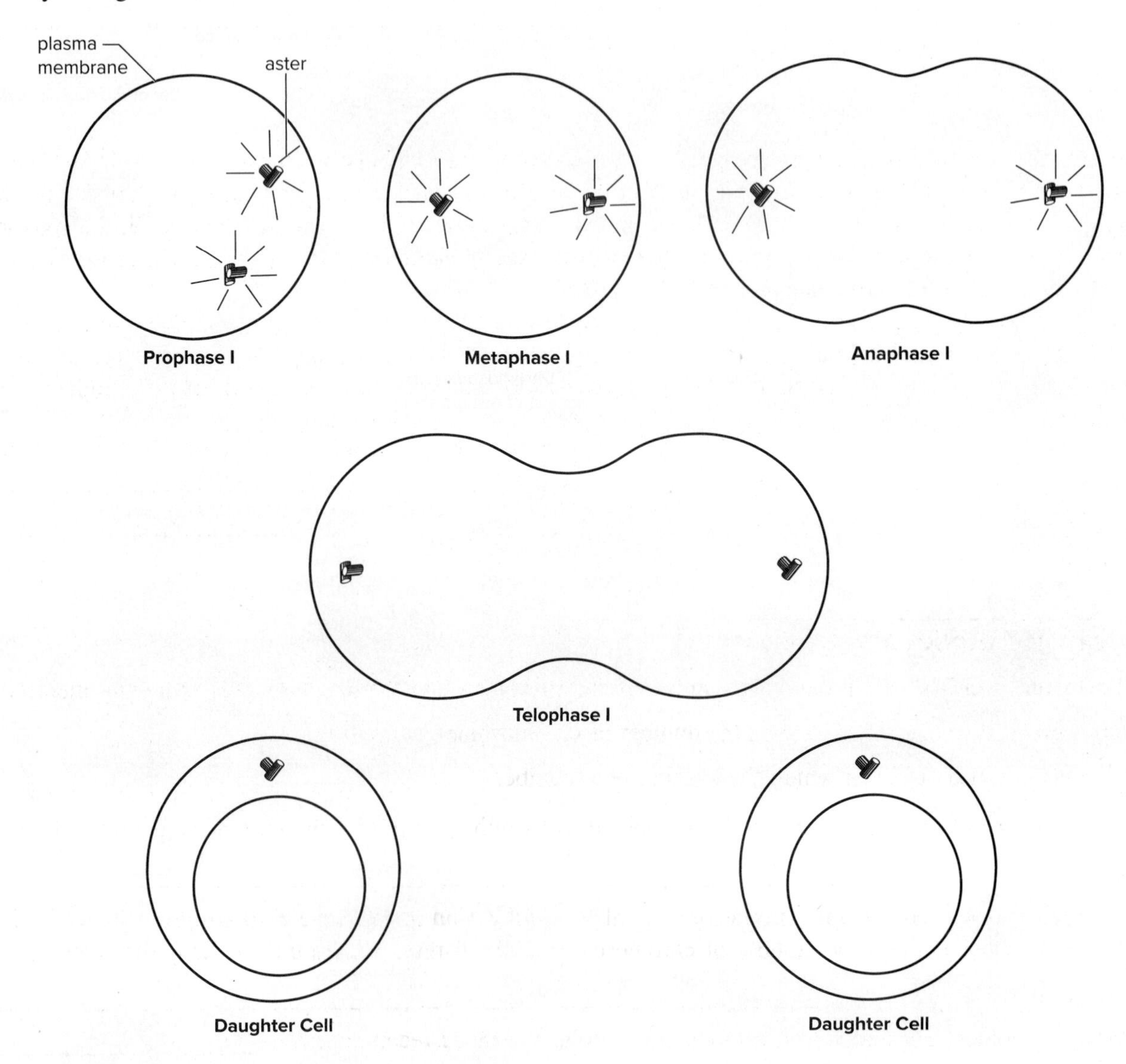

Figure 15.7 Meiosis I exercise art.

Conclusions: Meiosis I

- Do the chromosomes inherited from the mother (e.g., red) or father (e.g., blue) have to remain together following meiosis I? ____________________ Does the same genetic material have to remain together? ____________________
- Name two ways that meiosis contributes to genetic recombination:

  a. ____

  b. ____

Interkinesis

Interkinesis is the period between meiosis I and meiosis II. In some species, daughter cells do not form, and meiosis II follows right after meiosis I. Does DNA replication occur during interkinesis prior to meiosis II? ________________ Explain why not.____

____

Meiosis II

During prophase of meiosis II, a spindle appears. Each chromosome individually attaches to the spindle. During metaphase II, the chromosomes are lined up at the equator. During anaphase II, the centromeres divide and the chromatids separate, becoming daughter chromosomes that move toward the poles. In telophase II, the spindle disappears as the nuclear envelope reappears. Meiosis II is exactly like mitosis (see Fig. 15.4) except that the nuclei of the parent cell and the daughter cells are haploid.

If so directed by your instructor, use the space below to create a series of drawings showing the stages of meiosis II. Otherwise, simply answer these questions. In Figure 15.6, how many chromosomes are in the nucleus during each phase of meiosis II, whether prophase II, metaphase II, telophase II, or the daughter cells? ___________ Explain how this is possible. ____

____

____

Summary of Meiotic Cell Division

1. The parent cell has the diploid (2n) number of chromosomes, and the daughter cells following meiosis have the ____________________(n) number of chromosomes.
2. Meiosis is cell division in which the chromosome number ______________________________.
3. If a parent cell has 16 chromosomes, the daughter cells will have how many chromosomes following meiosis? ____
4. Whereas meiosis reduces the chromosome number, **fertilization** restores the chromosome number. A zygote contains the same number of chromosomes as the parent, but are these exactly the same chromosomes? ____
5. What is another way that sexual reproduction results in genetic recombination? ______________

   ____

   ____

15.3 Mitosis Versus Meiosis

In this section we will utilize Figure 15.8 to (1) compare the process of mitosis to the process of meiosis in general and then (2) to specifically compare mitosis to meiosis I and finally (3) to specifically compare mitosis to meiosis II.

General Differences

To fill in Table 15.2, review Figure 15.8 to see how many divisions were required for mitosis versus meiosis and the chromosome number (2n/n) of the daughter cells and the number of daughter cells for each process. Now fill in Table 15.2

Table 15.2 Differences Between Mitosis and Meiosis

	Mitosis	Meiosis
1. Number of divisions		
2. Chromosome number in daughter cells		
3. Number of daughter cells		

Figure 15.8 Meiosis I and meiosis II compared to mitosis.
Compare metaphase I of meiosis I to metaphase of mitosis. Only in metaphase I are the homologues paired at the equator. Members of each homologue pair separate during anaphase I, and therefore the daughter cells are haploid. The exchange of color between nonsister chromatids represents the crossing-over that occurred during meiosis I.

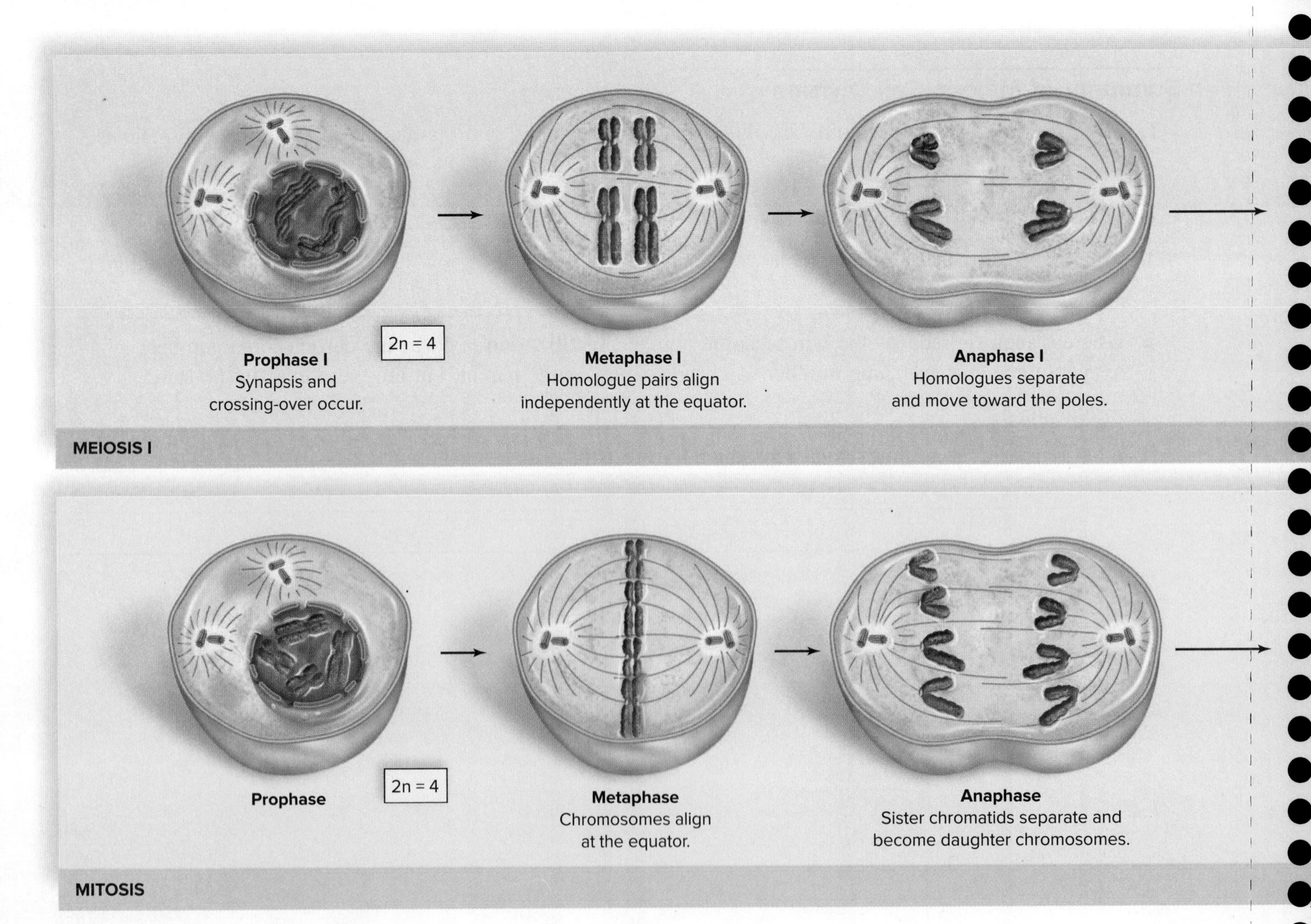

Specific Differences

To compare the specifics of mitosis to meiosis I, notice that phrases in the first column of Table 15.3 are correct for mitosis. If the phrase also applies to meiosis I, just repeat it in the second column. If it doesn't apply, correct it for meiosis I. To get you started, the correct phrase has been supplied for prophase I.

Table 15.3 Mitosis Compared with Meiosis I

Mitosis	Meiosis I
Prophase: no pairing of chromosomes.	Prophase I: Pairing of homologues
Metaphase: duplicated chromosomes at equator.	Metaphase I: ______________
Anaphase: sister chromatids separate.	Anaphase I: ______________
Telophase: chromosomes have one chromatid.	Telophase I: ______________

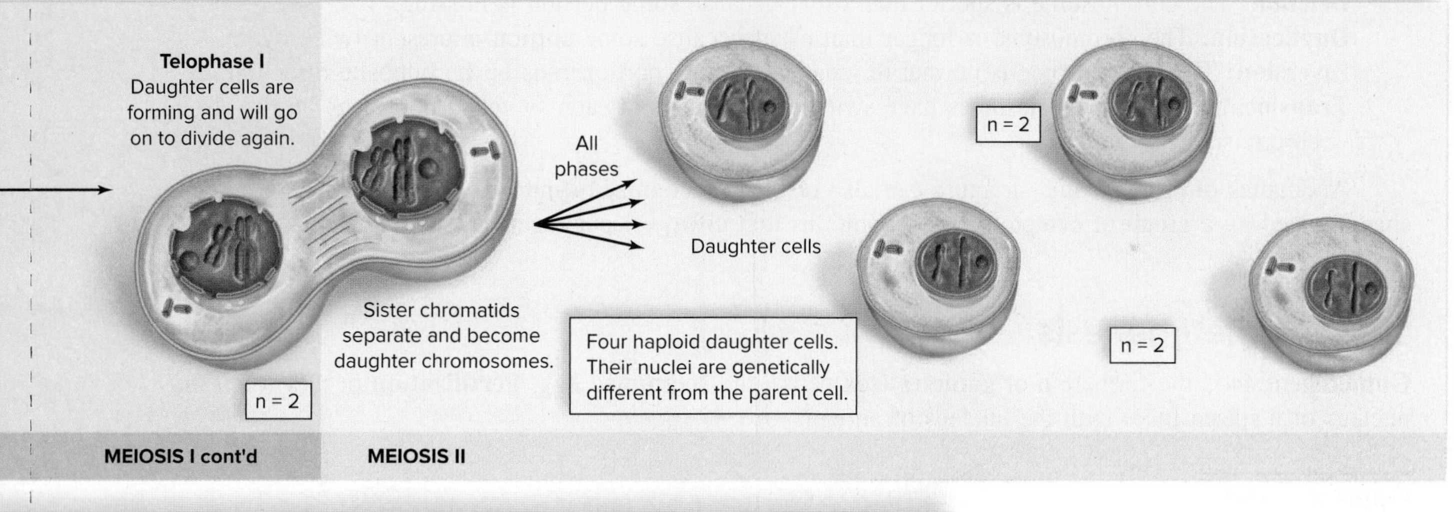

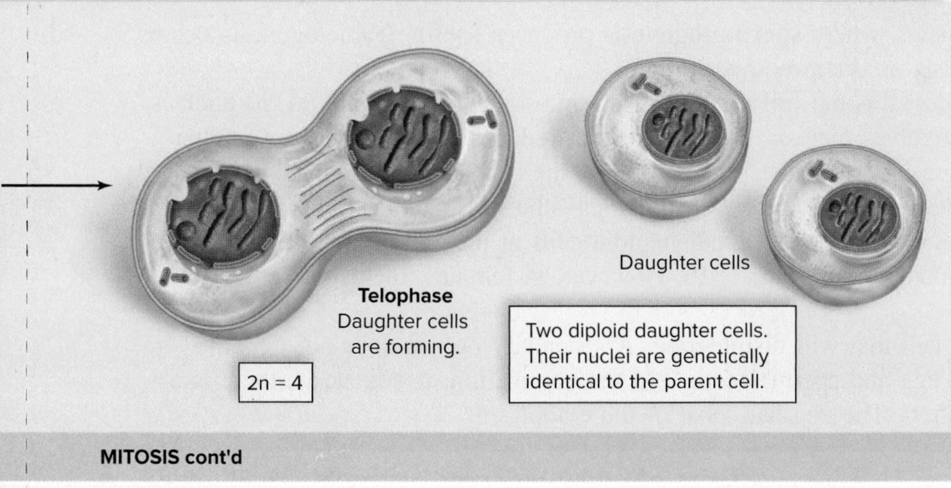

To compare mitosis to meiosis II, notice that the phrases in the first column of Table 15.4 are correct for mitosis. If the phrase also applies to meiosis II (see one daughter cell at the bottom of Fig. 15.6), just repeat it in the second column. To get you started, the correct phrase has been supplied for prophase II.

Table 15.4 Mitosis Compared with Meiosis II

Mitosis	Meiosis II
Prophase: no pairing of chromosomes.	Prophase II: No pairing of chromosomes
Metaphase: duplicated chromosomes at equator.	Metaphase II: ______________
Anaphase: sister chromatids separate.	Anaphase II: ______________
Telophase: two diploid daughter cells	Telophase II: ______________

Change in Chromosome Structure

Anomalies of chromosome structure can occur when cells divide, particularly if cells have been subject to environmental influences such as radiation or drug intake. Some of the more common structural anomalies that can be seen in a karyotype are

Deletion: The chromosome is shorter than usual because some portion is missing.
Duplication: The chromosome is longer than usual because some portion is present twice over.
Inversion: The chromosome is normal in length but some portion runs in the opposite direction.
Translocation: Two chromosomes have switched portions and each switched portion is on the wrong chromosome.

Anomalies of chromosome structure can also result in recognized syndromes. A **syndrome** is characterized by a group of symptoms and conditions that always occur together (see page 225).

15.4 Gametogenesis

Gametogenesis is the formation of **gametes** (sex cells), the sperm and egg. **Fertilization** occurs when the nucleus of a sperm fuses with the nucleus of an egg.

Gametogenesis

Gametogenesis occurs in the testes of males, where **spermatogenesis** produces sperm. Gametogenesis occurs in the ovaries of females, where **oogenesis** produces oocytes (eggs).

Recall that a diploid (2n) nucleus contains the full number of chromosomes, and a haploid (n) nucleus contains half as many. Gametogenesis involves meiosis, the process that reduces the chromosome number from 2n to n. In sexually reproducing species, if meiosis did not occur, the chromosome number would double with each generation. Because meiosis consists of two divisions—the first meiotic division (meiosis I) and the second meiotic division (meiosis II)—you expect four haploid cells at the end of the process. Indeed, there are four sperm as a result of spermatogenesis (Fig. 15.9). However, in females, meiosis I results in a secondary oocyte and one polar body.

A **polar body** is a nonfunctioning cell that will disintegrate. A secondary oocyte does not undergo meiosis II unless fertilization (fusion of egg and sperm) occurs. At the completion of oogenesis, there is a single egg. The polar bodies die (Fig. 15.9). The egg has most of the cytoplasm.

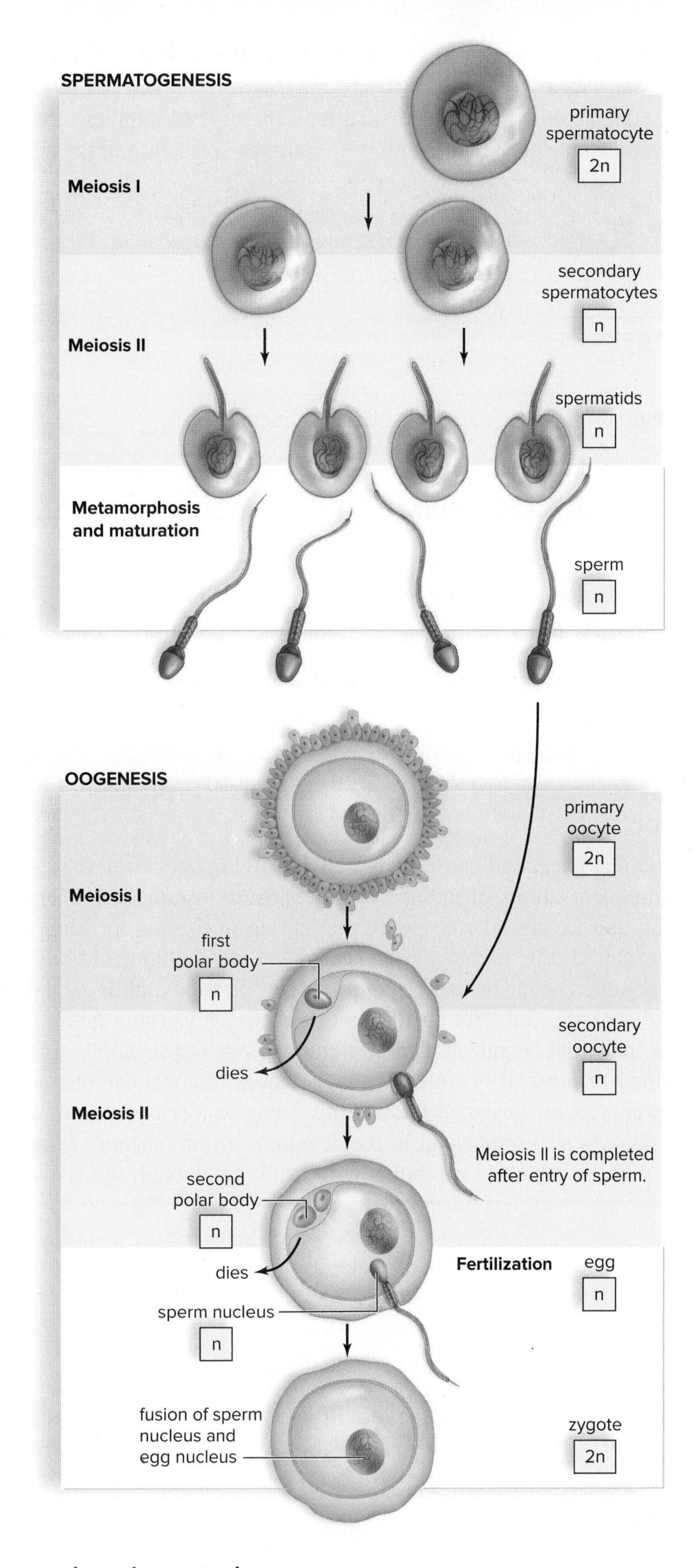

Figure 15.9 Spermatogenesis and oogenesis.
Spermatogenesis produces four viable sperm, whereas oogenesis produces one egg and two polar bodies. In humans, both sperm and egg have 23 chromosomes each; therefore, following fertilization, the zygote has 46 chromosomes.

Observation: Gametogenesis

Examine any available gametogenesis models and state here the diploid number of the parent cell and the haploid number of a gamete. Remember that counting the number of centromeres tells you the number of chromosomes.

Diploid number of parent cell: ______________________________

Haploid number of a gamete: ______________________________

Summary of Gametogenesis

1. What is gametogenesis? ______________________________

   In general, how many chromosomes are in a human gamete? ______________________________

2. What is spermatogenesis? ______________________________

   How many chromosomes does a human sperm have? ______________________________

3. What is oogenesis? ______________________________

   How many chromosomes does a human egg have? ______________________________

4. Following fertilization, how many chromosomes does the zygote, the first cell of the new individual, have? ______________________________

Application for Daily Living

Mitosis and Cancer

Mutations can upset the cell cycle and cause body cells to divide uncontrollably. Such mutations often lead to cancer. Two of the frequent causes of mutations are exposure to radiation or organic chemicals. Radiation is in sunlight, and we are all well aware that sitting in the sun for long hours can lead to skin cancer. Frequent X-rays can also be a matter of concern, and we should avoid any that are not medically necessary. Mutagenic organic chemicals can be found in certain food additives, industrial chemicals, and pesticides. That's why lawns sprayed with pesticides often carry a warning label. People sometimes show extreme concern about industrial chemicals and pesticides in our water supply, and yet they still smoke. Tobacco smoke contains a number of organic chemicals that are known carcinogens, and it is estimated that one-third of cancer deaths can be attributed to smoking. Lung cancer is the most frequently lethal cancer in the United States; smoking is also implicated in the development of cancers of the mouth, larynx, bladder, kidney, and pancreas. When smoking is combined with drinking alcohol, the risk of cancer increases.

Laboratory Review 15

_______________ **1.** During what stage of the cell cycle does DNA replication occur?

_______________ **2.** Name the phase of mitosis during which separation of sister chromatids occurs.

_______________ **3.** By what process does the cytoplasm of a human cell separate?

_______________ **4.** Name the phase of mitosis in which duplicated chromosomes first appear.

_______________ **5.** Where in humans would you expect to find meiosis taking place?

_______________ **6.** If there are 13 pairs of homologues at the start of spermatogenesis, how many chromosomes are there in a sperm?

_______________ **7.** What term refers to the production of an egg?

_______________ **8.** During which type of gametogenesis would you see polar bodies?

_______________ **9.** What do you call chromosomes that have the same length and carry genes for the same traits?

_______________ **10.** If homologues are separating, what phase is this?

_______________ **11.** If the parent cell has 24 chromosomes, how many does each daughter cell have at the completion of meiosis II?

_______________ **12.** Name the type of cell division during which homologues pair.

_______________ **13.** Name the type of cell division described by $2n \rightarrow 2n$.

_______________ **14.** Does metaphase of mitosis, meiosis I, or meiosis II have the haploid number of duplicated chromosomes at the equator of the spindle?

Thought Questions

15. Meiosis functions to reduce chromosome number. When, during the human life cycle, is the diploid number of chromosomes restored?

16. How does the alignment of chromosomes differ between metaphase of mitosis and metaphase of meiosis I?

17. A student is simulating meiosis I with homologues that are red-long and yellow-long. Describe the appearance of two nonsister chromatids following crossing-over.

LABORATORY

16

Human Genetics

Learning Outcomes

16.1 Determining the Genotype

- Determine the genotypes by observation of individuals and their relatives.

Prelab Question: Both Nancy and Jim are homozygous for freckles, a dominant trait. Using letters, what is their genotype?

16.2 Determining Inheritance

- Do genetic problems involving autosomal dominant, autosomal recessive, and X-linked recessive alleles.
- Do genetic problems involving multiple allele inheritance and use blood type to help determine paternity.

Prelab Question: What are the chances that Nancy and Jim will have children with freckles?

16.3 Genetic Counseling

- Analyze a karyotype to determine if a person's chromosomal inheritance is as expected or whether a chromosome anomaly has occurred.
- Analyze a pedigree to determine if the pattern of inheritance is autosomal dominant, autosomal recessive, or X-linked recessive.
- Construct a pedigree to determine the chances of inheriting a particular phenotype when provided with generational information.

Prelab Question: In a pedigree of Nancy's family and Jim's family, would you expect to find few or many people with freckles?

***Application for Daily Living:* Choosing the Gender of Your Child**

Introduction

In this laboratory, you will discover that the same principles of genetics apply to humans as they do to plants and fruit flies. A gene has two alternate forms, called **alleles,** for any trait, such as hairline, finger length, and so on. One possible allele, designated by a capital letter, is **dominant** over the **recessive** allele, designated by a lowercase letter. An individual can be **homozygous dominant** (two dominant alleles, *EE*), **homozygous recessive** (two recessive alleles, *ee*) or **heterozygous** (one dominant and one recessive allele, *Ee*). **Genotype** refers to an individual's alleles, and **phenotype** refers to an individual's appearance (Fig. 16.1). Homozygous dominant and also heterozygous individuals show the dominant phenotype; homozygous recessive individuals show the recessive phenotype.

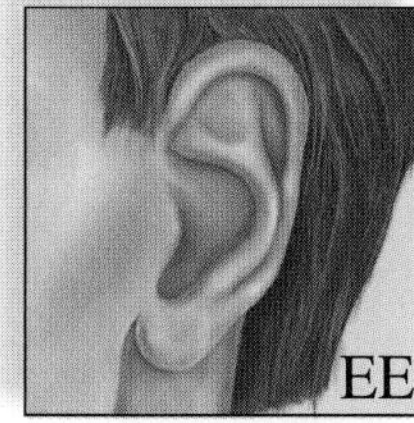

a. Unattached earlobe

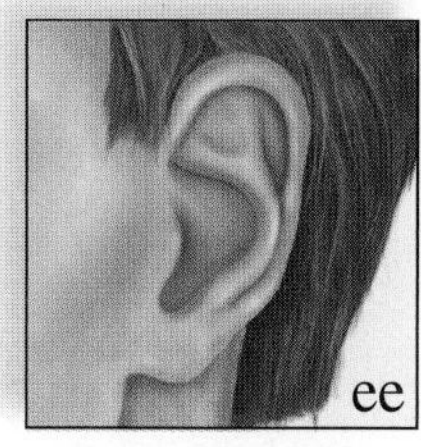

b. Attached earlobe

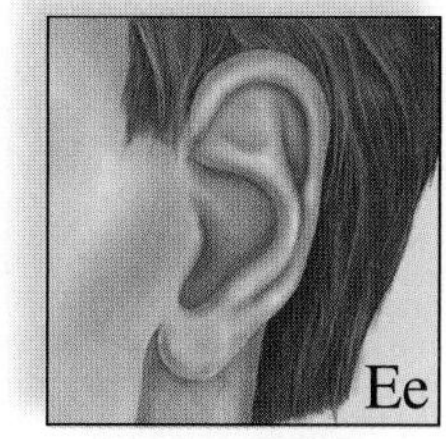

c. Unattached earlobe

Figure 16.1 Genotype versus phenotype. Unattached earlobes (*E*) are dominant over attached earlobes (*e*). **a.** Homozygous dominant individuals have unattached earlobes. **b.** Homozygous recessive individuals have attached earlobes. **c.** Heterozygous individuals have unattached earlobes.

21.1 Determining the Genotype

Humans inherit 46 chromosomes that occur in 23 pairs. Twenty-two of these pairs are called autosomes and one pair is the sex chromosomes. Autosomal traits are determined by alleles on the autosomal chromosomes.

Autosomal Dominant and Recessive Traits

Figure 16.2 shows a few human traits.

1. What is the homozygous dominant genotype for type of hairline? ______________ What is the phenotype? ____________________________

2. What is the homozygous recessive genotype for finger length? ______________ What is the phenotype? ____________________________

3. Why does the heterozygous individual *Ff* have freckles? ____________________________

Figure 16.2 Commonly inherited traits in humans.
The alleles indicate which traits are dominant and which are recessive. (a) © Superstock; (b) © Dynamic Graphics/PictureQuest RF; (c-f) © McGraw-Hill Education/Bob Coyle, photographer; (g) © Corbis RF; (h) © Creatas/PunchStock RF

a. Widow's peak: *WW* or *Ww*

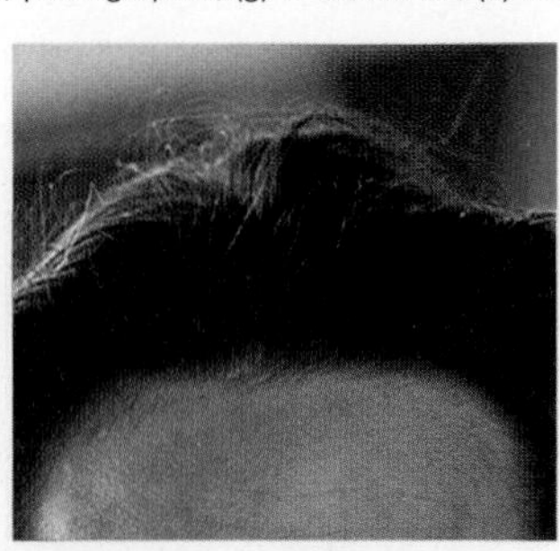

b. Straight hairline: *ww*

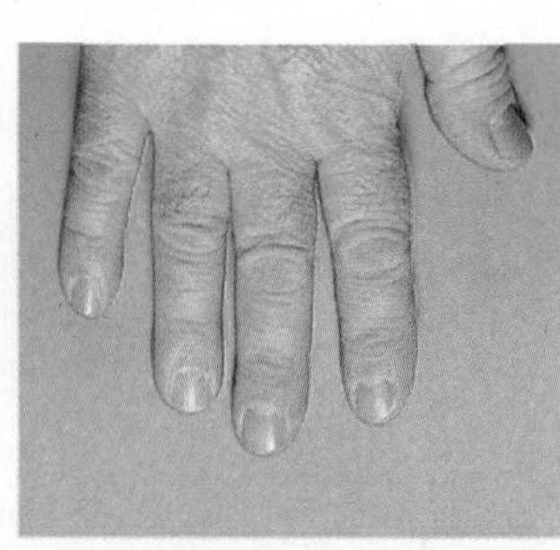

e. Short fingers: *SS* or *Ss*

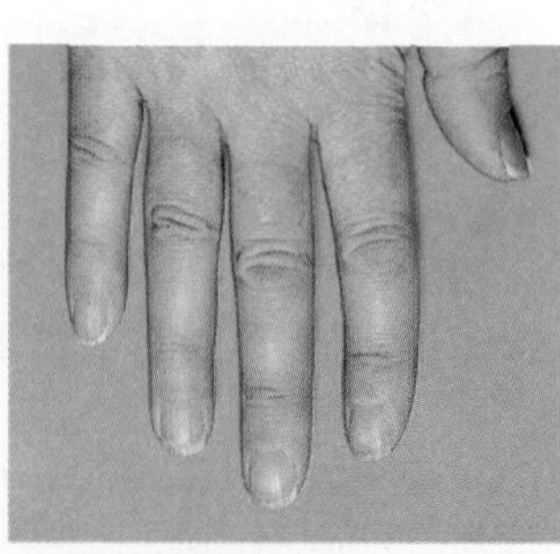

f. Long fingers: *ss*

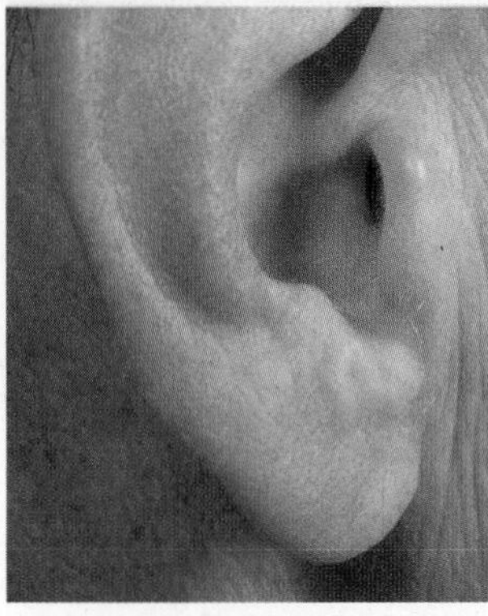

c. Unattached earlobes: *EE* or *Ee*

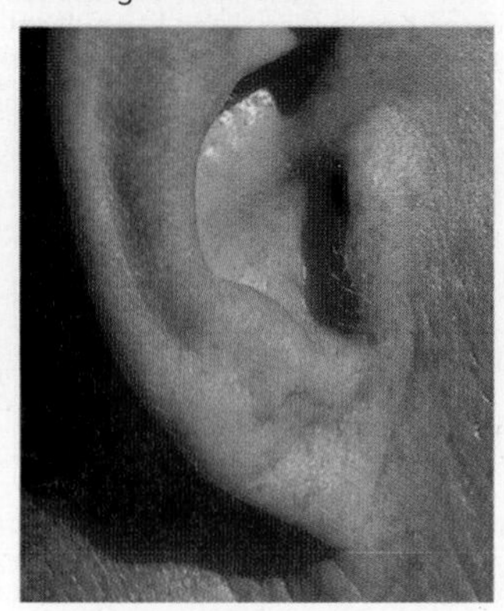

d. Attached earlobes: *ee*

g. Freckles: *FF* or *Ff*

h. No freckles: *ff*

These genetic problems use the alleles from Figure 16.2 and Table 16.1.

4. Maria and the members of her immediate family have attached earlobes. What is Maria's genotype? ______________ Her maternal grandfather has unattached earlobes. Deduce the genotype of her maternal grandfather. ______________ Explain your answer. ____________________________

5. Moses does not have a bent little finger, but his parents do. Deduce the genotype of his parents. ______________ of Moses. ______________ Explain your answer. ____________________________

6. Manny is adopted. He has hair on the back of his hand. Could both of his parents have had hair on the back of the hand? ______________ Could both of his parents have had no hair on the back of the hand? ______________ Explain your answer. ____________________________

Experimental Procedure: Human Traits

1. For this Experimental Procedure, you will need a lab partner to help you determine your phenotype for the traits listed in the first column of Table 16.1.
2. Determine your probable genotype. If you have the recessive phenotype, you know your genotype. If you have the dominant phenotype, you may be able to decide whether you are homozygous dominant or heterozygous by recalling the phenotype of your parents, siblings, or children. Circle your probable genotype in the second column of Table 16.1.
3. Your instructor will tally the class's phenotypes for each trait so that you can complete the third column of Table 16.1.
4. Complete Table 16.1 by calculating the percentage of the class with each trait. Are dominant phenotypes always the most common in a population? ____________ Explain your answer. ________________

____

Table 16.1 Autosomal Human Traits

Trait: d = Dominant r = Recessive	Probable Genotypes	Number in Class	Percentage of Class with Trait
Hairline:			
Widow's peak (d)	*WW* or *Ww*	4	______
Straight hairline (r)	*ww*	23	______
Earlobes:			
Unattached (d)	*UU* or *Uu*	9	______
Attached (r)	*uu*	18	______
Skin pigmentation:			
Freckles (d)	*FF* or *Ff*	3	______
No freckles (r)	*ff*	24	______
Hair on back of hand:			
Present (d)	*HH* or *Hh*	15	______
Absent (r)	*hh*	12	______
Thumb hyperextension—"hitchhiker's thumb":			
Last segment cannot be bent backward (d).	*TT* or *Tt*	10	______
Last segment can be bent back to 60° (r).	*tt*	17	______
Bent little finger:			
Little finger bends toward ring finger (d).	*LL* or *Ll*	1	______
Straight little finger (r)	*ll*	26	______
Interlacing of fingers:			
Left thumb over right (d)	*II* or *Ii*	9	______
Right thumb over left (r)	*ii*	18	______

16.2 Determining Inheritance

Recall that a Punnett square is a means to determine the genetic inheritance of offspring if the genotypes of both parents are known. In a **Punnett square,** all possible types of sperm are lined up vertically, and all possible types of eggs are lined up horizontally, or vice versa, so that every possible combination of gametes occurs within the square. Figure 16.3 shows how to construct a Punnett square when autosomal alleles are involved.

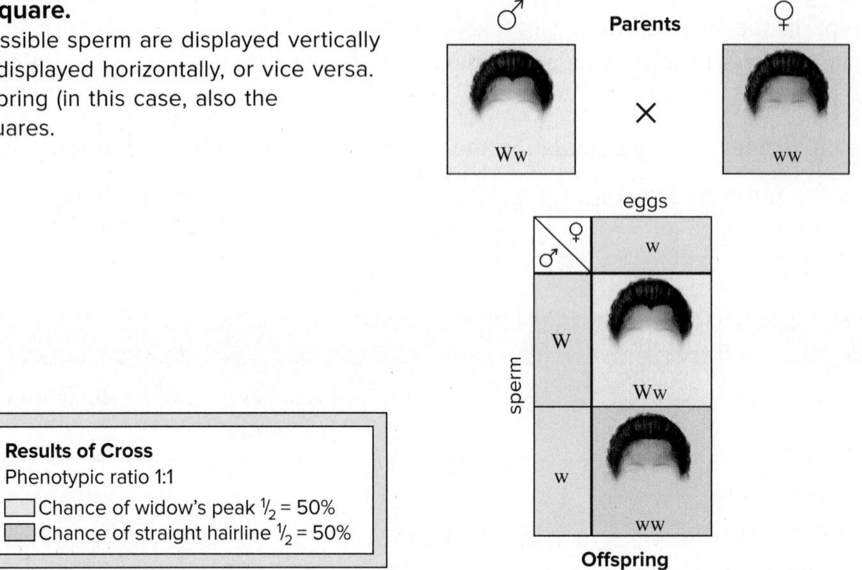

Figure 16.3 Punnett square.
In a Punnett square, all possible sperm are displayed vertically and all possible eggs are displayed horizontally, or vice versa. The genotypes of the offspring (in this case, also the phenotypes) are in the squares.

Inheritance of Genetic Disorders

Figure 16.4 can be used to learn the chances that a particular phenotype will occur.

In Figure 16.4*a,*

¼ of the offspring have the recessive phenotype = __________ % chance

¾ of the offspring have the dominant phenotype = __________ % chance

In Figure 16.4*b,*

½ of the offspring have the recessive or the dominant phenotype = __________ % chance

In all the following genetic problems, use letters to fill in the parentheses with the genotype of the parents.

1. **a.** With reference to Figure 16.4*a,* if a genetic disorder is recessive and both parents are heterozygous (__________), what are the chances that an offspring will have the disorder? __________

   b. With reference to Figure 16.4*a,* if a genetic disorder is dominant and both parents are heterozygous (__________), what are the chances that an offspring will have the disorder? __________

2. **a.** With reference to Figure 16.4*b,* if the parents are heterozygous (__________) by homozygous recessive (__________), and the genetic disorder is recessive, what are the chances that the offspring will have the disorder? __________

   b. With reference to Figure 16.4*b,* if the parents are heterozygous (__________) by homozygous recessive (__________), and the genetic disorder is dominant, what are the chances that an offspring will have the disorder? __________

Figure 16.4 Two common patterns of autosomal inheritance in humans.
a. Both parents are heterozygous. **b.** One parent is heterozygous and the other is homozygous recessive. The letter *A* stands for any trait that is dominant and the letter *a* stands for any trait that is recessive. Substitute the correct alleles for the problem you are working on. For example, *C* = normal; *c* = cystic fibrosis.

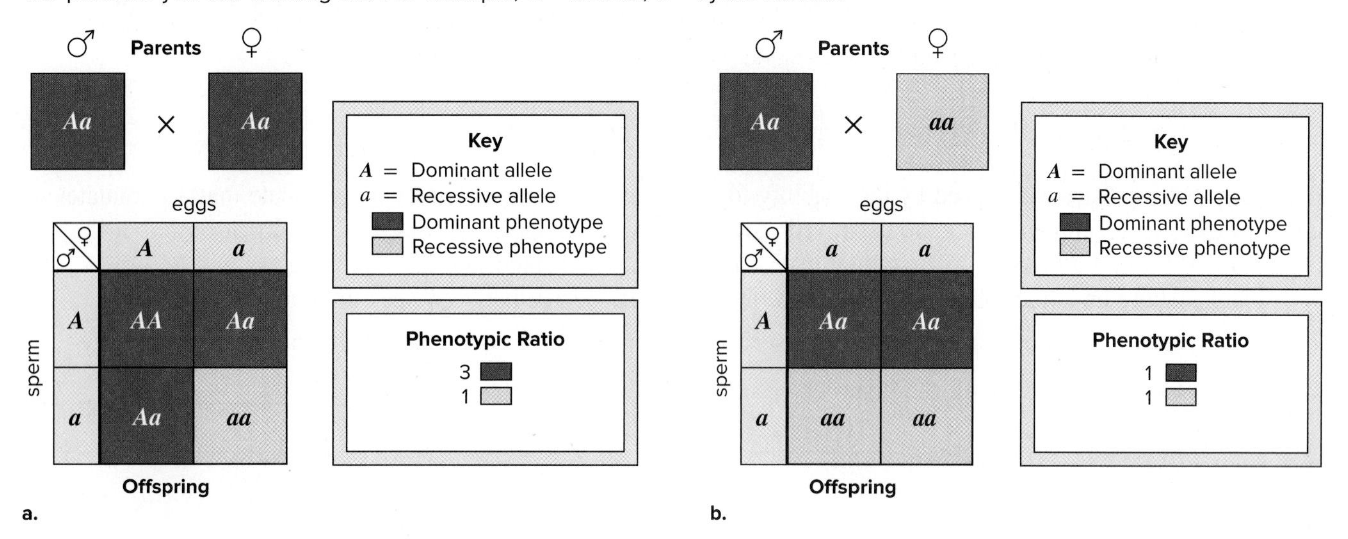

Autosomal Disorders

1. **Neurofibromatosis** (NF), sometimes called von Recklinghausen disease, is one of the most common genetic disorders. It affects roughly 1 in 3,000 people. It is seen equally in every racial and ethnic group throughout the world. At birth or later, the affected individual may have six or more large tan spots on the skin. Such spots may increase in size and number and become darker. Small benign tumors (lumps) called neurofibromas may occur under the skin or in the muscles. Neurofibromas are made up of nerve cells and other cell types.

   Neurofibromatosis is a dominant disorder. If a heterozygous woman reproduces with a homozygous normal man, what are the chances a child will have neurofibromatosis? ___________________________

2. **Cystic fibrosis** is due to abnormal mucus-secreting tissues. At first, the infant may have difficulty regaining the birth weight despite good appetite and vigor. A cough associated with a rapid respiratory rate but no fever indicates lung involvement. Large, frequent, and foul-smelling stools are due to abnormal pancreatic secretions. Whereas children previously died in infancy due to infections, they now often survive because of antibiotic therapy.

   Cystic fibrosis is a recessive disorder. A **carrier** is an individual that appears to be normal but carries a recessive allele for a genetic disorder. A man and a woman are both carriers (__________) for cystic fibrosis. What are the chances a child will have cystic fibrosis? ______________________________

3. **Huntington disease** does not appear until the 30s or early 40s. There is a progressive deterioration of the individual's nervous system that eventually leads to constant thrashing and writhing movements until insanity precedes death. Studies suggest that Huntington disease is due to a single faulty gene that has multiple effects, in which case there is now hope for a cure.

   People with Huntington disease seem to be more fertile than others. It is amazing that more than 1,000 of the cases in the United States in the past century can be traced to one man born in 1831.

   Huntington disease is a dominant disorder. Drina is 25 years old and as yet has no signs of Huntington disease. Her mother does have Huntington disease (__________), but her father is free (__________) of the disorder. What are the chances that Drina will develop Huntington disease? __________

   ____

4. **Phenylketonuria** (**PKU**) is characterized by severe intellectual impairment due to an abnormal accumulation of the common amino acid phenylalanine within cells, including neurons. The disorder takes its name from the presence of a breakdown product, phenylketone, in the urine and blood. Newborn babies are routinely tested at the hospital and, if necessary, are placed on a diet low in phenylalanine.

   Phenylketonuria (PKU) is a recessive disorder. Mr. and Mrs. Martinez appear to be normal, but they have a child with PKU. What are the genotypes of Mr. and Mrs. Martinez? ________________________

5. **Tay–Sachs disease** is caused by the inability to break down a certain type of fat molecule that accumulates around nerve cells until they are destroyed. Afflicted newborns appear normal and healthy at birth, but they do not develop normally. At first, they may learn to sit up and stand, but later they regress and become intellectually impaired, blind, and paralyzed. Death usually occurs between ages three and four.

   Tay–Sachs is an autosomal recessive disorder. Is it possible for two individuals who do not have Tay–Sachs to have a child with the disorder? Explain your answer. ________________________________

   ____

X-Linked Disorders

The sex chromosomes designated X and Y carry genes just like the autosomal chromosomes. Some genes, particularly on the X chromosome, have nothing to do with gender inheritance and are said to be X-linked. **X-linked recessive disorders** are due to recessive genes carried on the X chromosomes. Males are more likely to have an X-linked recessive disorder than females because the Y chromosome is blank for this trait. Does a color-blind male give his son a recessive-bearing X or a Y that is blank for the recessive allele? __________

The possible genotypes and phenotypes for an X-linked recessive disorder are as follows:

Females	Males
X^BX^B = normal vision	X^BY =normal vision
X^BX^b = normal vision (carrier)	X^bY = color blindness
X^bX^b = color blindness	

An X-linked recessive disorder in a male is always inherited from his mother. Most likely, his mother is heterozygous and therefore does not show the disorder. She is designated a carrier for the disorder. Figure 16.5 shows how females can become carriers. *Use the key to genotypes and phenotypes for X-linked recessive disorders to complete the parentheses of questions 2 and 3 with genotypes of the parents.*

1. **a.** What is the genotype for a color-blind female?__________ How many recessive alleles does a female inherit to be color-blind? __________

   b. What is the genotype for a color-blind male? __________ How many recessive alleles does a male inherit to be color-blind? __________

2. **a.** With reference to Figure 16.5*a,* if the mother is a carrier (__________) and the father has normal vision (__________), what are the chances that a daughter will be color blind? __________

   b. A daughter will be a carrier? ___________ **c.** A son will be color blind? ___________

3. **a.** With reference to Figure 16.5*b,* if the mother has normal vision (__________) and the father is color blind (__________), what are the chances that a daughter will be color blind? __________

   b. A daughter will be a carrier? ___________ **c.** A son will be color blind? ___________

Figure 16.5 Two common patterns of X-linked inheritance in humans.
a. The sons of a carrier mother have a 50% chance of being color blind. **b.** A color-blind father has carrier daughters.

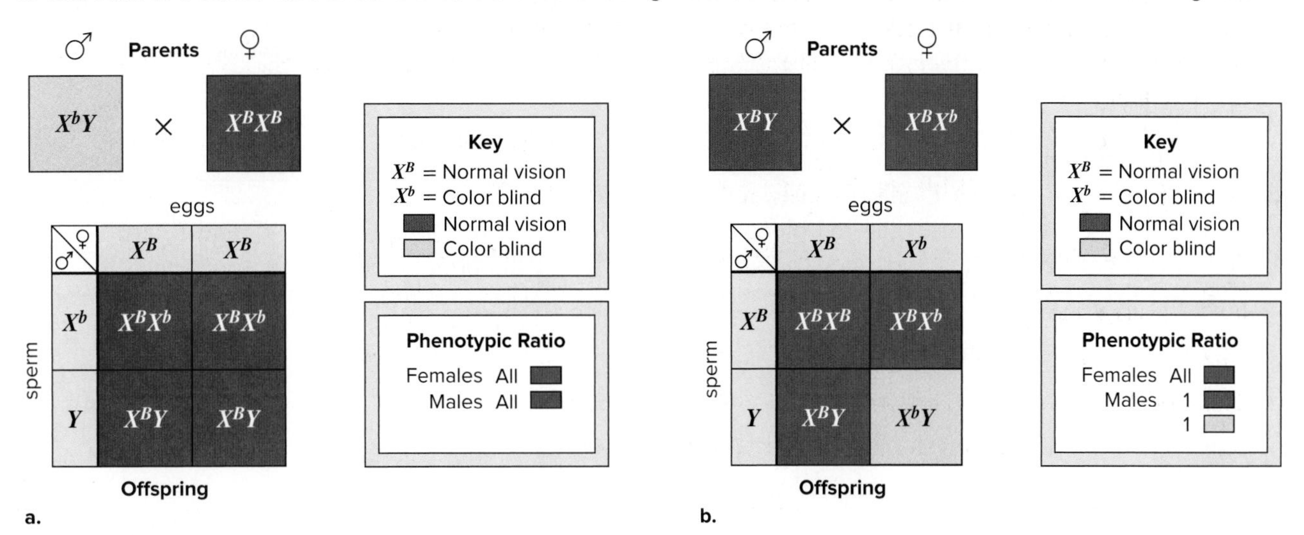

X-Linked Genetics Problems

For **color blindness,** there are two possible X-linked alleles involved. One affects the green-sensitive cones, whereas the other affects the red-sensitive cones. About 6% of men in the United States are color blind due to a mutation involving green perception, and about 2% are color blind due to a mutation involving red perception.

1. A woman with normal color vision (__________), whose father was color blind (__________), marries a man with normal color vision (__________). What genotypes could occur among their offspring? _______

____

What genotypes could occur if it was the normal-visioned man's father who was color blind? _________

____

2. Antonio's father is color blind (__________) but his mother is not color blind (__________).

Is Antonio necessarily color blind? _______________ Explain your answer. _______________

Could he be color blind? _______________ Explain your answer. _______________

____

Hemophilia is called the bleeder's disease because the affected person's blood is unable to clot. Although hemophiliacs do bleed externally after an injury, they also suffer from internal bleeding, particularly around joints. Hemorrhages can be checked with transfusions of fresh blood (or plasma) or concentrates of the clotting protein. The most common type of hemophilia is hemophilia A, due to absence or minimal presence of a particular clotting factor called factor VIII.

3. Make up a cross involving hemophilia that could be answered by a Punnett square, as in Figure 16.5*a* or *b*.

____

____

What is the answer to your genetics problem? _______________________________

Multiple Alleles

> **Protective clothing** Wear protective laboratory clothing, latex gloves, and goggles. If the chemicals touch the skin, eyes, or mouth, wash immediately. If inhaled, seek fresh air.

When a trait is controlled by **multiple alleles,** the gene has more than two possible alleles. But each person has only two of the possible alleles. For example, ABO blood type is determined by multiple alleles: I^A, I^B, i. Red blood cells have surface molecules called antigens that indicate they belong to the person. The I^A allele causes red blood cells to carry an A antigen, the I^B allele causes red blood cells to carry a B antigen, and the i allele causes the red blood cells to have neither of these antigens. I^A and I^B are dominant to i. Remembering that each person can have any two of the possible alleles, these are possible genotypes and phenotypes for blood types.

Genotypes	Antigens on Red Cells	Blood Types
I^AI^A, I^Ai	A	A
I^BI^B, I^Bi	B	B
I^AI^B	A and B	AB
ii	none	O

Blood type also indicates whether the person is Rh positive or Rh negative. If the genotype is *DD* or *Dd,* the person is Rh positive, and if the genotype is *dd,* the person is Rh negative. It is customary to simply attach a + or – superscript to the ABO blood type, as in A^-.

Blood Typing Problems

1. A man with type A blood reproduces with a woman who has type B blood. Their child has blood type O. Using I^A, I^B, and i, give the genotype of all persons involved. man _____________ woman _____________ child ___________
2. If a child has type AB blood and the father has type B blood, what could the genotype of the mother be? _____________ or _____________
3. If both mother and father have type AB blood, they cannot be the parents of a child who has what blood type? _____________
4. What blood types are possible among the children if the parents are $I^Ai \times I^Bi$? (*Hint:* Do a Punnett square using the possible gametes for each parent.)

Experimental Procedure: Using Blood Type to Help Determine Paternity

In this Experimental Procedure a mother, Wanda, is seeking the identity of the father of her child, Sophia. We will use blood typing to decide which of three men could possibly be the father.

1. Obtain three testing plates, each of which contains three depressions; vials of blood from possible fathers 1, 2, and 3 respectively; vials of anti-A serum, anti-B serum, and anti-Rh serum. (All of these are synthetic.)
2. Using a wax pencil, number the plates so you know which plate is for possible father #1, #2, or #3. Look carefully at a plate and notice the wells are designated as A, B, or Rh.
3. Being sure to close the cap to each vial in turn, do the following using plate #1:
   Add a drop of father #1 blood to all three wells—close the cap.
   Add a drop of anti-A (blue) to the well designated A—close the cap.
   Add a drop of anti-B (yellow) to the well designated B—close the cap.
   Add a drop of anti-Rh (clear) to the well designated Rh—close the cap.

Skip

Table 21.2 Blood Types of Involved Persons					
	Mother*	Child*		Father?	
	Wanda	Sophia	#1	#2	#3
Blood type					

*Record the blood types of the Mother and Child that your instructor supplies.

4. Stir the contents of each well with a mixing stick of the correct color. After a few minutes, examine the wells for agglutination, i.e., granular appearances that indicate the blood type. (It may take a few minutes for the reaction to occur.) If a person had AB^+ blood, which wells would show agglutination? _________
____
5. Repeat steps 3 and 4 for plates #2 and #3.
6. Record the blood type results for each of the men in Table 16.2.

Conclusion

Which man is Sophia's father based on the results of your blood typing experiment? Explain your reply based on what you've learned about the inheritance of alleles for blood type and Rh. _______________________
____
____
____

16.3 Genetic Counseling

Potential parents are becoming aware that many illnesses are caused by abnormal chromosomal inheritance or by gene mutations. Therefore they are seeking genetic counseling, which is available in many major hospitals. The counselor helps the couple understand the mode of inheritance for a condition of concern so that the couple can make an informed decision about how to proceed.

Determining Chromosomal Inheritance

If a genetic counselor suspects that a condition is due to a chromosome anomaly, he or she may suggest that the parents' chromosomes be examined. It is possible to view the chromosomes of an individual because cells can be microscopically examined and photographed just before cell division occurs. A computer is then used to arrange the chromosomes by pairs. The resulting pattern of chromosomes is called a **karyotype.**

A trisomy occurs when the individual has three chromosomes instead of two chromosomes at one karyotype location. **Trisomy 21** (Down syndrome) is the most common autosomal trisomy in humans. Survival to adulthood is common. Characteristic facial features include an eyelid fold, a flat face, and a large, fissured tongue. Some degree of intellectual impairment is common as is early-onset Alzheimer disease. Sterility due to sexual underdevelopment may be present.

Observation: Sex Chromosome Anomalies

A female with **Turner syndrome** (XO) has only one sex chromosome, an X chromosome; the O signifies the absence of the second sex chromosome. Because the ovaries never become functional, these females do not undergo puberty or menstruation, and their breasts do not develop. Generally, females with Turner syndrome have a short build, folds of skin on the back of the neck, difficulty recognizing various spatial patterns, and normal intelligence. With hormone supplements, they can lead fairly normal lives.

When an egg having two X chromosomes is fertilized by an X-bearing sperm, an individual with **poly-X syndrome** results. The body cells have three X chromosomes and therefore 47 chromosomes. Although they tend to have learning disabilities, poly-X females have no apparent physical anomalies, and many are fertile and have children with a normal chromosome count.

When an egg having two X chromosomes is fertilized by a Y-bearing sperm, a male with **Klinefelter syndrome** results. This individual is male in general appearance, but the testes are underdeveloped, and the breasts may be enlarged. The limbs of XXY males tend to be longer than average, muscular development is poor, body hair is sparse, and many XXY males have learning disabilities.

Jacob syndrome occurs in males who are usually taller than average, suffer from persistent acne, and tend to have speech and reading problems. At one time, it was suggested that XYY males were likely to be criminally aggressive, but the incidence of such behavior has been shown to be no greater than that among normal XY males.

Label each karyotype in Figure 16.6 as one of the syndromes just discussed. Explain your answers on the lines provided.

Figure 16.6 Sex chromosome anomalies.

(1-4) © CNRI/SPL/Science Source

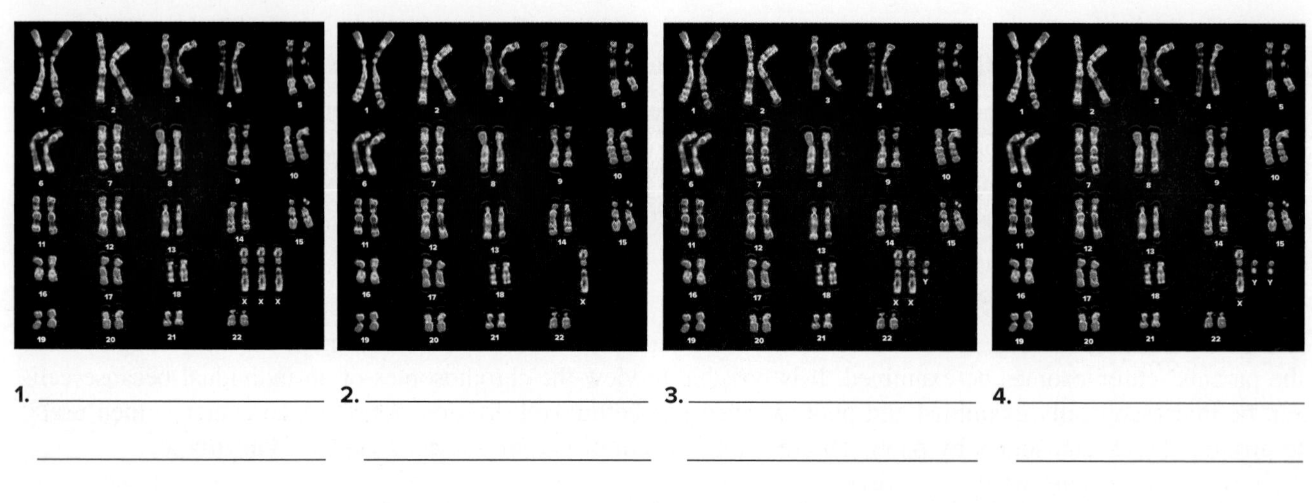

1. ______________________

2. ______________________

3. ______________________

4. ______________________

Determining the Pedigree

A pedigree shows the inheritance of a genetic disorder within a family and can help determine the inheritance pattern and whether any particular individual has an allele for that disorder. Then a Punnett square can be done to determine the chances of a couple producing an affected child.

The symbols used to indicate normal and affected males and females, reproductive partners, and siblings in a pedigree are shown in Figure 16.7.

For example, suppose you wanted to determine the inheritance pattern for straight hairline and you knew which members of a generational family had the trait (Fig. 16.8*a*).

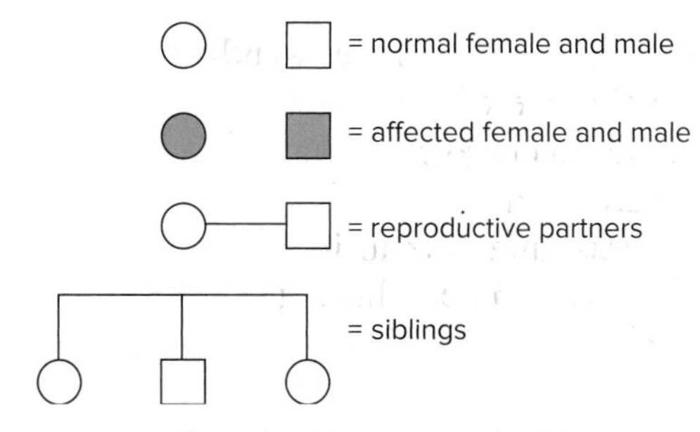

Figure 16.7 Pedigree symbols.

Figure 16.8 Autosomal pedigrees
a. Child with recessive phenotype can have parents without the recessive phenotype. **b.** Child with the dominant phenotype has parent(s) with the dominant phenotype; heterozygous parents can also have a child without the dominant phenotype.

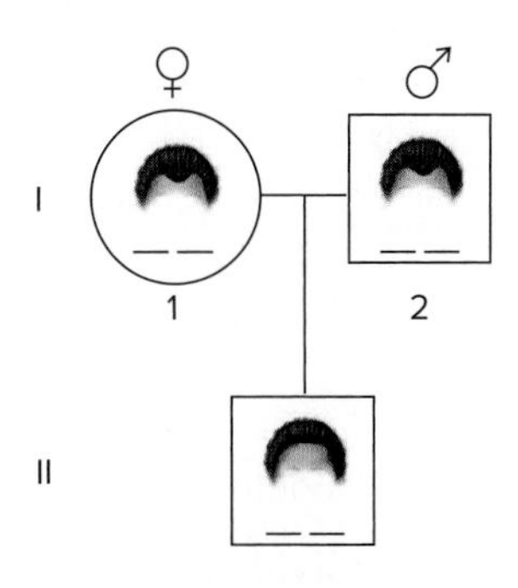

a. Straight hairline is recessive.

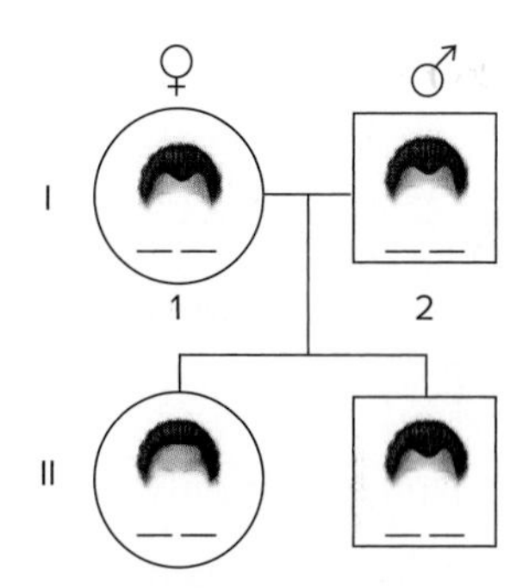

b. Widow's peak is dominant.

A pedigree allows you to determine that straight hairline is autosomal recessive because two parents without this phenotype have a child with the phenotype. This can happen only if the parents are heterozygous and straight hairline is recessive. Similarly, a pedigree allows you to determine that widow's peak is autosomal dominant (Fig. 16.8*b*): a child with this phenotype has at least one parent with the dominant phenotype, but again, heterozygous parents can produce a child without widow's peak. *Give each person in Figure 16.8*a *and* b *a genotype.*

Not shown is an X-linked recessive pedigree. An X-linked recessive phenotype occurs mainly in males, and it skips a generation because a female who inherits a recessive allele for the condition from her father may have a son with the condition.

Pedigree Analysis

For each of the following pedigrees, decide whether a trait is inherited as an autosomal dominant, autosomal recessive, or X-linked recessive. Then decide the genotype of particular individuals in the pedigree.

1. Study the following pedigree:

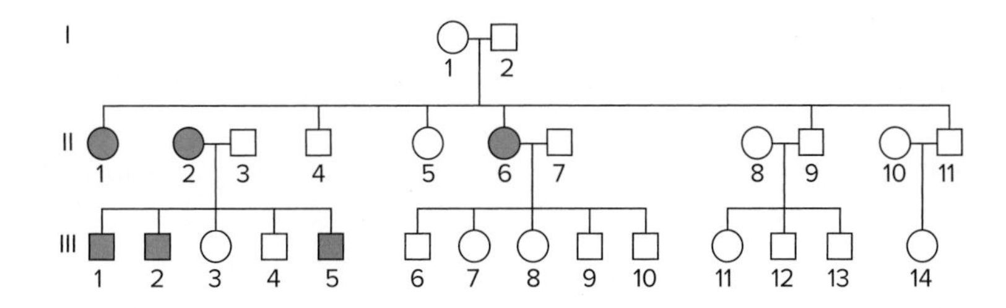

a. Notice that neither of the original parents is affected, but several children are affected. This could happen only if the trait were ______________________________.

b. What is the genotype of the following individuals? Use *A* for the dominant allele and *a* for the recessive allele and explain your answer.

Generation I, individual 1: ______________________________

Generation II, individual 1: ______________________________

Generation III, individual 8: ______________________________

2. Study the following pedigree:

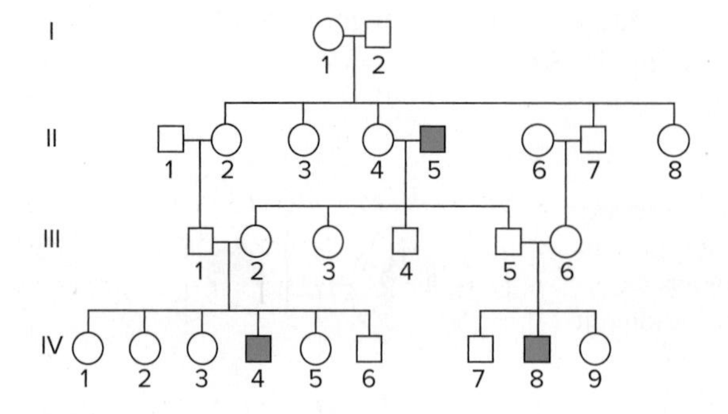

a. Notice that only males are affected. This could happen only if the trait were ______________.

b. What is the genotype of the following individuals? Explain your answer.

Generation I, individual 1: ______________________________

Generation II, individual 8: ______________________________

Generation III, individual 1: ______________________________

Construction of a Pedigree

You are a genetic counselor who has been given the following information from which you will construct a pedigree.

1. Your data: Henry has a double row of eyelashes, which is a dominant trait. Both his maternal grandfather and his mother have double eyelashes. Their spouses are normal. Henry is married to Isabella and their first child, Polly, has normal eyelashes. The couple wants to know the chances of any child having a double row of eyelashes.
2. What is your key for this trait?

   Key: __________ normal eyelashes __________ double row of eyelashes

3. *Construct two identical pedigrees with symbols (circles and squares) only for the underlined persons in #1.* The pedigrees start with the maternal grandfather and grandmother and end with Polly.

Pedigree 1 **Pedigree 2**

4. Pedigree 1: Try out a pattern of autosomal dominant inheritance by assigning appropriate genotypes for an autosomal dominant pattern of inheritance to each person in this pedigree. Pedigree 2: Try out a pattern of X-linked dominant inheritance by assigning appropriate genotypes for this pattern of inheritance to each person in your pedigree. Which pattern is correct? ____________________

5. Use correct genotypes to show a cross between Henry and Isabella and from experience with crosses, state the expected phenotypic ratio among the offspring:

Cross	Henry		Isabella	**Phenotypic ratio:**
	__________	×	__________	__________

6. What are the percentage chances that Henry and Isabella will have a child with double eyelashes? __________ "Chance has no memory," and each child has the same chance for double eyelashes. Explain why.

____

____

Application for Daily Living

Choosing the Gender of Your Child

The gender of a child depends upon whether an X-bearing sperm or a Y-bearing sperm enters the egg. A new technology that can separate X-bearing and Y-bearing sperm offers prospective parents the opportunity to choose the sex of their child. The results are not perfect. Following artificial insemination, there's about an 85% success rate for a girl and about a 65% success rate for a boy.

Proponents of sex-selection technology argue that there are many instances in which the ability to choose the sex of a child may benefit society. For example, if a mother is a carrier of an X-linked genetic disorder, such as hemophilia or Duchenne muscular distrophy, this would be the simplest way to ensure a healthy child. Previously, a pregnant woman with these concerns had to wait for the results of an amniocentesis test and then decide whether to abort the pregnancy if it was a boy. In such cases, is it better to ensure that the child will not have a specific genetic disorder than to take the risk?

Also, it's possible for parents to use this technology simply because they choose to have a child of one particular gender. Do you think it is acceptable to select the sex of your child? Some are concerned that allowing parents to select the sex of their children could lead to selecting specific traits for their children. Should this be a concern?

Laboratory Review 16

Homozygous dominant 1. What phrase describes an individual with two dominant alleles?

Genotype 2. What word refers to the alleles possessed by an individual?

Twenty-two 3. How many autosomes do humans have?

Dominant trait 4. An individual will have a widow's peak if the genotype is *WW* or *Ww*. Is widow's peak an autosomal dominant or autosomal recessive trait?

Punnett Square 5. What technique can be used to determine the genetic inheritance of offspring when the parents have known genotypes?

Carriers 6. Cystic fibrosis is an autosomal recessive disease. What word refers to an individual who has one recessive allele for cystic fibrosis?

hh 7. What is the genotype of individuals who will not develop Huntington disease (autosomal dominant)?

________ 8. From whom does a male child inherit color blindness—his mother or his father?

________ 9. What is the genotype of a carrier for hemophilia?

________ 10. How many dominant alleles does someone with AB blood type have?

________ 11. Can a child with AB blood type have a parent with O blood?

________ 12. What is the presence of three chromosomes (instead of two) called, and what human chromosome is most commonly affected in this fashion?

________ 13. What is created when chromosomes are paired by size and shape?

Pedigree 14. What technique is used to determine the inheritance pattern of specific traits within a family?

Thought Questions

15. The blood types of parents and their two children are determined. The father has type A blood. The mother has type O blood. One child has type A blood and the other has type O blood. What are the genotypes of all people involved? Explain your reply.

16. What are the chances that a color-blind man will have a color-blind grandson (the child of the color-blind man's normal vision daughter and a normal vision male)? What are the genotypes of all people involved?

17. Recall what you learned about the fate of homologous chromosomes during meiosis I and the sister chromatids during meiosis II, and explain how trisomies like trisomy 21 or poly-X syndrome could occur.

LABORATORY

17

DNA Biology and Technology

Learning Outcomes

17.1 DNA Structure and Replication

- Explain how the structure of DNA facilitates replication.
- Explain how DNA replication is semiconservative.

Prelab Question: What does semiconservative replication mean?

17.2 RNA Structure

- List the ways in which RNA structure differs from DNA structure.

Prelab Question: The R in RNA stands for what?

17.3 DNA and Protein Synthesis

- Describe how DNA is able to store information.
- State the function of transcription and translation during protein synthesis.

Prelab Question: Which one, transcription or translation, produces a protein?

17.4 Isolation of DNA and Biotechnology

- Describe the procedure for isolating DNA.
- Describe the process of DNA gel electrophoresis.

Prelab Question: Give an example to show that DNA can be manipulated in the laboratory like any other chemical.

17.5 Detecting Genetic Disorders

- Understand the relationship between an abnormal DNA base sequence and a genetic disorder.
- Suggest two ways to detect a genetic disorder.

Prelab Question: If the DNA base sequence changes, why does the protein change?

Application for Daily Living: **Personal DNA Sequencing**

Introduction

This laboratory pertains to molecular genetics and biotechnology. Molecular genetics is the study of the structure and function of **DNA** (**deoxyribonucleic acid**), the genetic material. Biotechnology refers to the use of natural biological systems to create a product or achieve some other end desired by humans. **Genetic engineering**, or modification of an organism's genes, is a form of biotechnology.

First we will study the structure of DNA and see how that structure facilitates DNA replication in the nucleus of cells. DNA replicates prior to mitosis; following mitosis, each daughter nucleus has a complete copy of the genetic material. DNA replication also precedes the production of gametes.

Then we will study the structure of **RNA** (**ribonucleic acid**) and how it differs from that of DNA before examining how DNA, with the help of RNA, specifies protein synthesis. The linear construction of DNA, in which nucleotide follows nucleotide, is paralleled by the linear construction of the primary structure of protein, in which amino acid follows amino acid. Essentially, we will see that the sequence of nucleotides in DNA codes for the sequence of amino acids in a protein. We will also review the role of three types of RNA in protein synthesis. DNA's code is passed to messenger RNA (mRNA), which moves to the ribosomes containing ribosomal RNA (rRNA). Transfer RNA (tRNA) brings the amino acids to the ribosomes, and they become linked by peptide bonds in the order directed by mRNA. In this way, a particular polypeptide forms.

We now understand that a mutated gene has an altered DNA base sequence, and altered sequences can cause genetic disorders. You will have an opportunity to carry out a laboratory procedure that detects whether an individual is normal, has sickle-cell disease, or is a carrier.

17.1 DNA Structure and Replication

The structure of DNA lends itself to **replication,** the process that makes a copy of a DNA molecule. DNA replication is a necessary part of chromosome duplication, which precedes mitosis and meiosis. Therefore, DNA replication is needed for growth and repair and for sexual reproduction.

DNA Structure

DNA is a polymer of nucleotide monomers (Fig. 17.1). Each nucleotide is composed of three molecules: deoxyribose (a 5-carbon sugar), a phosphate, and a nitrogen-containing base.

Figure 17.1 Overview of DNA structure.
Diagram of DNA double helix shows that the molecule resembles a twisted ladder. Sugar–phosphate backbones make up the sides of the ladder, and hydrogen-bonded bases make up the rungs of the ladder. Complementary base pairing dictates that A is bonded to T and G is bonded to C and vice versa. *Label the boxed nucleotide pair as directed in the next Observation.*

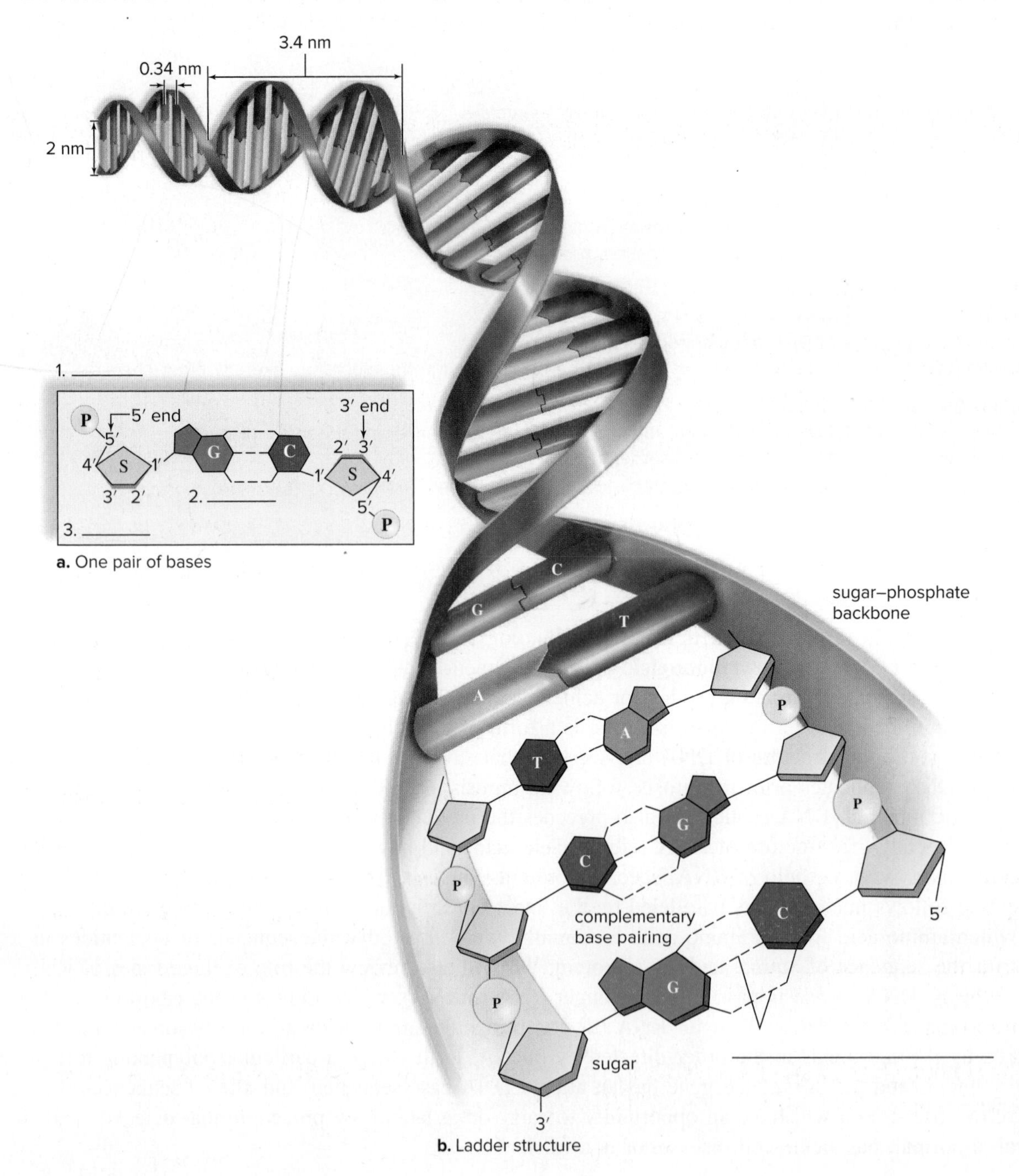

a. One pair of bases

b. Ladder structure

Observation: DNA Structure

1. A boxed nucleotide pair is shown in Figure 17.1. If you are working with a kit, draw a representation of one of your nucleotides here. *Label phosphate, base pair, and deoxyribose in your drawing and in spaces 1–3 in Figure 17.1a.*

2. Notice the four types of bases: cytosine (C), thymine (T), adenine (A), and guanine (G). What is the color of each of the four types of bases in Figure 17.1? In your kit? Complete Table 17.1 by writing in the colors of the bases.

Table 17.1 Base Colors

	In Figure 17.1	In Your Kit
Cytosine	red	blue
Thymine	purple / blue	yellow
Adenine	orange	red
Guanine	green	green.

3. Using Figure 17.1 as a guide, join several nucleotides together. Observe the entire DNA molecule. What types of molecules make up the backbone (uprights of ladder) of DNA (Fig. 17.1*b*)? Sugar-phosphate and hydrogen bonded In the backbone, the phosphate of one nucleotide is bonded to a sugar of the next nucleotide.
4. Using Figure 17.1 as a guide, join the bases together with hydrogen bonds. *Label a hydrogen bond in Figure 17.1.* Dashes are used to represent hydrogen bonds in Figure 17.1*b* because hydrogen bonds are (strong or weak) Strong.
5. Notice in Figure 17.1*b* and in your model that the base A is always paired with the base T, and the base C is always paired with the base G. This is called complementary base pairing.
6. In Figure 17.1*b,* what molecules make up the rungs of the ladder? hydrogen bonded bases
7. Each half of the DNA molecule is a DNA strand. Why is DNA also called a double helix (Fig. 17.1*b*)?

____

DNA Replication

During replication, the DNA molecule is duplicated so that there are two identical DNA molecules. We will see that complementary base pairing makes replication possible.

Observation: DNA Replication

1. Before replication begins, DNA is unzipped. Using Figure 17.2*a* as a guide, break apart your two DNA strands. What bonds are broken in order to unzip the DNA strands? ___________
2. Using Figure 17.2*b* as a guide, attach new complementary nucleotides to each strand using complementary base pairing.
3. Show that you understand complementary base pairing by completing Table 17.2.
   You now have two DNA molecules (Fig. 17.2*c*). Are your molecules identical?

   ____
4. Because of complementary base pairing, each new double helix is composed of an ___________ strand and a ___________ strand. *Write old or new in 1–10, Figure 17.2*a, b, *and* c.
   Conservative means to save something from the past. Why is DNA replication called semiconservative?

   ____

   ____

   ____

Figure 17.2 DNA replication.
Use of the ladder configuration better illustrates how replication takes place. **a.** The parental DNA molecule. **b.** The "old" strands of the parental DNA molecule have separated. New complementary nucleotides available in the cell are pairing with those of each old strand. **c.** Replication is complete.

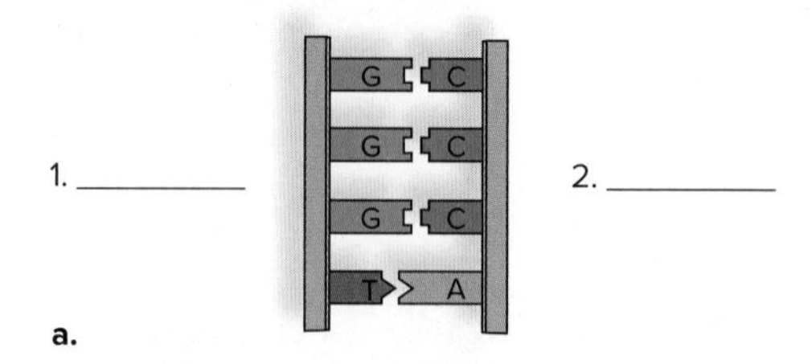

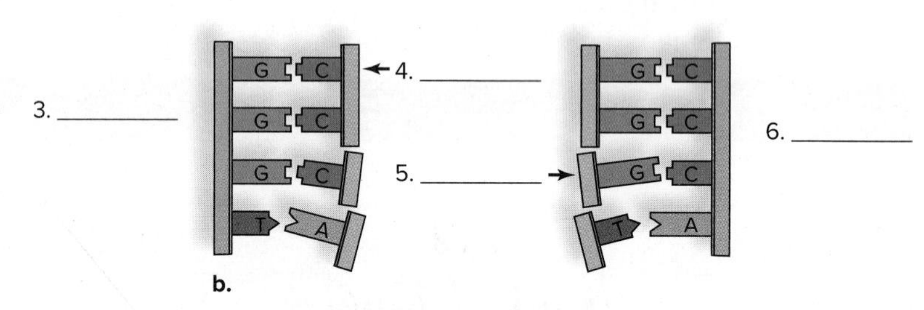

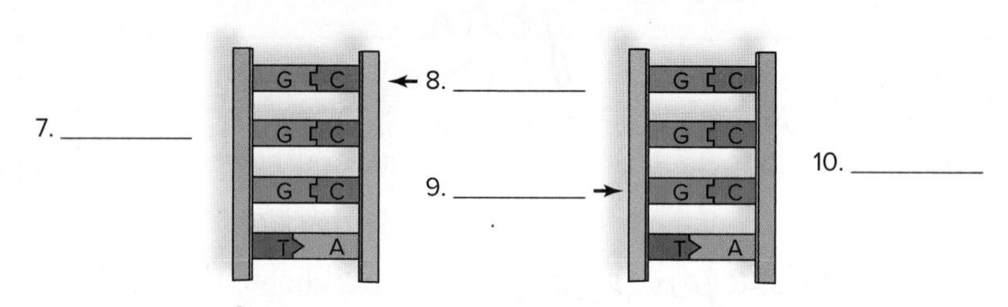

5. Genetic material has to be inherited from cell to cell and organism to organism. Consider that because of DNA replication, a chromosome is composed of two chromatids and each chromatid is a DNA double helix. The chromatids separate during cell division so that each daughter cell receives a copy of each chromosome. In your own words, how does replication provide a means for passing DNA from cell to cell and organism to organism? ___________________________________

   ____

Table 17.2 DNA Replication

Old strand	G	G	G	T	T	C	C	A	T	T	A	A	A	T	T	C	C	A	G	A	A	A	T	C	A	T	A
New strand																											

17.2 RNA Structure

Like DNA, RNA is a polymer of nucleotides (Fig. 17.3). In an RNA nucleotide, the sugar ribose is attached to a phosphate molecule and to a nitrogen-containing base, C, U, A, or G. In RNA, the base uracil replaces thymine as one of the pyrimidine bases. RNA is single stranded, whereas DNA is double stranded.

Figure 17.3 Overview of RNA structure.
RNA is a single strand of nucleotides. Label the boxed nucleotide as directed in the next Observation.

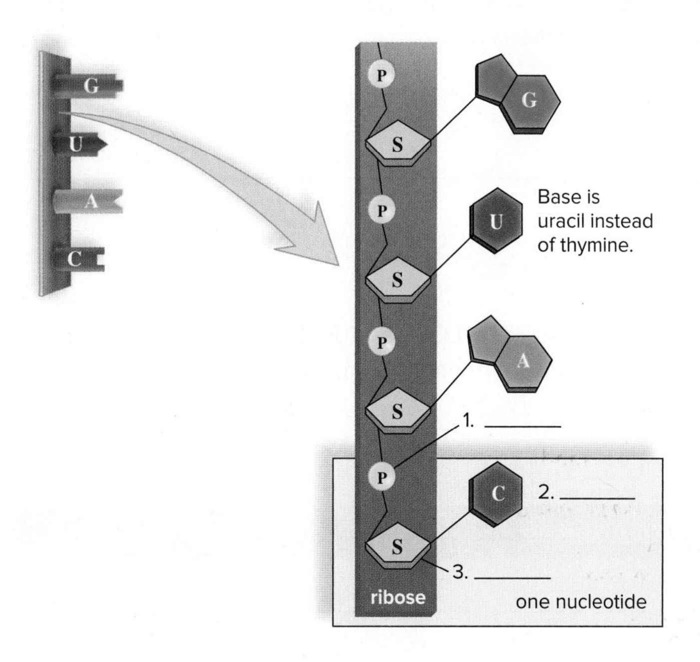

1. Describe the backbone of an RNA molecule. ______________________________

2. Where are the bases located in an RNA molecule? ______________________________

Observation: RNA Structure

1. If you are using a kit, draw a nucleotide for the construction of mRNA. *Label the ribose (the sugar in RNA), the phosphate, and the base in your drawing and in 1–3, Figure 17.3.*
2. Complete Table 17.3 by writing in the colors of the bases for Figure 17.3 and for your kit.

Table 17.3 Base Colors

	In Figure 17.3	In Your Kit
Cytosine	Red	Blue
Uracil	Purple	white
Adenine	Yellow	Red
Guanine	Green	Green

3. The base uracil substitutes for the base thymine in RNA. Complete Table 17.4 to show the several other ways RNA differs from DNA.

Table 17.4 DNA Structure Compared with RNA Structure

	DNA	RNA
Sugar	Deoxyribose	
Bases	Adenine, guanine, thymine, cytosine	
Strands	Double stranded with base pairing	
Helix	Yes	

Complementary Base Pairing

Complementary base pairing occurs between DNA and RNA. The RNA base uracil pairs with the DNA base adenine; the other bases pair as shown previously. Complete Table 17.5 to show the complementary DNA bases for the RNA bases.

Table 17.5 DNA and RNA Bases

RNA bases	C	U	A	G
DNA bases				

17.3 DNA and Protein Synthesis

Protein synthesis requires the processes of transcription and translation. During **transcription,** which takes place in the nucleus, an RNA molecule called **messenger RNA (mRNA)** is made complementary to one of the DNA strands. This mRNA leaves the nucleus and goes to the ribosomes in the cytoplasm. Ribosomes are composed of **ribosomal RNA (rRNA)** and proteins in two subunits. They provide a location for protein synthesis to occur.

During **translation,** RNA molecules called **transfer RNA (tRNA)** bring amino acids to the ribosome, and the amino acids join in the order prescribed by mRNA. In this way, the sequence of amino acids in a new polypeptide was originally specified by DNA, demonstrating that DNA carries the codes to make all of the proteins necessary for a cell to function.

Explain the role DNA, mRNA, and tRNA have in protein synthesis here:

DNA ______________________________

mRNA ______________________________

tRNA ______________________________

Transcription

During transcription, complementary RNA is made from a DNA template (Fig. 17.4). A portion of DNA unwinds and unzips at the point of attachment of the enzyme **RNA polymerase.** A strand of mRNA is produced when complementary nucleotides join in the order dictated by the sequence of bases in DNA. Transcription occurs in the nucleus, and the mRNA passes out of the nucleus to enter the cytoplasm.

Label Figure 17.4. For number 1, note the name of the enzyme that carries out mRNA synthesis. For number 2, note the name of this polynucleotide molecule.

Figure 17.4 Messenger RNA (mRNA).
Messenger RNA, which is complementary to a section of DNA, forms during transcription.

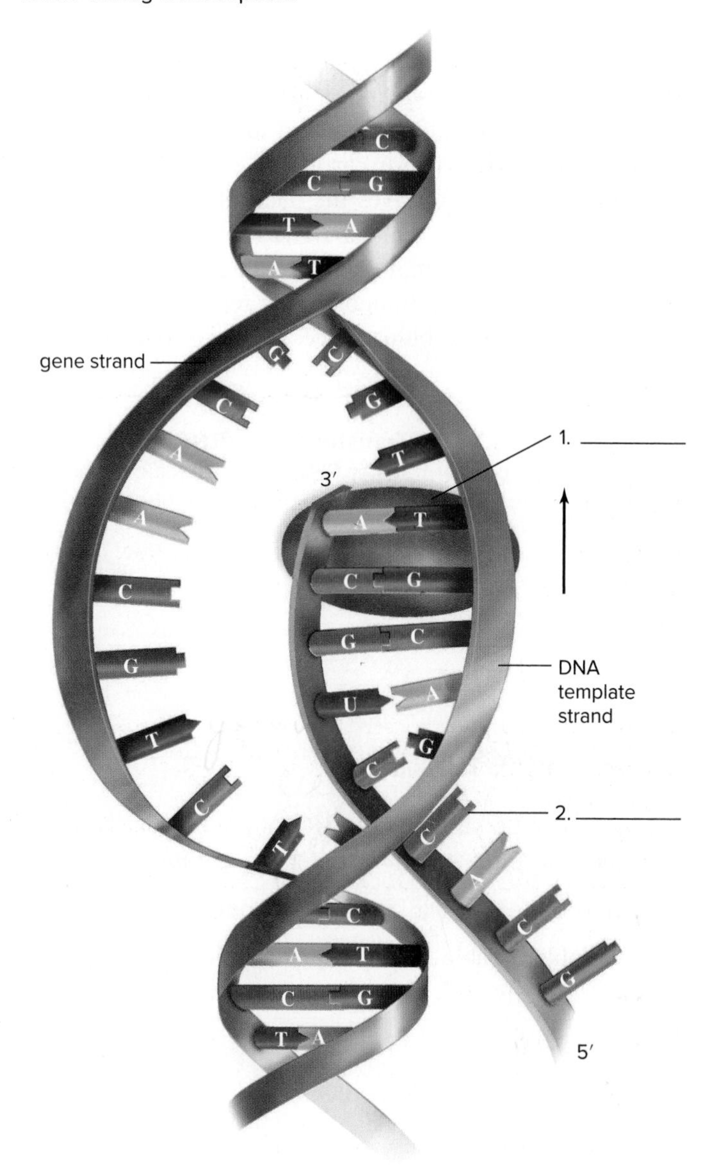

Observation: Transcription

1. If you are using a kit, unzip your DNA model so that only one strand remains. This strand is called the **template strand*** because it will be used as a template to construct an mRNA molecule. The **gene strand** is preferred terminology for the complementary DNA strand you discarded because as you can verify in Figure 17.4, it has the same sequence of bases as the mRNA molecule, except that uracil substitutes for thymine.
2. Using Figure 17.4 as a guide, construct a messenger RNA (mRNA) molecule by first lining up RNA nucleotides complementary to the template strand of your DNA molecule. Join the nucleotides together to form mRNA.
3. A portion of DNA has the sequence of bases shown in Table 17.6. *Complete Table 17.6 to show the sequence of bases in mRNA.*
4. If you are using a kit, unzip mRNA transcript from the DNA. Locate the end of the strand that will move to the ________________ in the cytoplasm.

Table 17.6 Transcription

DNA	T A C A C G A G C A A C T A A C A T
mRNA	UT

*Sometimes the template strand is called the sense strand and the gene strand is called the antisense strand.

Translation

During translation, a polypeptide is made. DNA specifies the sequence of amino acids in a polypeptide because every three bases code for an amino acid. Therefore, DNA is said to have a **triplet code.** The bases in mRNA are complementary to the bases in DNA. Every three bases in mRNA are called a **codon.** One codon of mRNA represents one amino acid. Thus, the sequence of DNA bases serves as the blueprint for the sequence of amino acids assembled to make a protein. The correct sequence of amino acids in a polypeptide is the message that mRNA carries.

Messenger RNA leaves the nucleus and proceeds to the ribosomes, where protein synthesis occurs. As previously mentioned, transfer RNA (tRNA) molecules transfer amino acids to the ribosomes. Each RNA has one particular tRNA amino acid at one end and a specific **anticodon** at the other end (Fig. 17.5). *Label Figure 17.5,* where the amino acid is represented as a colored ball, the tRNA is green, and the anticodon is the sequence of three bases. (The anticodon is complementary to the mRNA codon.)

Figure 17.5 Transfer RNA (tRNA).
Each type of transfer RNA with a specific anticodon carries a particular amino acid to the ribosomes.

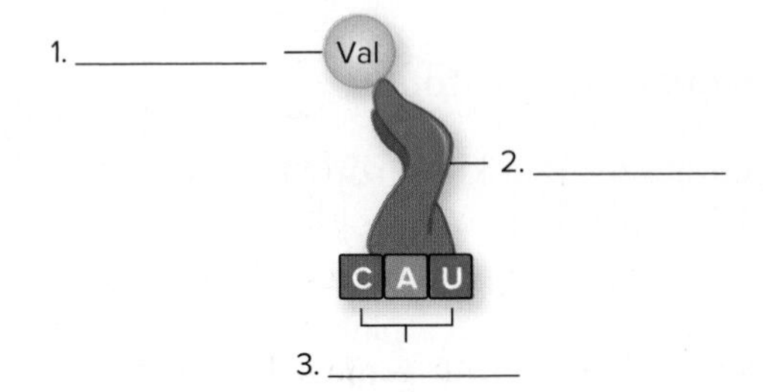

Observation: Translation

1. Figure 17.6 shows seven tRNA–amino acid complexes. Every amino acid has a name; in the figure, only the first three letters of the name are inside the ball.
2. If you are using a kit, arrange your tRNA–amino acid complexes in the order consistent with Table 17.7. Complete Table 17.7. Why are the codons and anticodons in groups of three? ____________________

____

Figure 17.6 Transfer RNA diversity.
Each type of tRNA carries only one particular amino acid, designated here by the first three letters of its name.

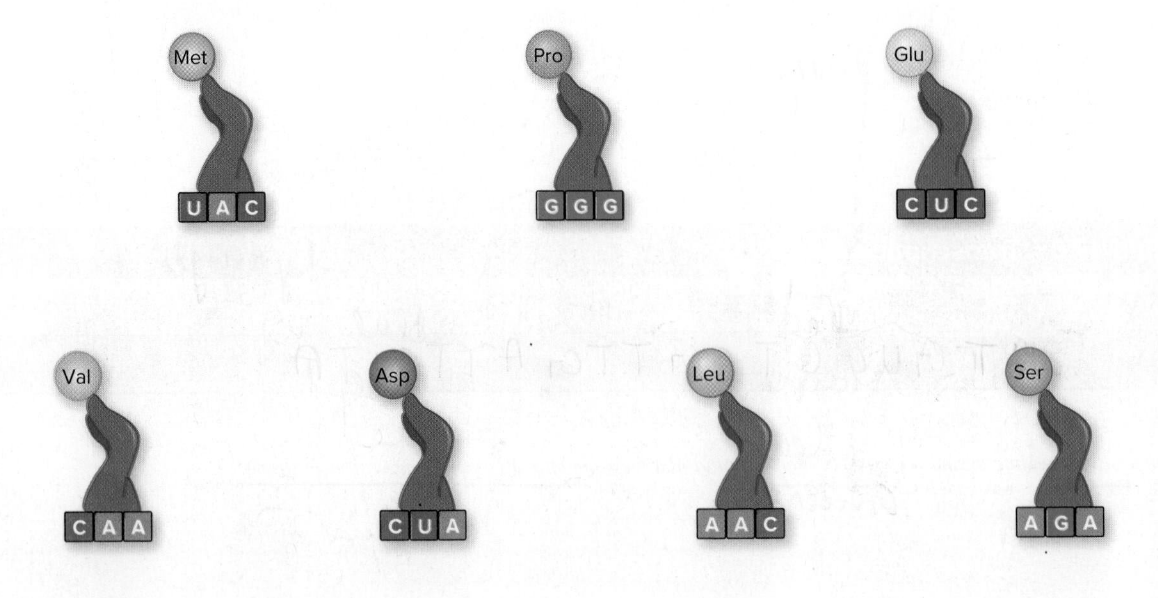

Table 17.7 Translation							
mRNA codons	AUG	CCC	GAG	GUU	GAU	UUG	UCU
tRNA anticodons	TAC	GGG	CUC				
Amino acid*							

*Use three letters only. See Table 17.8 for the full names of these amino acids.

Table 17.8 Names of Amino Acids

Abbreviation	Name
Met	methionine
Pro	proline
Asp	aspartate
Val	valine
Glu	glutamic acid
Leu	leucine
Ser	serine

3. Figure 17.7 shows the manner in which the polypeptide grows. A ribosome has three binding sites. They are the A (amino acid) site, the P (peptide) site, and the E (exit) site. A tRNA leaves from the E site after it has passed its amino acid or peptide to the newly arrived tRNA–amino acid complex. Then the ribosome moves forward, making room for the next tRNA–amino acid. This sequence of events occurs over and over until the entire polypeptide is borne by the last tRNA to come to the ribosome. Then a release factor releases the polypeptide chain from the ribosome. *In Figure 17.7, label the ribosome, the mRNA, and the peptide.*

Figure 17.7 Protein synthesis.
During protein synthesis, amino acids are added one at a time to the growing chain based on the mRNA codons. This process occurs at ribosomes in the cell.

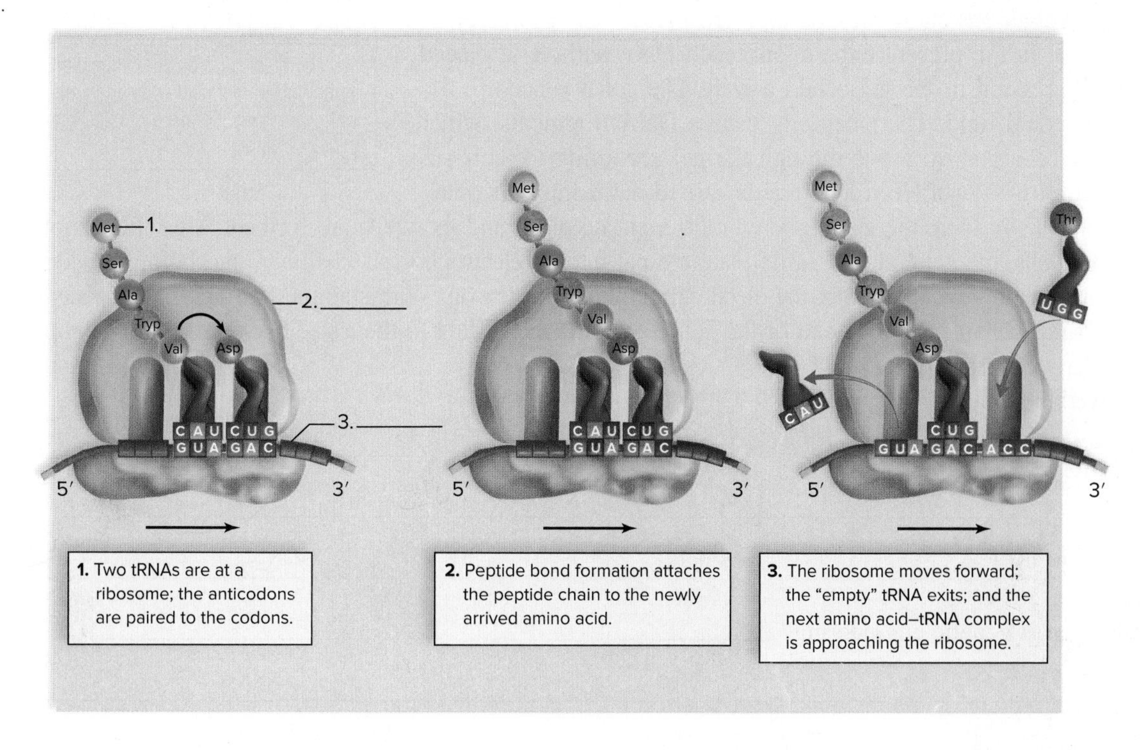

17.4 Isolation of DNA and Biotechnology

In the following Experimental Procedure, you will isolate DNA from the cells of a fruit or vegetable. It will only be necessary to expose the cells to an agent (dishwasher detergent) that emulsifies membrane in order to "free" the DNA from its enclosures (plasma membrane and nuclear envelope). When transferred to a tube, the presence of NaCl allows DNA to precipitate as a sodium salt. The precipitate forms at the interface between ethanol and the salt solution, and then it floats to the top of the tube where it may be collected.

Experimental Procedure: Isolation of DNA

1. You will need a slice of fruit or vegetable (i.e., tomato, onion, or apple), a mortar and pestle, and a large, clean glass test tube on ice.
2. Crush and grind a slice of fruit or vegetable in a mortar and pestle. Remove the pestle and set aside. Add enough 0.9% NaCl solution to achieve a "soupy" consistency.
3. Add two drops of dishwasher detergent (such as Blue Dawn) to the mixture in the mortar. Swirl until the color of the detergent disappears. Wait 3 to 5 minutes. The solution becomes clear and viscous as the DNA escapes its enclosures.
4. Check the viscosity of the solution in the mortar, and if it is too viscous for you to pipet, add a few more ml of NaCl solution. Use a transfer pipet to move 2 to 3 ml of the DNA from the mortar to the glass tube. Try to pick up clear zones of the solution WITHOUT CELL DEBRIS.
5. Slowly add 5 to 7 ml of ice-cold ethanol down the side of the tube on ice. As you do so, DNA will precipitate *between* the ethanol and the NaCl solutions. Then it will float to the top, where it can be picked up by a fresh transfer pipet. Any "white dust" you see consists of RNA molecules.
6. Use a transfer pipet to place the DNA in a small, clean test tube. Pipet away any extra water/ethanol and air-dry the DNA for a few minutes. Dissolve the DNA in 3 to 4 ml distilled H_2O.
7. Add five drops of phenol red, a pH indicator. The resulting dark pink color confirms the presence of nucleic acid (i.e., the DNA).

Experimental Procedure: Gel Electrophoresis

For experimental procedures, wear safety goggles, gloves, and protective clothing. If any chemical spills on your skin, wash immediately with mild soap and water; flood eyes with water only. Report any spills immediately to your instructor.

During gel electrophoresis, charged DNA molecules migrate across a span of gel (gelatinous slab) because they are placed in a powerful electrical field. In the present experiment, each DNA sample is placed in a small depression in the gel called a well. The gel is placed in a powerful electrical field. The electricity causes DNA fragments, which are negatively charged, to move through the gel according to their size.

Almost all DNA gel electrophoresis is carried out using horizontal gel slabs (Fig. 17.8). First, the gel is poured onto a plastic plate, and the wells are formed. After the samples are added to the wells, the gel and the plastic plate are put into an electrophoresis chamber, and buffer is added. The DNA samples begin to migrate after the electrical current is turned on. With staining, the DNA fragments appear as a series of bands spread from one end of the gel to the other according to their size because smaller fragments move faster than larger fragments.

Answer the following questions:

a. What is biotechnology? (See lab introduction) ____________________

____

____

b. What is genetic engineering? ____________________

____

Figure 17.8 Equipment and procedure for gel electrophoresis.

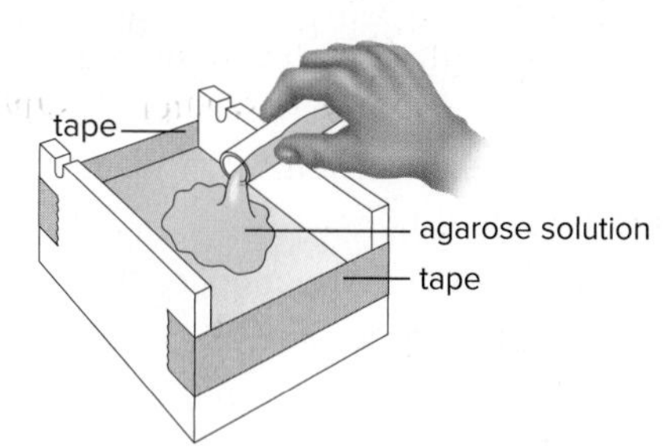

a. Agarose solution poured into casting tray

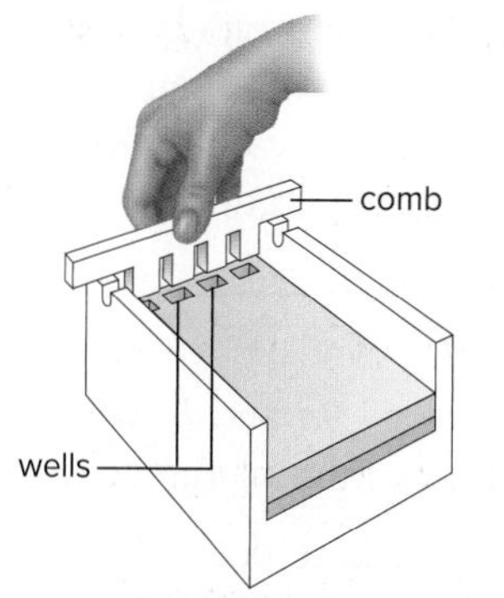

b. Comb that forms wells for samples

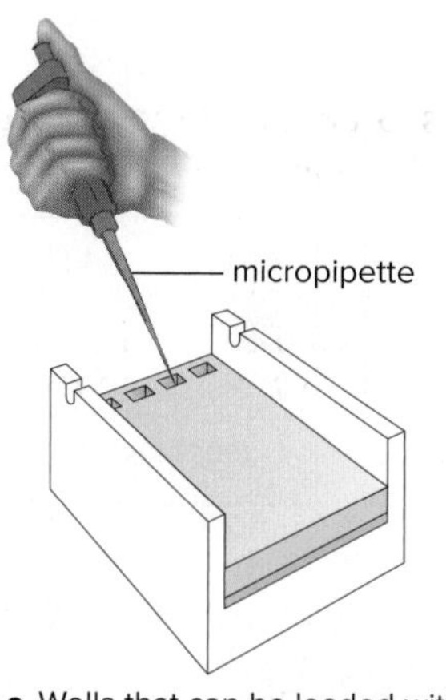

c. Wells that can be loaded with samples

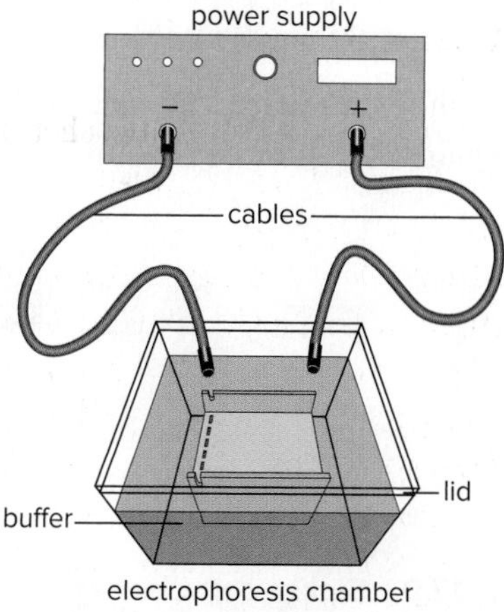

d. Electrophoresis chamber and power supply

c. Speculate how the ability to isolate DNA and run gel electrophoresis of DNA relates to biotechnology.

___

___

___

d. Name a biotechnology product someone you know is now using or taking as a medicine.

___

17.5 Detecting Genetic Disorders

The base sequence of DNA in all the chromosomes is an organism's genome. The Human Genome Project was a monumental effort to determine the normal order of all the 3.6 billion nucleotide bases in the human genome. Someday it will be possible to sequence anyone's genome within a relatively short time; this will enable us to determine what particular base sequence alterations signify that he or she has a disorder or will have one in the future. In this laboratory, you will study the alteration in base sequence that causes a person to have sickle-cell disease.

In persons with sickle-cell disease, the red blood cells aren't biconcave disks like normal red blood cells—they are sickle shaped. Sickle-shaped cells can't pass along narrow capillary passageways. They clog the vessels and break down, causing the person to suffer from poor circulation, anemia, and poor resistance to infection. Internal hemorrhaging leads to further complications, such as jaundice, episodic pain in the abdomen and joints, and damage to internal organs.

Sickle-shaped red blood cells are caused by an abnormal hemoglobin (Hb^S). Individuals with the Hb^AHb^A genotype are normal; those with the Hb^SHb^S genotype have sickle-cell disease, and those with the Hb^AHb^S have sickle-cell trait. Persons with sickle-cell trait do not usually have sickle-shaped cells unless they experience dehydration or mild oxygen deprivation.

Genetic Sequence for Sickle-Cell Disease

Examine Figure 17.9*a* and *b,* which show the DNA base sequence, the mRNA codons, and the amino acid sequence for a portion of the gene for Hb^A and the same portion for Hb^S.

1. In what one base does Hb^A differ from Hb^S? Hb^A ________ Hb^S ________
2. What are the codons that contain this base? Hb^A ________ Hb^S ________
3. What is the amino acid difference? Hb^A ________ Hb^S ________

This amino acid difference causes the polypeptide chain in sickle-cell hemoglobin to pile up as firm rods that push against the plasma membrane and deform the red blood cell into a sickle shape:

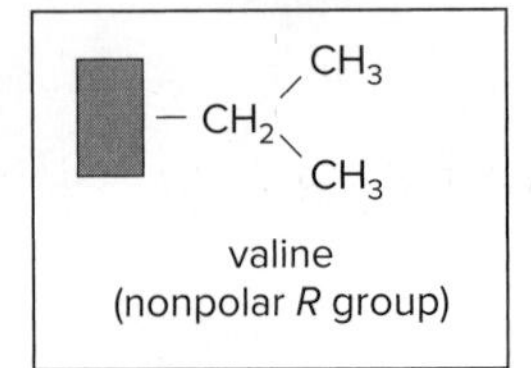

Figure 17.9 Sickle-cell disease.
a. When red blood cells are normal, the base sequence (in one location) for Hb^A alleles is CTC. **b.** In sickle-cell disease at these locations, it is CAC.

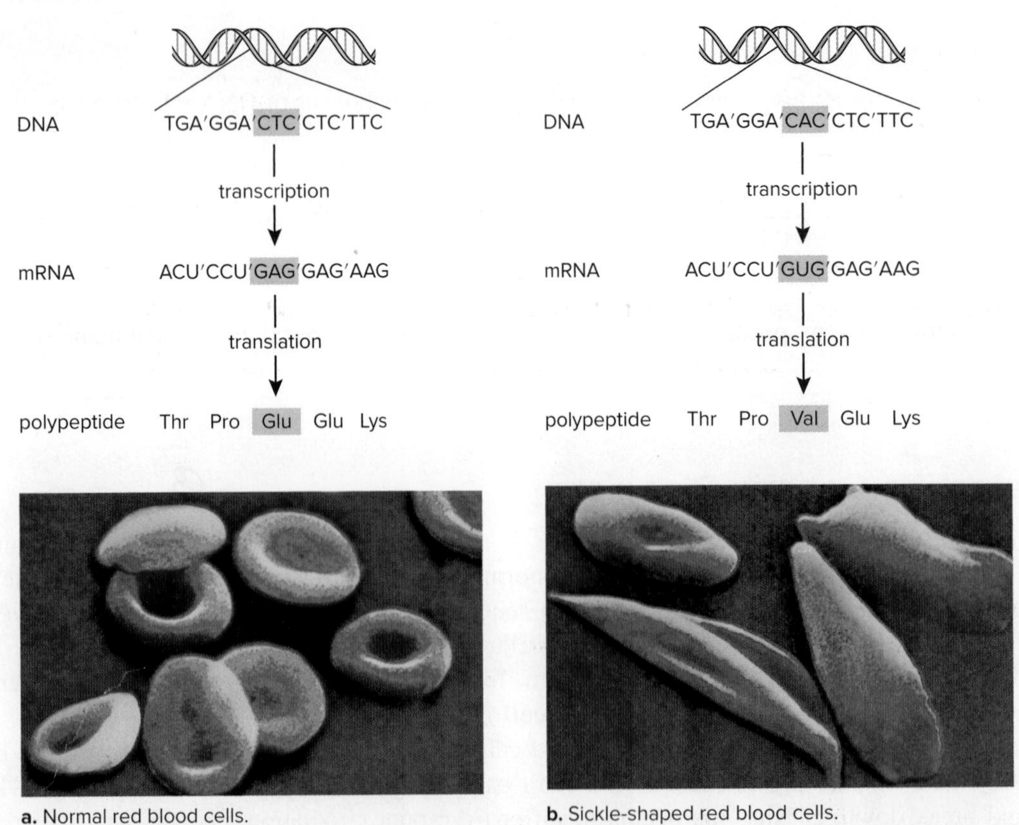

a. Normal red blood cells.

b. Sickle-shaped red blood cells.

Detection of Sickle-Cell Disease by Gel Electrophoresis

Three samples of hemoglobin have been subjected to protein gel electrophoresis. Protein gel electrophoresis (Fig. 17.10) is carried out in the same manner as DNA gel electrophoresis (see Fig. 17.8) except the gel has a different composition.

1. Sickle-cell hemoglobin (Hb^S) migrates more slowly toward the positive pole than normal hemoglobin (Hb^A) because the amino acid valine has no polar *R* groups, whereas the amino acid glutamic acid does have a polar *R* group.
2. In Figure 17.10, which lane contains only Hb^S, signifying that the individual is Hb^SHb^S?

   ____

3. Which lane contains only Hb^A, signifying that the individual is Hb^AHb^A? ________
4. Which lane contains both Hb^S and Hb^A, signifying that the individual is Hb^AHb^S?

   ____

Figure 17.10 Gel electrophoresis of hemoglobins.

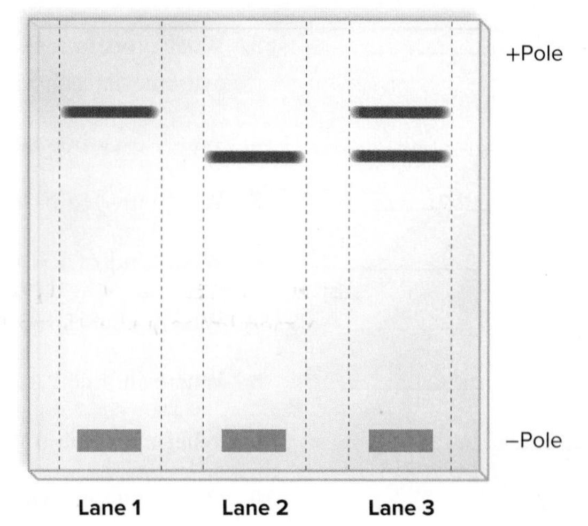

Detection by Genomic Sequencing

You are a genetic counselor. A young couple seeks your advice because sickle-cell disease occurs among the family members of each. You order DNA base sequencing to be done. The results come back that at one of the loci for normal hemoglobin, each has the abnormal sequence CAC instead of CTC. The other locus is normal. What are the chances that this couple will have a child with sickle-cell disease? ________________

Conclusion: Detecting Genetic Disorders

- What two methods of detecting sickle-cell disease were described in this section? ________________

  ____

- Which method is more direct and probably requires more expensive equipment to do? ________________

  ____

- Which method probably preceded the other method as a means to detect sickle-cell disease? ________________

  ____

Application for Daily Living

Personal DNA Sequencing

You may know that it took many years to get the human genome sequenced, but now personal DNA base sequencing can be done within days, and it is relatively affordable. The information can allow a trained geneticist to predict which genetic disorders you are susceptible to, and then you can choose either prevention and/or early detection to avoid a serious illness. DNA sequencing may also tell you which drugs might be especially useful to you in case of illness.

The move is on to require people to go through their personal physician to get their DNA sequenced. Why? It is always a good idea to have it done by a reputable company and to have someone help you sift through the data that might be made available. It's not always easy to learn that you may come down with a disorder, and you would want to know what your options are right away.

Laboratory Review 17

_______________ **1.** What biomolecule is composed of deoxyribose sugars, phosphates, and the nitrogen-containing bases adenine, guanine, cytosine, and thymine?

_______________ **2.** What holds the nitrogen bases together in DNA?

_______________ **3.** What process makes a copy of DNA so that daughter cells that result from mitosis and cytokinesis each have a copy of DNA?

_______________ **4.** Which 5-carbon sugar is characteristic of RNA?

_______________ **5.** What nitrogen base is associated with RNA but not with DNA?

_______________ **6.** What kind of RNA is made when transcription occurs?

_______________ **7.** Which strand of DNA is used to make mRNA?

_______________ **8.** Where in a cell does protein synthesis occur?

_______________ **9.** What molecule transports amino acids to the ribosomes?

_______________ **10.** How many nitrogen bases make up a codon, and what does each codon represent?

_______________ **11.** What connects one amino acid to another in the polypeptide chain?

_______________ **12.** Which size DNA fragments, smaller or larger, move more quickly through a gel during electrophoresis?

_______________ **13.** How are many genetic disorders now identified?

_______________ **14.** What protein is affected when an individual has sickle-cell disease?

Thought Questions

15. Students sometimes get the impression that DNA must be replicated prior to protein synthesis. Explain why DNA is only replicated prior to the start of mitosis and meiosis and what the result would be if DNA was replicated each time a new protein needed to be made.

16. If the portion of DNA to be transcribed has 66 bases, how many amino acids will the resulting protein have? Explain your reply.

17. a. What concerns should high school and college athletic coaches have about the potential for their players to have sickle-cell trait?

b. How could players be identified as having sickle-cell trait or not?

LABORATORY

18

Human Evolution

Learning Outcomes

18.1 Evidence from the Fossil Record

- Use the geologic timescale to trace the evolution of life in broad outline.
- Describe several types of fossils and explain how fossils help establish the sequence of evolution.
- Explain how scientists use fossils to establish that organisms are related by common descent.

Prelab Question: Explain why the order of fossils in rock matters.

18.2 Evidence from Comparative Anatomy

- Explain how comparative anatomy provides evidence that humans are related to other groups of vertebrates.
- Compare the human skeleton with the chimpanzee skeleton, and explain differences based on lifestyle.
- Compare hominid skulls and hypothesize a possible evolutionary sequence.

Prelab Question: What anatomical features are similar between a chimpanzee and human skeleton?

18.3 Molecular Evidence

- Explain how molecular evidence aids the study of how humans are related to all other groups of organisms.
- Explain how molecular evidence also helps show how humans are related to other vertebrates and primates.

Prelab Question: What are the advantages of using molecular evidence to determine relatedness?

Introduction

Evolution is the process by which organisms are related by **common descent:** All organisms can trace their ancestry to the first cells. The process of evolution is amazingly simple. A group of organisms changes over time because the members of the group most suited to the natural environment have more offspring than the others in the group. So, for example, among bacteria, those which can withstand an antibiotic produce more offspring, and with time the entire group of bacteria becomes resistant to the antibiotic. Reproduction and therefore evolution have been going on since the first cells appeared on Earth. By studying (1) the fossil record, (2) anatomical and embryological comparative anatomy, and (3) molecular evidence, science is able to show that all organisms are related to one another.

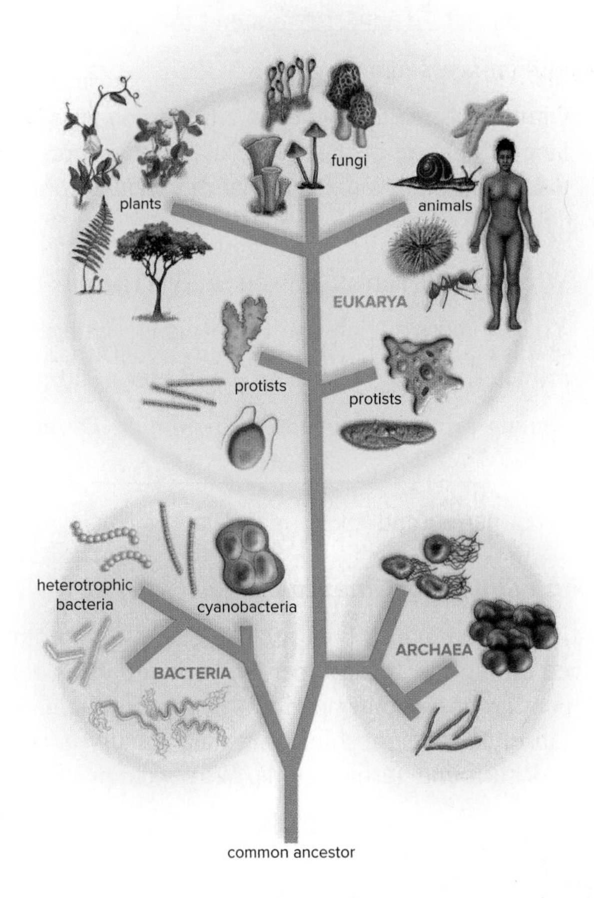

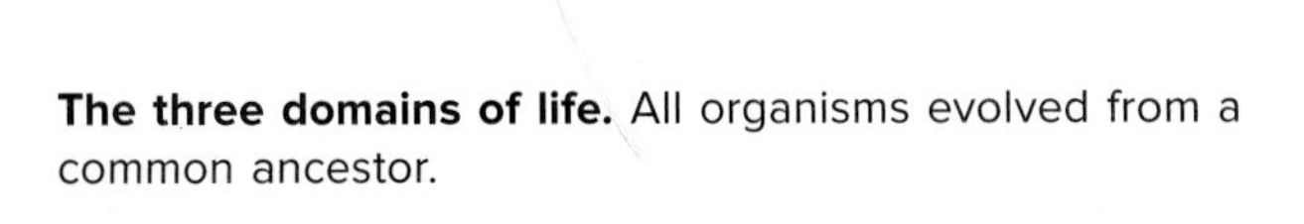

The three domains of life. All organisms evolved from a common ancestor.

18.1 Evidence from the Fossil Record

The geologic timescale, which was developed by both geologists and paleontologists, depicts the history of life based on the fossil record (Table 18.1). A **fossil** is any evidence of the existence of an organism in ancient times as opposed to modern times. In this section, we will study the geologic timescale and then examine some fossils.

Geologic Timescale

Divisions of the Timescale

Notice that the timescale divides the history of Earth into eras, then periods, and then epochs. The four eras span the greatest amounts of time, and the epochs are the shortest time frames. Notice that only the periods of the Cenozoic era are divided into epochs, meaning that more attention is given to the evolution of primates and flowering plants than to the earlier evolving organisms. List the four eras in the timescale starting with Precambrian time: Cenozoic, Mesozoic, Paleozoic and Precambrian time

1. Using the geologic timescale, you can trace the history of life by beginning with Precambrian time at the bottom of the timescale. The timescale indicates that the first cells (the prokaryotes) arose some 3,500 million years ago (MYA). The prokaryotes evolved before any other group. Why do you read the timescale starting at the bottom? Because it's the oldest.
2. The Precambrian time was very long, lasting from the time the Earth first formed until 542 MYA. The fossil record during the Precambrian time is meager, but the fossil record from the Cambrian period onward is rich (for reasons still being determined). This helps explain why the timescale usually does not show any periods until the Cambrian period of the Paleozoic era. You can also use the timescale to check when certain groups evolved and/or flourished.
   Example: During the Ordovician period, the nonvascular plants appear on land, and the first jawless and jawed fishes appear in the seas.
   During the Mesozoic era and the Jurassic period, the first flowering plants appear. How many million years ago was this? 199.6 – 145.5
3. On the timescale, note the Carboniferous period. During this period great swamp forests covered the land. These are also called coal-forming forests because, with time, they became the coal we burn today. How do you know that the plants in this forest were not flowering trees as most of our trees are today? Because there was no human influence on plant life
   What type of animal was diversifying at this time? Amphibians
4. You should associate the Cenozoic era with the evolution of humans. Among mammals, humans are primates. During what period and epoch did primates appear? Tertiary
   Among primates, humans are hominids. During what period and epoch did hominids appear? Paleocene
   __________ The scientific name for humans is *Homo sapiens.*
   What period and epoch is the age of *Homo sapiens?* Holocene

Dating within the Timescale

The timescale provides both relative dates and absolute dates. When you say, for example, "Flowering plants evolved during the Jurassic period," you are using relative time, because flowering plants evolved earlier or later than groups in other periods. If you use the dates that are given in millions of years (MYA), you are using absolute time. Absolute dates are usually obtained by measuring the amount of a radioactive isotope in the rocks surrounding the fossils. Why wouldn't you expect to find human fossils and dinosaur fossils together in rocks dated similarly? __________

Table 18.1 The Geologic Timescale: Major Divisions of Geological Time and Some of the Major Evolutionary Events That Occurred

Era	Period	Epoch	Millions of Years Ago	Plant Life	Animal Life
Cenozoic*		Holocene	0.01–present	Human influence on plant life	Age of *Homo sapiens*
				Significant Mammalian Extinction	
	Quaternary	Pleistocene	1.8–0.01	Herbaceous plants spread and diversify.	Presence of Ice Age mammals Modern humans appear.
	Tertiary	Pliocene	5.3–1.8	Herbaceous angiosperms flourish.	First hominids appear.
		Miocene	23–5.3	Grasslands spread as forests contract.	Apelike mammals and grazing mammals flourish; insects flourish.
		Oligocene	33.9–23	Many modern families of flowering plants evolve.	Browsing mammals and monkeylike primates appear.
		Eocene	55.8–33.9	Subtropical forests with heavy rainfall thrive.	All modern orders of mammals are represented.
		Paleocene	65.5–55.8	Flowering plants continue to diversify.	Primitive primates, herbivores, carnivores, and insectivores appear.
				Mass Extinction: Dinosaurs and Most Reptiles	
Mesozoic	Cretaceous		145.5–65.5	Flowering plants diversify; conifers persist.	Placental mammals appear; modern insect groups appear.
	Jurassic		199.6–145.5	Flowering plants appear.	Dinosaurs flourish; birds appear.
				Mass Extinction	
	Triassic		251–199.6	Forests of conifers and cycads dominate.	First mammals appear; first dinosaurs appear; corals and molluscs dominate seas.
				Mass Extinction	
Paleozoic	Permian		299–251	Gymnosperms diversify.	Reptiles diversify; amphibians decline.
	Carboniferous		359.2–299	Age of great coal-forming forests; Ferns, club mosses, and horsetails flourish.	Amphibians diversify; first reptiles appear; first great radiation of insects.
				Mass Extinction	
	Devonian		416–359.2	First seed plants appear. Seedless vascular plants diversify.	Jawed fishes diversify and dominate the seas; first insects and first amphibians appear.
	Silurian		443.7–416	Seedless vascular plants appear.	First jawed fishes appear.
				Mass Extinction	
	Ordovician		488.3–443.7	Nonvascular land plants appear. Marine algae flourish.	Invertebrates spread and diversify; jawless fishes (first vertebrates) appear.
	Cambrian		542–488.3	First plants appear on land. Marine algae flourish.	All invertebrate phyla present; first chordates appear.
Precambrian Time			630	Oldest soft-bodied invertebrate fossils	
			1,000–700	Protists evolve and diversify.	
			2,100	Oldest eukaryotic fossils	
			2,700	O_2 accumulates in atmosphere.	
			3,500	Oldest known fossils (prokaryotes)	
			4,570	Earth forms.	

*Many authorities divide the Cenozoic era into the Paleogene period (contains the Paleocene, Eocene, and Oligocene epochs) and the Neogene period (contains the Miocene, Pliocene, Pleistocene, and Holocene epochs).

Limitations of the Timescale

Because the timescale tells when various groups evolved and flourished, it might seem that evolution has been a series of events leading only from the first cells to humans. This is not the case; for example, prokaryotes (bacteria and archaea) never declined and are still the most abundant and successful organisms on Earth. Even today, they constitute up to 90% of the total weight of organisms.

Then, too, the timescale lists mass extinctions, but it doesn't tell when specific groups became extinct. **Extinction** is the total disappearance of a species or a higher group; **mass extinction** occurs when a large number of species disappear in a few million years or less. For lack of space, the geologic timescale can't depict in detail what happened to the members of every group mentioned. Figure 18.1 does show how mass extinction affected a few groups of animals. Which of the animals shown in Figure 18.1 suffered the most during the **P-T extinction** (Permian-Triassic extinction)? ______________________

The **K-T extinction** occurred between the Cretaceous and the Tertiary periods. Which animals shown in Figure 18.1 became extinct during the K-T extinction? ______________________

Figure 18.1 shows only periods and no eras. *Fill in the eras on the lines provided in the figure.*

Figure 18.1 Mass extinctions.
Five significant mass extinctions and their effects on the abundance of certain forms of marine and terrestrial life. The width of the horizontal bars indicates the varying number of each life-form considered.

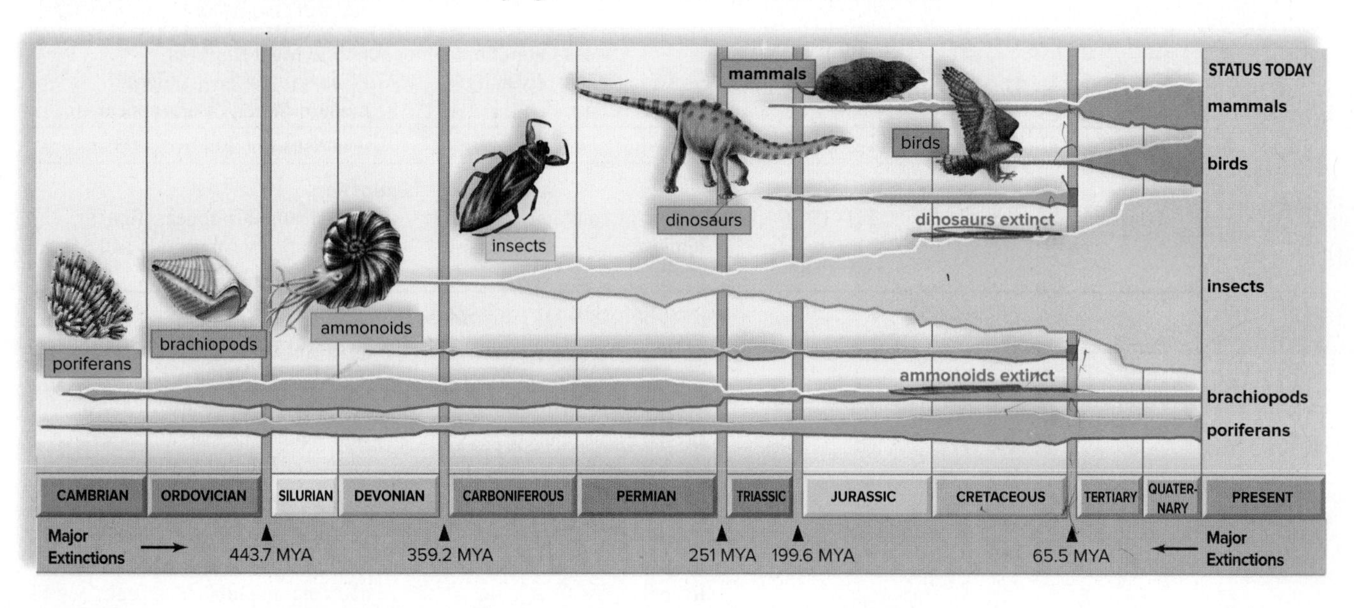

Era ______________ ______________ ______________

Fossils

The fossil record depends heavily on anatomical data to show the evolutionary changes that have occurred as one group gives rise to another group. Why would that be? ______________________

____

Invertebrate Fossils

The **invertebrates** are animals without a backbone (Fig. 18.2). Even the chordates, which is the phylum that contains the **vertebrates** (animals with a backbone), contain a couple of insignificant groups of invertebrates.

In order to familiarize yourself with possible invertebrate fossils, a few of the most common fossilized invertebrate groups are depicted in Figure 18.2.

Figure 18.2 Invertebrate fossils.

Fossilized ammonite, a mollusc

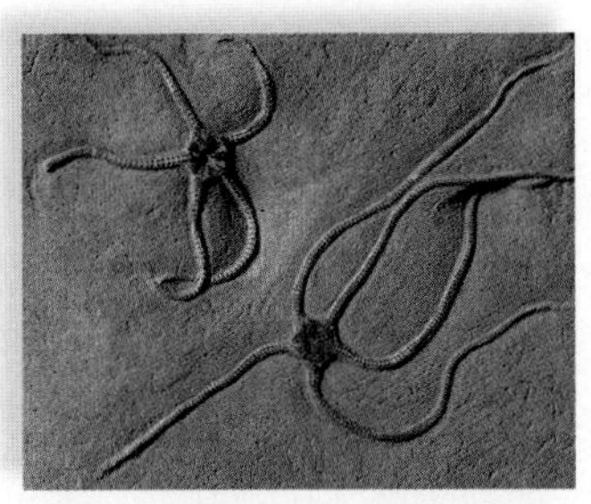

Fossilized brittle stars, echinoderms

Fossilized mayfly, an arthropod

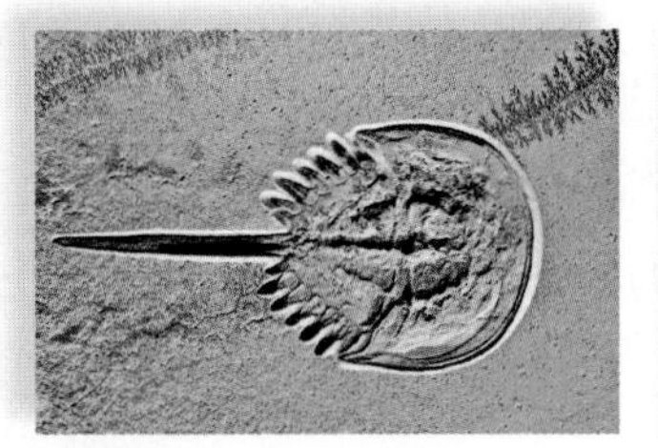

Fossilized horseshoe crab, an arthropod

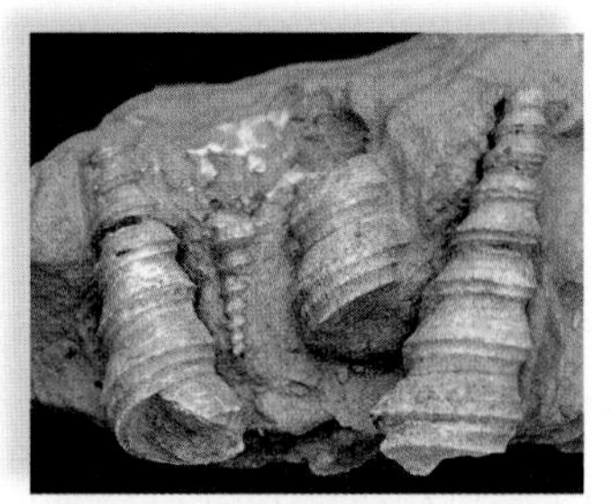

Fossilized snails, molluscs

Fossilized trilobites, arthropods

Observation: Invertebrate Fossils

1. Obtain a box of selected fossils. If the fossils are embedded in rocks, examine the rock until you have found the fossil. Fossils are embedded in rocks because the sediment that originally surrounded them hardened over time. Most fossils consist of hard parts such as shells, bones, or teeth because these parts are not consumed or decomposed over time. One possible reason the Cambrian might be rich in fossils is that organisms now had ______________________________ whereas before they did not.
2. The kit you are using, or your instructor, will identify which of the fossils are invertebrate animals. List the names of these fossils in Table 18.2 and give a description of the fossil. Use the Geologic Timescale (Table 18.1) to tell the era and period they were most abundant; list the fossils from the latest (*top*) to earliest (*bottom*).

Table 18.2 Invertebrate Fossils

Type of Fossil	Era, Period	Description

Observation: Vertebrate Fossils

The various groups of vertebrates are shown in Figure 18.3. Today it is generally agreed that birds are reptiles rather than being a separate group. That means that the major groups of vertebrates are (1) various types of fishes (jawless, jawed, cartilaginous, and bony fishes), (2) amphibians such as frogs and salamanders, (3) reptiles such as lizards, crocodiles, and birds, and (4) mammals. There are many types of mammals, from whales to mice to humans. The fossil record can rely on skeletal differences, such as limb structure, to tell if an animal is a mammal.

In order to familiarize yourself with possible vertebrate fossils, a few of the most common fossilized vertebrate groups are depicted in Figure 18.3. Which of the fossils available to you are vertebrates? __________

____

Use Table 18.1 to associate each fossil with the particular era and period when this type of animal was most abundant. Fill in Table 18.3 according to sequence of the time frames from the latest (*top*) to earliest (*bottom*).

Fossilized bird, a reptile

Fossilized bony fish

Fossilized deerlike mammal

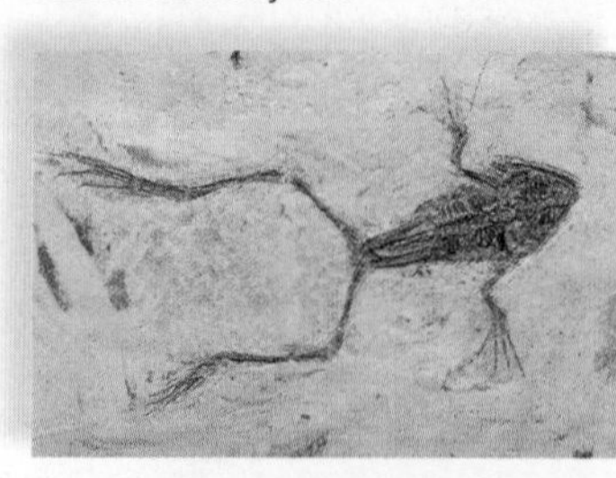

Fossilized frog, an amphibian

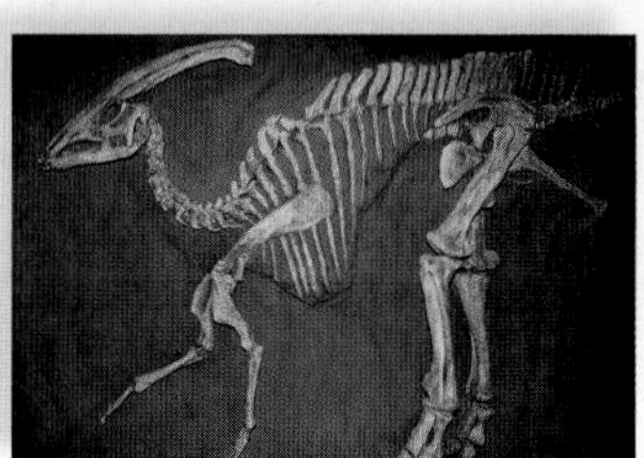

Fossilized duckbill dinosaur, a reptile

Fossilized turtle, a reptile

Figure 18.3 Vertebrate fossils.

Table 18.3 Vertebrate Fossils

Type of Fossil	Era, Period	Description

Observation: Plant Fossils

See Figure 18.4, which shows the evolution of plants, including the bryophytes, ferns and their allies, gymnosperms, and angiosperms. The fossil record for plants is not as good as that for invertebrates and vertebrates because plants have no hard parts that are easily fossilized. In order to familiarize you with possible plant fossils, a few of the most common fossilized plant groups are depicted in Figure 18.4.

Fossilized early seed plant leaves

Figure 18.4 Plant fossils.

(fern) © Ekaterina Andreeva/123RF; (flower) © Barbara Strnadova/Science Source; (maple leaf) © Biophoto Associates/Science Source; (gingko leaf) © Corbin17/Alamy; (sassafras leaf) © Jonathan Blair/Getty Images; (seed plant leaves) © Sinclair Stammers/Science Source

Plants that have no hard parts become fossils when their impressions are filled in by minerals. Use Table 18.1 to associate each fossilized plant in your kit with a particular era and period. Assume trees are flowering plants and associate them with the era and period when flowering plants were most abundant. Fill in Table 18.4 with a description of each fossil in sequence from the latest (*top*) to earliest (*bottom*).

Table 18.4 Plant Fossils

Type of Fossil	Era, Period	Description

Summary of Evidence from the Fossil Record

In this section, you studied the geologic timescale and various fossils represented in the record. The geologic timescale gives powerful evidence of evolution because

1. Fossils are ______________________.
2. Fossils can be arranged ______________________.
3. Younger fossils and not older fossils are more like ______________________.
4. In short, the fossil record shows that ______________________.

18.2 Evidence from Comparative Anatomy

In the study of evolutionary relationships, parts of organisms are said to be **homologous** if they exhibit similar basic structures and embryonic origins. If parts of organisms are similar in function only, they are said to be **analogous.** Only homologous structures indicate an evolutionary relationship and are used to classify organisms.

Comparison of Adult Vertebrate Forelimbs

The limbs of vertebrates are homologous structures (Fig. 18.5). The similarity of homologous structures is explainable by descent from a common ancestor.

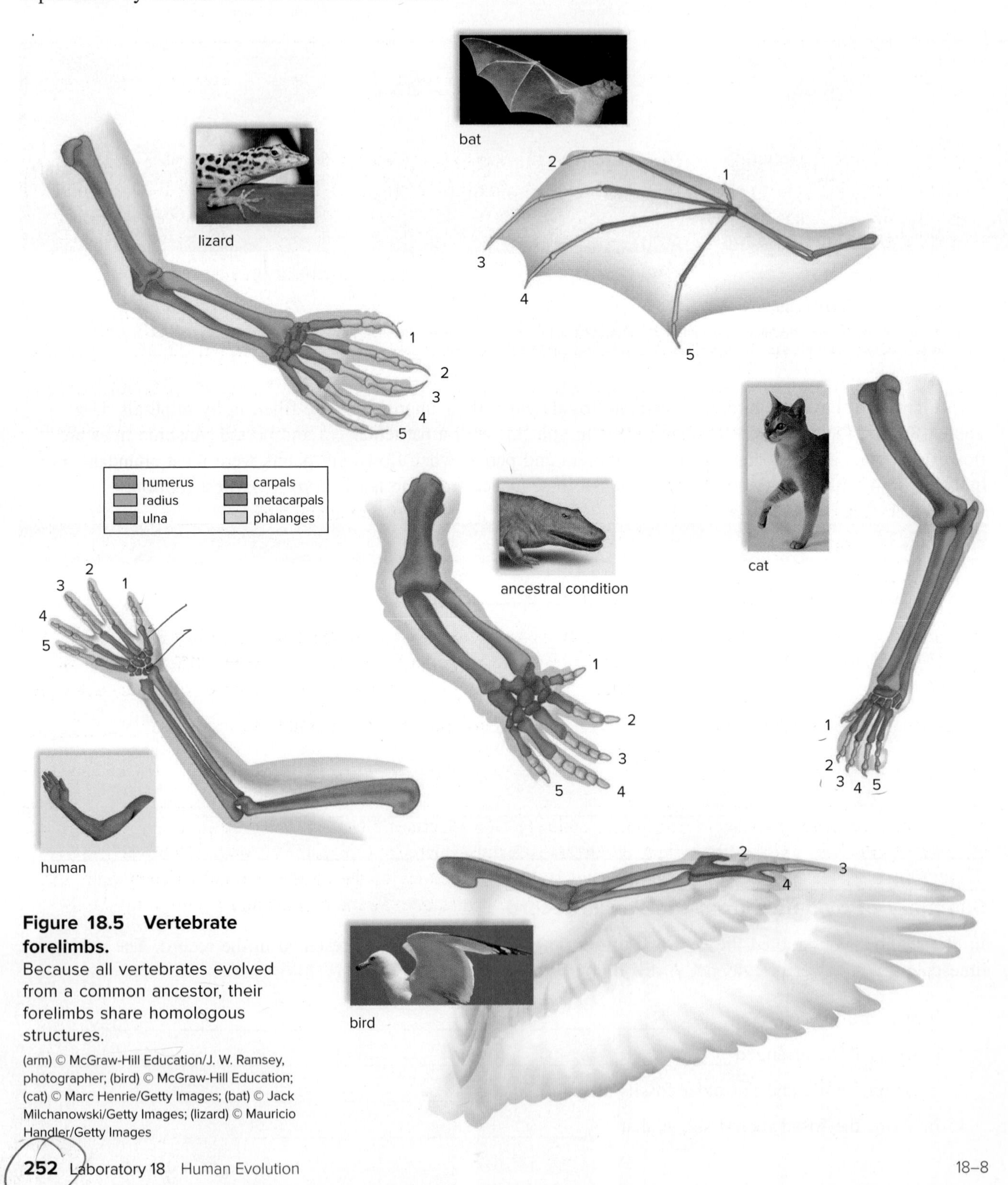

Figure 18.5 Vertebrate forelimbs.
Because all vertebrates evolved from a common ancestor, their forelimbs share homologous structures.

Observation: Vertebrate Forelimbs

1. Find the forelimb bones of the ancestral vertebrate in Figure 18.5. The basic components are the humerus (h), ulna (u), radius (r), carpals (c), metacarpals (m), and phalanges (p) in the five digits.
2. *Label the corresponding forelimb bones of the lizard, the bird, the bat, the cat, and the human.*
3. Fill in Table 18.5 to indicate which bones in each specimen appear to most resemble the ancestral condition and which most differ from the ancestral condition.
4. Adaptation to a way of life can explain the modifications that have occurred. Relate the change in bone structure to mode of locomotion in two examples.

   Example 1: ______________________________

   Example 2: ______________________________

Table 18.5 Comparison of Vertebrate Forelimbs

Animal	Bones That Resemble Common Ancestor	Bones That Differ from Common Ancestor
Lizard	Metacarpals. Carpals	Phalanges.
Bird	None	Phalanges, Metacarpals, Carpals
Bat	Metacarpals,	Phalanges.
Cat	Metacarpals, and all	except phalanges.
Human	All	

Conclusion: Vertebrate Forelimbs

- Vertebrates are descended from a ________________, but they are adapted to ________________.

Comparison of Vertebrate Embryos

The anatomy shared by vertebrates extends to their embryological development. During early developmental stages, all animal embryos resemble each other closely, but as development proceeds the different types of vertebrates take on their own shape and form. In Figure 18.6, the reptile and bird embryo resemble each other more than either resembles a fish. What does that tell you about their evolutionary relationship? ____________

____

____

In this observation, you will see that as embryos all vertebrates have a postanal tail, a dorsal spinal cord, pharyngeal pouches, and various organs. In aquatic animals, pharyngeal pouches become functional gills (Fig. 18.6). In humans, the first pair of pouches becomes the cavity of the middle ear and auditory tube, the second pair becomes the tonsils, and the third and fourth pairs become the thymus and parathyroid glands.

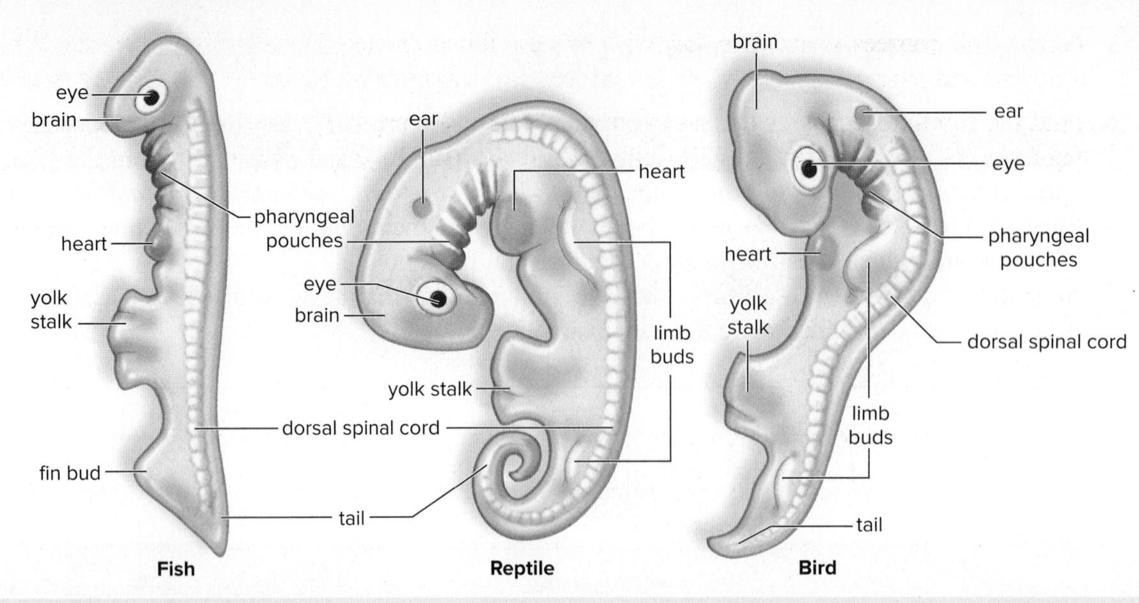

Figure 18.6 Vertebrate embryos.

Observation: Vertebrate Embryos

1. Obtain prepared slides of vertebrate embryos at comparable stages of development. Observe each of the embryos using a stereomicroscope.
2. List five similarities of the embryos:
   - **a.** ______________________________
   - **b.** ______________________________
   - **c.** ______________________________
   - **d.** ______________________________
   - **e.** ______________________________

Conclusion: Vertebrate Embryos

Vertebrate embryos resemble one another because ______________________________
______________________________.

Comparison of Chimpanzee and Human Skeletons

Chimpanzees and humans are closely related, as is apparent from an examination of their skeletons. However, they are adapted to different ways of life. Chimpanzees are adapted to living in trees and are herbivores—they eat mainly plants. Humans are adapted to walking on the ground and are omnivores—they eat both plants and meat.

Observation: Chimpanzee and Human Skeletons

Posture

Chimpanzees are arboreal and climb in trees. While on the ground, they tend to knuckle-walk, with their hands bent. Humans are terrestrial and walk erect. In Table 18.6, compare:

1. **Head and torso:** Where are the head and trunk with relation to the hips and legs—thrust forward over the hips and legs or balanced over the hips and legs (Fig. 18.7)? Move forward.
2. **Spine:** Which animal has a long and curved lumbar region, and which has a short and stiff lumbar region? How does this contribute to an erect posture in humans? ______________________________
____

3. **Pelvis:** Chimpanzees sway when they walk because lifting one leg throws them off balance. Which animal has a narrow and long pelvis, and which has a broad and short pelvis? Record your observations in Table 18.6.

Table 18.6 Comparison of Chimpanzee and Human Postures

Skeletal Part	Chimpanzee	Human
1. Head and torso	Broad and short	Narrow and long
2. Spine	Broad and short	Narrow and long
3. Pelvis	Narrow and long	Broad and short
4. Femur	Broad and short	Narrow and long
5. Knee joint	Broad and short	Narrow and long
6. Foot: Opposable toe	Chimpanzee	
Arch	Human	

4. **Femur:** In humans, the femur better supports the trunk. In which animal is the femur angled between articulations with the pelvic girdle and the knee? In which animal is the femur straight with no angle? Record your observations in Table 18.6.
5. **Knee joint:** In humans, the knee joint is modified to support the body's weight. In which animal is the femur larger at the bottom and the tibia larger at the top? Record your observations in Table 18.6.
6. **Foot:** In humans, the foot is adapted for walking long distances and running with less chance of injury.

   In which animal is the big toe opposable? ________ How does an opposable toe assist chimpanzees?

   ____

   Which foot has an arch? ____________
   How does an arch assist humans?

   ____

   Record your observations in Table 18.6.
7. How does the difference in the position of the foramen magnum, a large opening in the base of the skull for the spinal cord, correlate with the posture and stance of the two organisms?

   ____

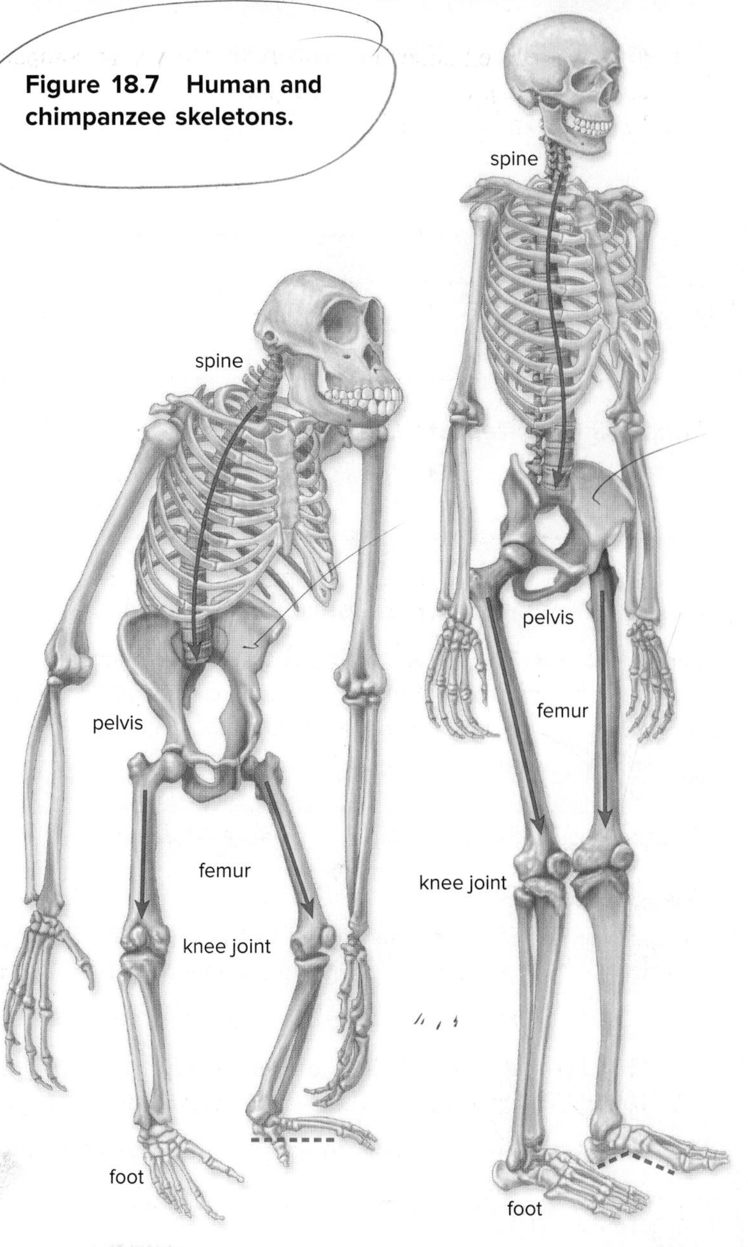

Figure 18.7 Human and chimpanzee skeletons.

Skull Features

Humans are omnivorous. A diet rich in meat does not require strong grinding teeth or well-developed facial muscles. Chimpanzees are herbivores, and a vegetarian diet requires strong teeth and strong facial muscles that attach to bony projections. Compare the skulls of the chimpanzee and the human in Figure 18.8 and answer the following questions:

1. **Supraorbital ridge:** For which skull is the supraorbital ridge (the region of frontal bone just above the eye socket) thicker? Record your observations in Table 18.7.
2. **Sagittal crest:** Which skull has a sagittal crest, a projection for muscle attachments that runs along the top of the skull? Record your observation in Table 18.7.
3. **Frontal bone:** Compare the slope of the frontal bones of the chimpanzee and human skulls. How are they different? Record your observations in Table 18.7.
4. **Teeth:** Examine the teeth of the adult chimpanzee and adult human skulls. Are the incisors (two front teeth) vertical or angled? Do the canines overlap the other teeth? Are the molars larger or moderate in size? Record your observations in Table 18.7.
5. **Chin:** What is the position of the mouth and chin in relation to the profile for each skull? Record your observations in Table 18.7.

Figure 18.8 Chimpanzee and human skulls.

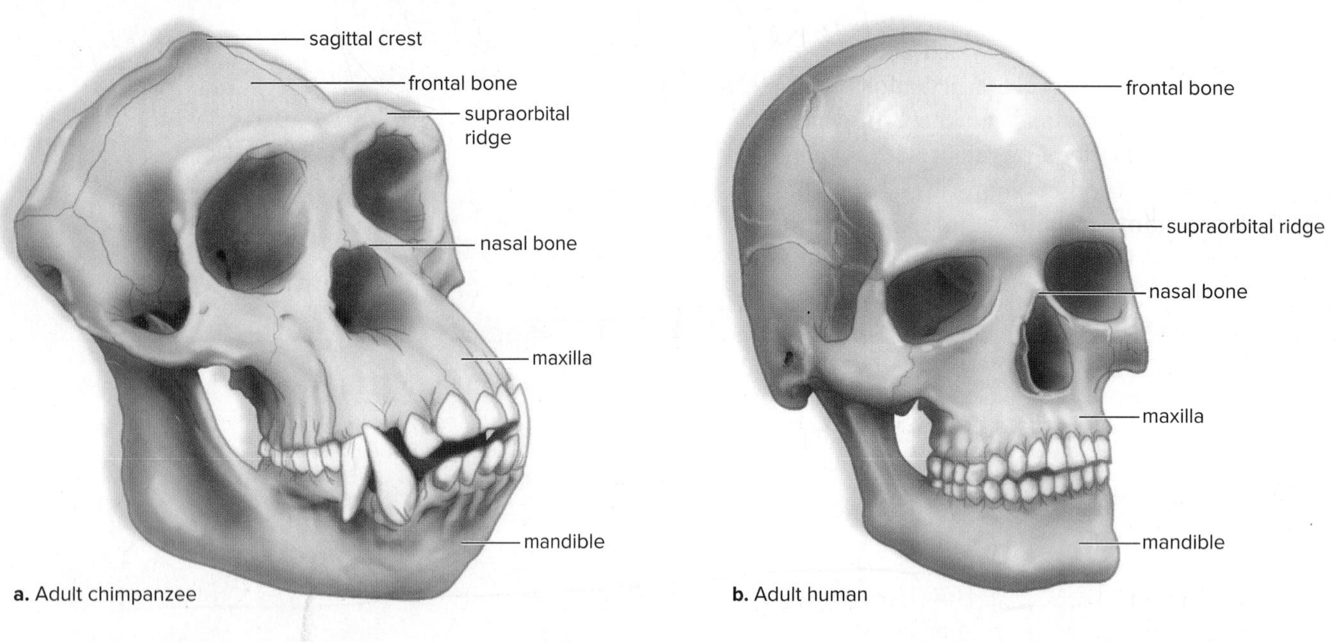

Table 18.7 Skull Features of Chimpanzees and Humans

Feature	Chimpanzee	Human
1. Supraorbital ridge		
2. Sagittal crest		
3. Slope of frontal bone		
4. Teeth		
5. Chin		

Conclusion: Chimpanzee and Human Skeletons

- Do your observations show that the skeletal differences between chimpanzees and humans can be related to posture? Yes Explain your answer. chimpanzee were moving forward
- Do your observations show that diet can be related to the skull features of chimpanzees and humans? Yes Explain your answer. they gonna eat less and are herbivorous eat mainly plants - Human are omnivorous they eat both plants and meat

Comparison of Hominid Skulls

The designation *hominid* includes humans and primates that are humanlike. Paleontologists have uncovered several fossils dated from 7.5 MYA to 30,000 years BP (before present), when humans called Cro-Magnons arose that are virtually identical to modern humans (*Homo sapiens*). (Fig. 18.9).

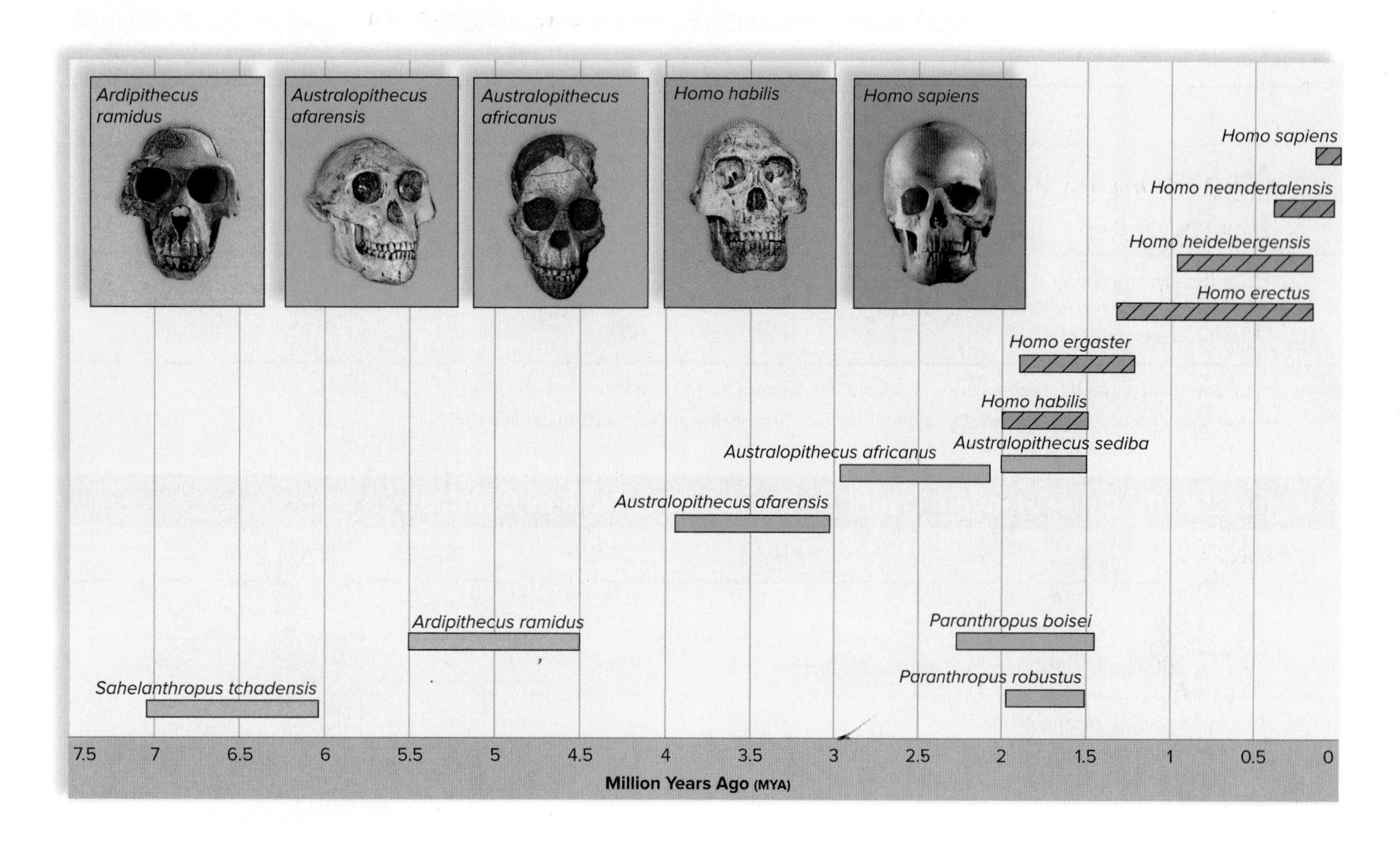

Figure 18.9 Human evolution.

(A. ramidus) © Richard T. Nowitz/Science Source; *(A. afarensis)* © Scott Camazine/Alamy; *(A. africanus)* © Philippe Plailly/Science Source; *(H. habilis)* © Kike Calvo VWPics/Superstock; *(H. sapiens)* © Kenneth Garrett/Getty Images

Observation of Hominid Skulls

Several of the skulls noted in Figure 18.9 are on display. Use Tables 18.8, 18.9, and 18.10 to record data pertaining to the cranium (or braincase), the face, and the teeth. Compare the early skulls on display with a modern human skull. For example, is the forehead like or more flat than the human skull? List at least three skulls you examined for each feature.

Table 18.8 Other Hominid Craniums Compared to Human Cranium			
Feature		**Skulls**	
	1.	2.	3.
a. Frontal bone (like or more flat?)			
b. Supraorbital ridge (divided or continuous?)			
c. Sagittal crest (present?)			
d. Mastoid process (flat or projecting?)			

Table 18.9 Other Hominid Faces Compared to Human Face			
Feature		**Skulls**	
	1.	2.	3.
a. Nasal bones (raised or flat?)			
b. Nasal opening (larger?)			
c. Chin (projecting forward?)*			
d. Width of face (wider?)**			

*If your instructor directs you to, measure from the edge of the foramen magnum to between the incisors.

**If your instructor directs you to, measure the width of the face from mid-zygomatic arch to the other mid-arch.

Figure 18.10 Other Hominid Dentition Compared to Human Dentition			
Feature		**Skulls**	
	1.	2.	3.
a. Teeth rows (parallel or diverging from each other?)			
b. Incisors (vertical or angled?)			
c. Canine teeth (overlapping other teeth?)			
d. Molars (more massive?)			

Conclusions

- Do your data appear to be consistent with the evolutionary sequence of the hominids in Figure 18.9? Explain your answer. ____________________
- Report here any data you collected that would indicate a particular hominid was closer in time to humans than indicated in Figure 18.9. ____________________

  ____
- Report here any data you collected that would indicate a particular hominid was more distant in time from humans than indicated in Figure 18.9. ____________________

  ____

18.3 Molecular Evidence

Molecular data substantiate the comparative and developmental data that biologists have accumulated over the years. The activities of *Hox* genes differ, and this can account, for example, for why vertebrates have a dorsally placed nerve cord while it is ventrally placed in invertebrates. Sequencing DNA data show which organisms are closely related. Molecular data among primates is of extreme interest because it can help determine which of the primates we are most closely related to. Chromosomal and genetic data allow us to conclude that we are more closely related to chimpanzees than to other types of apes.

In this section, we note that scientists can compare the amino acid sequence in proteins to determine the degree to which any two groups of organisms are related. The sequence of amino acids in cytochrome *c,* a carrier of electrons in the electron transport chain found in mitochondria, has been determined per a variety of organisms. On the basis of the number of amino acid *differences* reported in Figure 18.10, it is concluded that the evolutionary relationship between humans and these organisms decreases in the order stated: monkeys, pigs, ducks, turtles, fishes, moths, and yeast. This conclusion agrees with the sequence of dates these organisms are found in the fossil record. Why can comparing amino acid data lead to the same conclusions as comparing DNA data? ___________

____

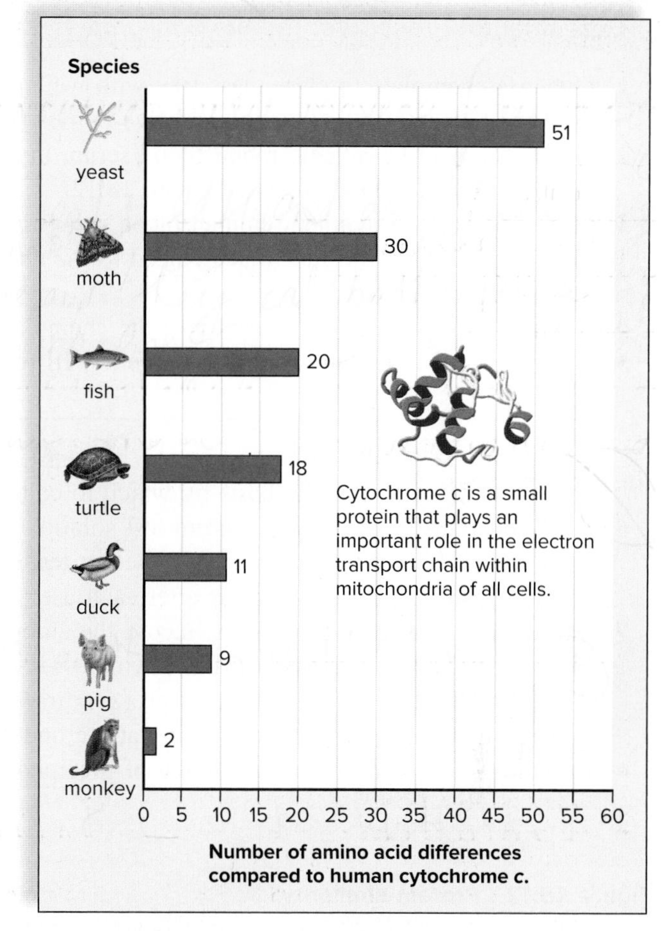

Figure 18.10 Amino acid differences in cytochrome *c*.
The few differences found in cytochrome c between monkeys and humans show that of these organisms, humans are most closely related to monkeys.

Protein Similarities

The immune system makes **antibodies** (proteins) that react with foreign proteins, termed **antigens.** Antigen-antibody reactions are specific. An antibody will react only with a particular antigen. In today's procedure, it is assumed rabbit antibodies to human antigens are in rabbit serum (Fig. 18.11). When these antibodies are allowed to react against the antigens of other animals, the stronger the antibody-antigen reaction (determined by the amount of precipitate), the more closely related the animal is to humans.

Figure 18.11 Antigen-antibody reaction.
When antibodies react to antigens, a precipitate appears.

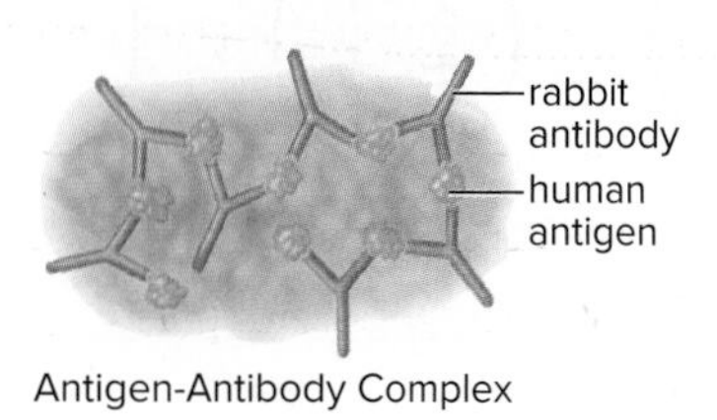

Experimental Procedure: Protein Similarities

1. Obtain a chemplate (a clear glass tray with wells), one bottle of synthetic human blood serum, one bottle of synthetic rabbit blood serum, and five bottles (I–V) of blood serum test solution.
2. Put two drops of synthetic rabbit blood serum in each of the six wells in the chemplate. *Label the wells 1–6.* See yellow circles in Figure 18.12.
3. Add two drops of synthetic human blood serum to each well. See red circles in Figure 18.12. Stir with the plastic stirring rod that was attached to the chemplate. The rabbit serum now contains antibodies against human antigens.
4. Rinse the stirrer. (The large cavity of the chemplate may be filled with water to facilitate rinsing.)
5. Add four drops of blood serum test solution III (contains human antigens) to well 6. Describe what you see. ____

   ____

   This well will serve as the basis by which to compare all the other samples of test blood serum.
6. Now add four drops of blood serum test solution I to well 1. Stir and observe. Rinse the stirrer. Do the same for each of the remaining blood serum test solutions (II–V)—adding II to well 2, III to well 3, and so on. Be sure to rinse the stirrer after each use.
7. At the end of 10 and 20 minutes, record the amount of precipitate in each of the six wells in Figure 18.12. Well 6 is recorded as having ++++ amount of precipitate after both 10 and 20 minutes. Compare the other wells with this well (+ = trace amount; 0 = none). Holding the plate slightly above your head at arm's length and looking at the underside toward an overhead light source will allow you to more clearly determine the amount of precipitate.

Figure 18.12 Protein similarity.
The greater the amount of precipitate, the more closely related an animal is to humans.

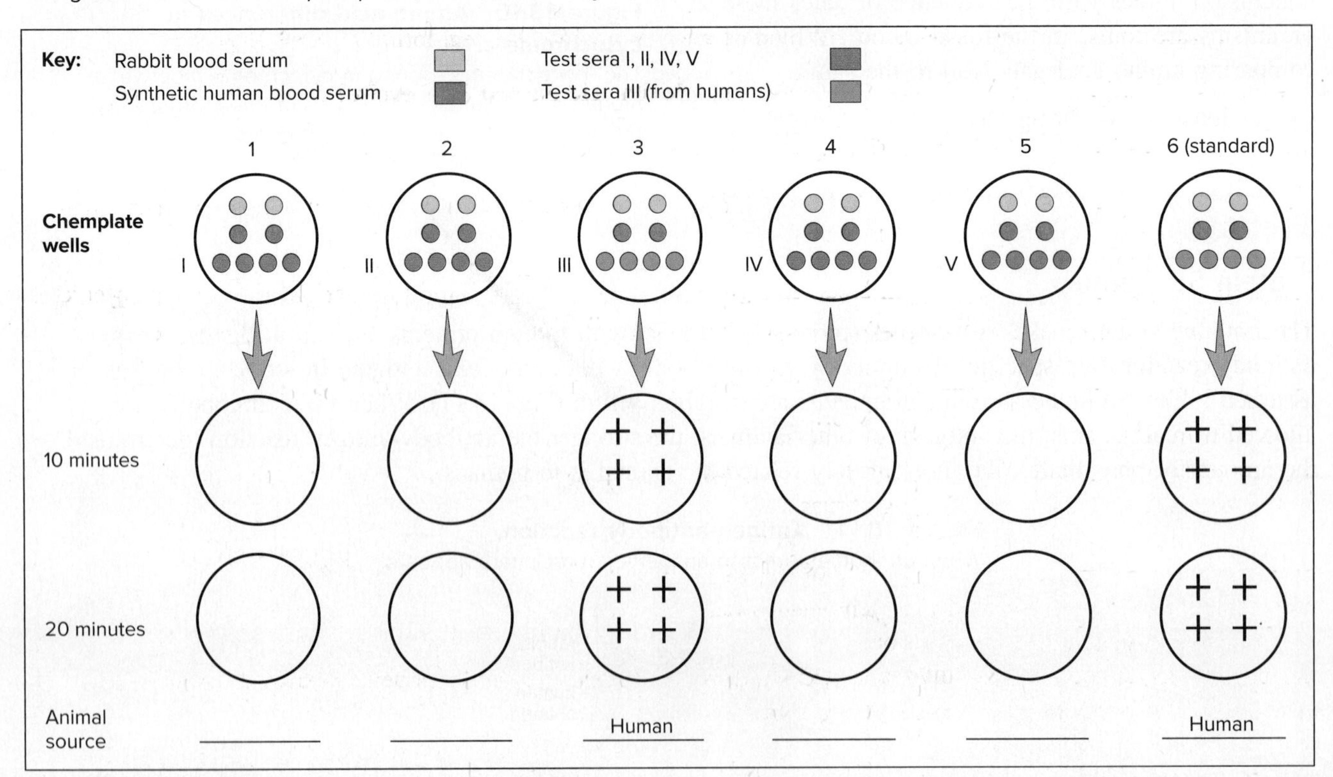

19 Effects of Pollution on Ecosystems

Learning Outcomes

19.1 Studying the Effects of Pollutants

- Using a hay infusion culture as the experimental material, predict the effect of oxygen deprivation on species composition and diversity of ecosystems.
- Using a hay infusion culture as the experimental material, predict the effect of acid deposition on species composition and diversity of ecosystems.
- Using a hay infusion culture as the experimental material, predict the effect of enrichment on species composition and diversity of ecosytems.
- Using data from a seed germination experiment, predict the effect of acid rain on crop yield.
- Using *Gammarus* as the subject, predict the effect of acid rain on food chains involving animals.

Prelab Question: How is thermal pollution related to oxygen deprivation?

19.2 Studying the Effects of Cultural Eutrophication

- Predict the effect of cultural eutrophication on food chains so that pollution results.

Prelab Question: Why do fish die off as a result of cultural eutrophication?

***Application for Daily Living:* Climate Change**

Introduction

This laboratory will consider three causes of aquatic pollution: thermal pollution, acid pollution, and cultural eutrophication. **Thermal pollution** occurs when water temperature rises above normal. As water temperature rises, the amount of oxygen dissolved in water decreases, possibly depriving organisms and their cells of an adequate supply of oxygen. Deforestation, soil erosion, and the burning of fossil fuels contribute to thermal pollution, but the chief cause is use of water from a lake or the ocean as a coolant for the waste heat of a power plant.

When sulfur dioxide and nitrogen oxides enter the atmosphere, usually from the burning of fossil fuels, they are converted to acids, which return to Earth as **acid deposition** (acid rain or snow). Acid deposition kills plants, aquatic invertebrates, and also decomposers, threatening the entire ecosystem.

Figure 19.1 Cultural eutrophication. Eutrophic lakes tend to have large populations of algae and rooted plants.

Cultural eutrophication, or overenrichment, is due to runoff from agricultural fields, wastewater from sewage treatment plants, and even excess detergents. These sources of excess nutrients cause an algal bloom seen as a green scum on a lake (Fig. 19.1). When algae overgrow and die, decomposition robs the lake of oxygen, causing a fish die-off.

19.1 Studying the Effects of Pollutants

We are going to study the effects of pollution by observing its effects on hay infusion organisms, on seed germination, and on an animal called *Gammarus.*

Study of Hay Infusion Culture

A hay infusion culture (hay soaked in water) contains various microscopic organisms, but we will be concentrating on how the pollutants in our study affect the protozoan populations in the culture. We will consider both of these aspects:

species composition: number of different types of microorganisms.

species diversity: composition and the abundance of each type of microorganism.

Experimental Procedure: Effect of Pollutants on a Hay Infusion Culture

During this Experimental Procedure you will examine, by preparing a wet mount, hay infusion cultures that have been treated in the following manner.

1. **Control culture:** This culture simulates the species composition and diversity of an untreated culture. Prepare a wet mount and answer the following questions:
   With the assistance of Figure 19.2 and any guides available in the laboratory, identify as many different types of microorganisms as possible in the hay infusion culture. State whether species composition is high, medium, or low. Record your estimation in the second column of Table 19.1. Do you judge species diversity to be high, medium, or low? Record your estimation in the third column of Table 19.1.
2. **Oxygen-deprived culture:** Thermal pollution causes water to be oxygen deprived; therefore, when we study the effects of low oxygen on a hay infusion culture, we are studying an effect of thermal pollution. Prepare a wet mount of this culture and determine if there is a change in species composition and diversity. Again record the species composition and species diversity as high, medium, or low in Table 19.1.

Figure 19.2 Microorganisms in hay infusion cultures.
Organisms are not to size.

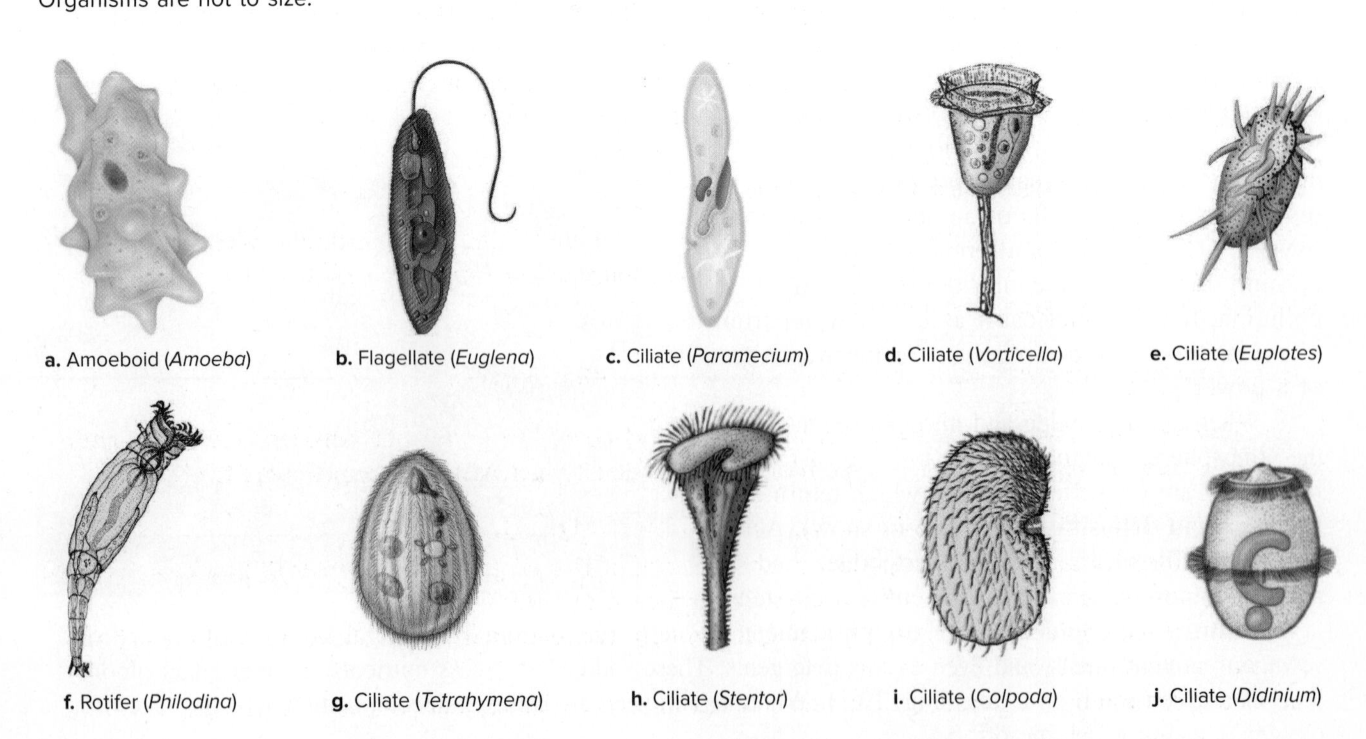

a. Amoeboid (*Amoeba*) **b.** Flagellate (*Euglena*) **c.** Ciliate (*Paramecium*) **d.** Ciliate (*Vorticella*) **e.** Ciliate (*Euplotes*)

f. Rotifer (*Philodina*) **g.** Ciliate (*Tetrahymena*) **h.** Ciliate (*Stentor*) **i.** Ciliate (*Colpoda*) **j.** Ciliate (*Didinium*)

Table 19.1 Effect of Pollution on a Hay Infusion Culture

Type of Culture	Species Composition (High, Medium, or Low)	Species Diversity (High, Medium, or Low)	Explanation
Control			
Oxygen-deprived			
Acidic			
Enriched			

3. **Acidic culture:** In this culture, the pH has been adjusted to 4 with sulfuric acid (H_2SO_4). This simulates the effect of acid rain on a hay infusion culture. Prepare a wet mount of this culture and determine if there is a change in species composition and diversity. Again record the species composition and species diversity as high, medium, or low in Table 19.1.
4. **Enriched culture:** More organic nutrients have been added to this culture. These nutrients will cause the algae population, which is food for most protozoans, to increase. In the short term, their species composition should increase. Eventually, as the algae die off, decomposition will rob the water of oxygen and the protozoans may start to die off. Prepare a wet mount of this culture and determine if there is a change in species composition and diversity. Again record the species composition and species diversity as high, medium, or low in Table 19.1.

Conclusions

- What could be a physiological reason for the adverse effects of oxygen deprivation on a hay infusion culture? If consistent with your results, enter this explanation in the last column of Table 19.1.
- What could be a physiological reason for the adverse effects of a low pH on a hay infusion culture? If consistent with your results, enter this explanation in the last column of Table 19.1.
- What could be an environmental reason for the adverse effects of an enriched culture? If consistent with your results, enter this explanation in the last column of Table 19.1.

Effect of Acid Rain on Seed Germination

Seeds depend on favorable environmental conditions of temperature, light, and moisture to germinate, grow, and reproduce. Like any other biological process, germination requires enzymatic reactions that can be adversely affected by an unfavorable pH.

Experimental Procedure: Effect of Acid Rain on Seed Germination

In this Experimental Procedure we will test whether there is a negative correlation between acid concentration and germination. In other words, it is hypothesized that as acidity increases, it becomes more likely seeds will ______________________________.

Your instructor has placed 20 sunflower seeds in each of five containers with water of increasing acidity: 0% vinegar (tap water), 1% vinegar, 5% vinegar, 20% vinegar, and 100% vinegar.

1. Test and record the pH of solutions having the vinegar concentrations noted above. Record the pH of each solution in Table 19.2.
2. Count the number of germinated sunflower seeds in each container, and complete Table 19.2.

Table 19.2 Effect of Increasing Acidity on Germination of Sunflower Seeds

Concentration of Vinegar	pH	Number of Seeds that Germinated	Percent Germination
0%			
1%			
5%			
20%			
100%			

Conclusions: Effect of Increasing Acidity on Germination of Sunflower Seeds

- As you know, each enzyme has an optimum pH. Explain why acid rain is expected to inhibit metabolism, and therefore, seedling development. ________________
- Do the data support or falsify your hypothesis? ________________

Study of *Gammarus*

A small crustacean called *Gammarus* lives in ponds and streams (Fig. 19.3) where it feeds on debris, algae, or anything smaller than itself, such as some of the microorganisms in Figure 19.2. In turn, fish like to feed on *Gammarus*.

Experimental Procedure: Gammarus

- Add 25 ml of spring water to a beaker and record the pH of the water. ________________ pH
- Add four *Gammarus* to the container. Do they all use their legs in swimming? ________________
- Which legs are used in jumping and climbing? ________________
- What do *Gammarus* do when they "bump" into each other? ________________

Control Sample

After observing *Gammarus,* decide what behaviors are most often observed. During a 5-minute time span, total the amount of time spent doing each of these behaviors.

Behaviors	Amount of Time	Total Time
1. ________	________	________
2. ________	________	________
3. ________	________	________

Test Sample

If so directed by your instructor, put a *Gammarus* in a beaker of spring water adjusted to pH 4 by adding vinegar. During a 5-minute time span, total the amount of time spent doing each of these behaviors.

Behaviors	Amount of Time	Total Time
1. ________	________	________
2. ________	________	________
3. ________	________	________

Figure 19.3 ***Gammarus.*** *Gammarus* is a type of crustacean classified in a subphylum that also includes shrimp.

Conclusion

- Draw a conclusion from this study: ______________________________

____

- Create a food chain that shows who eats whom when the food chain includes algae, protozoans, *Gammarus,* fish, and humans. ______________________________

  a. What would happen to this food chain if the water were oxygen deprived? ______________________________

  ____

  Acidic? ______________________________

  Enriched with inorganic nutrients (short term and long term)? ______________________________

  ____

Conclusions: Studying the Effects of Pollutants

- Give an example to show that the hay infusion study pertains to real ecosystems. ______________________________

  ____

- What are the potential consequences of acid rain on crops that reproduce by seeds? ______________________________

  ____

  On the food chains of the ocean? ______________________________

- How does the addition of nutrients affect species composition and species diversity of an ecosystem over time? ______________________________

19.2 Studying the Effects of Cultural Eutrophication

Chlorella, the green alga used in this study, is considered to be representative of algae in bodies of fresh water. The crustacean *Daphnia* feeds on green algae such as *Chlorella* (Fig. 19.4). First, you will observe how *Daphnia* feeds, and then you will determine the extent to which *Daphnia* could keep the effects of cultural eutrophication from occurring in a hypothetical example. Keep in mind that this case study is an oversimplification of a generally complex problem.

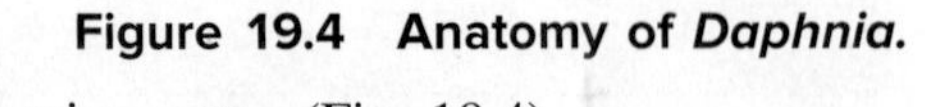

Figure 19.4 Anatomy of *Daphnia*.

Observation: Daphnia *Feeding*

1. Place a small pool of petroleum jelly in the center of a small petri dish.
2. Use a dropper to take a *Daphnia* from the stock culture, place it on its back (covered by water) in the petroleum jelly, and observe it under the stereomicroscope (Fig. 19.4).
3. Note the insectlike carapace and the legs waving rapidly as the *Daphnia* filters the water.
4. Add a drop of *carmine solution,* and observe how the *Daphnia* filters the "food" from the water and passes it through the gut. The gut is more visible if you push the animal onto its side. In this position, you may also observe the heart beating in the region above the gut and just behind the head.
5. Allow the *Daphnia* to filter-feed for up to 30 minutes, and observe the progress of the carmine particles through the gut. Does the carmine travel completely through the gut in 30 minutes? ________________

____

Experimental Procedure: Daphnia *Feeding on* Chlorella

This exercise requires the use of a spectrophotometer. Absorbance will be a measure of the algal population level; the greater the number of algal cells, the greater the absorbance. The higher the absorbance, the greater the amount of light absorbed and *not* passed through the solution.

1. Obtain two spectrophotometer tubes (cuvettes) and a Pasteur pipet.
2. Fill one of the cuvettes with distilled water, and use it to zero the spectrophotometer. Save this tube for step 6.
3. Use the Pasteur pipet to fill the second cuvette with *Chlorella.* Gently aspirate and expel the sample several times (without creating bubbles) to give a uniform dispersion of the algae.
4. Add ten hungry *Daphnia,* and following your instructor's directions, immediately measure the absorbance with the spectrophotometer. If a *Daphnia* swims through the beam of light, a strong deflection should occur; do not use any such higher readings—instead, use the lower figure for the absorbance. Record your reading in the first column of Table 19.3.
5. Remove the cuvette with the *Daphnia* to a safe place in a test tube rack. Allow the *Daphnia* to feed for 30 minutes.
6. Rezero the spectrophotometer with the distilled water cuvette.
7. Measure the absorbance of the experimental cuvette again. Record your data in the second column of Table 19.3, and explain your results in the third column.

Table 19.3 Spectrophotometer Data/*Daphnia* Feeding on *Chlorella*

Absorbance Before Feeding	Absorbance After Feeding	Explanation

Experimental Procedure: Case Study in Cultural Eutrophication

The following problem will test your understanding of the ecological value of a single species—in this case, *Daphnia.* Please realize that this is an oversimplification of a generally complex problem.

1. Assume that developers want to build condominium units on the shores of Silver Lake. Homeowners in the area have asked the regional council to determine how many units can be built without altering the nature of the lake. As a member of the council, you have been given the following information:

   The current population of *Daphnia,* 10 animals/liter, presently filters 24% of the lake per day, meaning that it removes this percentage of the algal population per day. This is sufficient to keep the lake essentially clear. Predation—the eating of the algae—will allow the *Daphnia* population to increase to no more than 50 animals/liter. Therefore, 50 *Daphnia*/liter will be available for feeding on the increased number of algae that would result from building the condominiums.

   Using this information, complete Table 19.4.

Table 19.4 *Daphnia* Filtering

Number of *Daphnia*/Liter	Percent of Lake Filtered
10	24%
50	

2. The sewage system of the condominiums will add nutrients to the lake. Phosphorus output will be 1 kg per day for every 10 condominiums. This will cause a 30% increase in the algal population. Using this information, complete Table 19.5.

Table 19.5 Cultural Eutrophication

Number of Condominiums	Phosphorus Added	Increase in Algal Population
10	1 kg	30%
20		
30		
40		
50		

Conclusion: Cultural Eutrophication

- Assume that phosphorus is the only nutrient that will cause an increase in the algal population and that *Daphnia* is the only type of zooplankton available to feed on the algae. How many condominiums would you allow the developer to build? ______________________________

  ____

- What other possible impacts could condominium construction have on the condition of the lake? __________

  ____

  ____

Application for Daily Living

Climate Change

Climate change includes an expected rise in temperature of the Earth's surface and oceans due to the burning of fossil fuels, such as oil and gasoline. When we burn fossil fuels, carbon dioxide enters the atmosphere and traps heat from the sun's rays near the Earth.

Any significant rise in the Earth's temperature can lead to all sorts of disastrous results (Fig. 19.5). The sea level can rise and flood cities from Miami to New York on the East Coast and San Diego and San Francisco on the West Coast. Rain patterns can change, and some areas that were dry before can become wet and wet areas can become dry. The favorable temperature for farming can move northward to regions that don't have good soil; plants and animals that can't migrate will become extinct. The soil in Alaska is called a permafrost because it is always frozen. If the glaciers and permafrost melt, animals such as the polar bear are threatened, and so is the Alaskan pipeline that helps supply American cities.

Agricultural land parched by lack of rain

Houses threatened by rising water

Human fatigued by exertion in hot weather

Figure 19.5 Some effects of climate change.

(lack of rain) © AP Photo/Seth Periman; (rising water) © AP Photo/HO/Mike Page, www.mike-page.co.uk; (exertion) © T. S. Corrigan/Alamy

Even if you don't believe climate change is imminent, you should do all you can to reduce consumption of oil. When we burn less oil, air pollution lessens, we are not as dependent on less-than-friendly foreign countries, and it may even help reduce the national debt.

Laboratory Review 19

_______________ 1. What results when sulfur dioxide and nitrogen oxides, formed by burning fossil fuels, enter the atmosphere?

_______________ 2. What kind of pollution results when water from a lake or the ocean is used to disperse waste heat?

_______________ 3. Does the amount of oxygen dissolved in water increase or decrease when the temperature of the water increases?

_______________ 4. What occurs to a body of water when there is runoff from agricultural fields?

_______________ 5. Which population experiences growth referred to as a "bloom" when excess nutrients enter a body of water?

_______________ 6. After excess nutrients cause algae to overgrow and then die, what causes the fish to die?

_______________ 7. What do we call the number of different microorganisms observed and their relative abundance in the hay infusion culture?

_______________ 8. What chemical was used to simulate the effect of acid rain on a hay infusion culture?

_______________ 9. What does an unfavorable pH adversely affect that inhibits seed germination?

_______________ 10. In the Experimental Procedure on the effect of acid rain on seed germination, if 18 of the sunflower seeds germinated in the 1% vinegar solution, what percent germinated?

_______________ 11. What kind of animal is *Gammarus?*

_______________ 12. What will *Daphnia* do that might prevent cultural eutrophication from occurring?

_______________ 13. Are there more or fewer algal cells present in a water sample as the absorption of light decreases?

_______________ 14. What element from the Silver Lake condominiums' sewage output contributes to the increased algal population?

Thought Questions

15. A "dead zone" forms where the Mississippi River empties into the Gulf of Mexico due to nutrient runoff into the river that is carried to the Gulf. Explain how the dead zone forms and what impact it may have on humans.

16. Pollutants often affect the producers in an ecosystem first and then have far-ranging effects. Use acid rain and a food chain to illustrate your understanding of the impact pollutants have after they enter an ecosystem.

17. Describe how the cultural eutrophication case study illustrates the need to have a balance of population sizes in an ecosystem.

Metric System

Unit and Symbol	Metric Equivalent	English Equivalents	
Length			
nanometer (nm)	= 10^{-9} m (10^{-3} µm)		
micrometer (µm)	= 10^{-6} m (10^{-3} mm)		
millimeter (mm)	= 0.001 (10^{-3}) m		
centimeter (cm)	= 0.01 (10^{-2}) m	1 inch	= 2.54 cm
		1 foot	= 30.5 cm
meter (m)	= 100 (10^{2}) cm	1 foot	= 0.30 m
	= 1,000 mm	1 yard	= 0.91 m
kilometer (km)	= 1,000 (10^{3}) m	1 mile	= 1.6 km
Weight (mass)			
nanogram (ng)	= 10^{-9} g		
microgram (µg)	= 10^{-6} g		
milligram (mg)	= 10^{-3} g		
gram (g)	= 1,000 mg	1 ounce	= 28.3 g
		1 pound	= 454 g
kilogram (kg)	= 1,000 (10^{3}) g		= 0.45 kg
metric ton (t)	= 1,000 kg	1 ton	= 0.91 t
Volume			
microliter (µl)	= 10^{-6} l (10^{-3} ml)		
milliliter (ml)	= 10^{-3} l	1 tsp	= 5 ml
	= 1 cm^3 (cc)	1 fl oz.	= 30 ml
	= 1,000 mm^3		
liter (l)	= 1,000 ml	1 pint	= 0.47 l
		1 quart	= 0.95 l
		1 gallon	= 3.79 l
kiloliter (kl)	= 1,000 l		

Units of Temperature

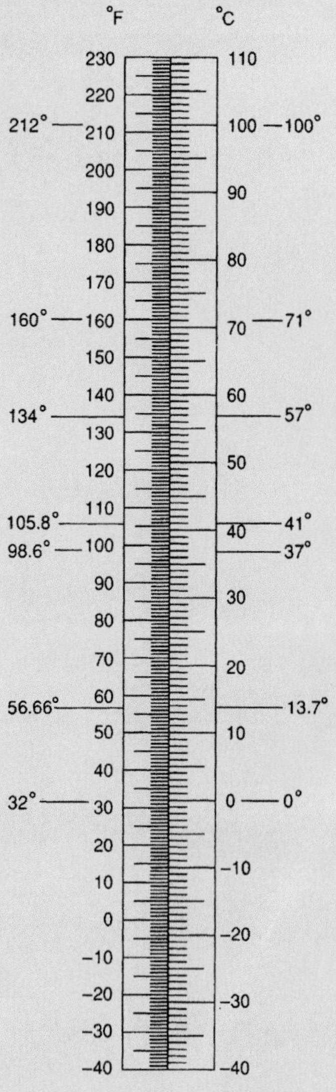

°F | °C

230 | 110
220 |
212° — 210 | 100 — 100°
200 |
190 | 90
180 | 80
170 |
160° — 160 | 70 — 71°
150 |
140 | 60
134° | 57°
130 |
120 | 50
105.8° — 110 | 40 — 41°
98.6° — 100 | 37°
90 |
80 | 30
70 | 20
56.66° — 60 | 13.7°
50 | 10
40 |
32° — 30 | 0 — 0°
20 |
10 | −10
0 | −20
−10 |
−20 | −30
−30 |
−40 | −40

To convert temperature scales:
°C = (°F − 32)/1.8
°F = 1.8(°C) + 32

Common Temperatures*

°C	°F	
100	212	Water boils at standard temperature and pressure.
57	134	Highest recorded temperature in the United States, Death Valley, July 10, 1913
41	105.8	Average body temperature of a marathon runner in hot weather
37	98.6	Human body temperature
0	32.0	Water freezes at standard temperature and pressure.

* See temperature scale

HINTS FOR USING THE METRIC SYSTEM

1. Use decimals, not fractions (e.g., 2.5 m, not 2½ m). Express measurements in units requiring only a few decimal places. For example, 0.3 m is more easily manipulated and understood than 300000000 nm.
2. When measuring pure water, the metric system offers an easy and common conversion from volume measured in liters to volume measured in cubic meters to mass measured in grams: 1 ml = 1 cm^3 = 1 g.
3. Do not place a period after metric symbols (e.g., 1 g, not 1 g.). Use a period after a symbol only at the end of a sentence. Do not mix units or symbols (e.g., 9.2 m, not 9 m 200 mm). Metric symbols are always singular (e.g., 10 km, not 10 kms).
4. Except for degrees Celsius, always leave a space between a number and a metric symbol (e.g., 20 mm, not 20mm; 10°C, not 10° C).
5. Use a zero before a decimal point when the number is less than one (e.g., 0.42 m, not .42 m).

Index

Note: Page numbers followed by *f* refer to figures; page numbers followed by *t* refer to tables.

D

E